QUATERNARY ENVIRONMENTAL MICROPALAEONTOLOGY

Edited by
Simon K. Haslett

Quaternary Research Unit, Department of Geography, Bath Spa University College

ARNOLD

A member of the Hodder Headline Group
LONDON
Co-published in the United States of America by
Oxford University Press Inc., New York

First published in Great Britain in 2002 by
Arnold, a member of the Hodder Headline Group
338 Euston Road, London NW1 3BH
http://www.arnoldpublishers.com

Co-published in the United States of America by
Oxford University Press Inc.,
198 Madison Avenue, New York, NY10016

British Library Cataloguing in Publication Data
A catalogue record for this book is available from the British Library

Library of Congress Cataloging-in-Publication Data
A catalog record for this book is available from the Library of Congress

ISBN 0 340 76197 0 (hb)
ISBN 0 340 76198 9 (pb)

1 2 3 4 5 6 7 8 9 10

Production Editor: Anke Ueberberg
Production Controller: Iain McWilliams
Cover Design: Terry Griffiths

Typeset in 10/12 pt Palatino by Integra Software Services Pvt Ltd, Pondicherry, India;
www.integra-india.com
Printed and bound in Malta by Gutenberg Press

What do you think about this book? Or any other Arnold title?
Please send your comments to feedback.arnold@hodder.co.uk

In memory of
Professor Brian M. Funnell
(1933–2000)
A friend and valued colleague

CONTENTS

LIST OF CONTRIBUTORS

Elisabeth Alve, Department of Geology, University of Oslo, Oslo, Norway.

Ian Boomer, Department of Geography, University of Newcastle, Newcastle-upon-Tyne, United Kingdom.

Frank M. Chambers, Centre for Environmental Change and Quaternary Research, GEMRU, University of Gloucestershire, Cheltenham, United Kingdom.

Amy L. Dale, GeoResearch Consulting, Bakli, Skarnes, Norway.

Barrie Dale, Geology Department, University of Oslo, Oslo, Norway.

W. Roland Gehrels, Quaternary Environments Research Group, Department of Geographical Sciences, University of Plymouth, Plymouth, United Kingdom.

Simon K. Haslett, Quaternary Research Unit, Department of Geography, Bath Spa University College, Newton Park, Bath, United Kingdom.

Richard W. Jordan, Department of Earth and Environmental Sciences, Faculty of Science, Yamagata University, Yamagata, Japan.

Kevin Kennington, Port Erin Marine Laboratory, School of Biological Sciences, University of Liverpool, Port Erin, Isle of Man, United Kingdom.

John W. Murray, School of Ocean and Earth Science, Southampton Oceanography Centre, European Way, Southampton, United Kingdom.

Christopher W. Smart, Department of Geological Sciences, University of Plymouth, Plymouth, United Kingdom.

ACKNOWLEDGEMENTS

The editor and publishers would like to thank the following for permission to use copyright material in this book:

American Geophysical Union and Thomas, E., Booth, L., Maslin, M. and Shackleton, N.J.: Northeastern Atlantic benthic foraminifera during the last 45,000 years: changes in productivity seen from the bottom up, *Paleoceanography*, 10, 1995, pp. 545–562; Geological Society of America and Nees, S., Altenbach, A.V., Kassens, H. and Thiede, J.: High-resolution record of foraminiferal response to late Quaternary sea-ice retreat in the Norwegian–Greenland Sea, *Geology*, 25, 1997, pp. 659–662; Elsevier Science and Schmiedl, G. and Mackensen, A.: Late Quaternary paleoproductivity and deep water circulation in the eastern South Atlantic Ocean: evidence from benthic foraminifera, *Palaeogeography, Palaeoclimatology, Palaeoecology*, 130, 1997, pp. 43–80; American Geophysical Union and Oppo, D.W., Fairbanks, R.G., Gordon, A.L. and Shackleton, N.J.: Late Pleistocene Southern Ocean ^{13}C variability, *Paleoceanography*, 5, 1990, pp. 43–54; Elsevier Science and Raymo, M.E., Ruddiman, W.F., Shackleton, N.J. and Oppo, D.W.: Evolution of Atlantic–Pacific ^{13}C gradients over the last 2.5 m.y., *Earth and Planetary Science Letters*, 5, 1990, pp. 353–368; American Geophysical Union and Hodell, D.A.: Late Pleistocene paleoceanography of the South Atlantic sector of the Southern Ocean: Ocean Drilling Program Hole 704A, *Paleoceanography*, 8, 1993, pp. 47–67; Springer-Verlag and Bickert, T. and Wefer, G.: Late Quaternary deep water circulation in the South Atlantic: reconstruction from carbonate dissolution and benthic stable isotopes, The South Atlantic: present and past circulation, Wefer, G., Berger, W.H., Siedler, G. and Webb, D.J. (eds), 1996, pp. 599–620; Elsevier Science and Ohkushi, K., Thomas, E. and Kawahata, H.: Abyssal benthic foraminifera from the north-western Pacific (Shatsky Rise) during the last 298 kyr, *Marine Micropaleontology*, 38, 2000, pp. 119–147; Elsevier Science and Den Dulk, M., Reichart, G.J., Memon, G.M., Roelofs, E.M.P., Zachariasse, W.J., Van der Zwaan, G.J.: Benthic foraminiferal response to variations in surface water productivity and oxygenation in the northern Arabian Sea, *Marine Micropaleontology*, 35, 1998, pp. 43–66; American Geophysical Union and Schmiedl, G., Hemleben, C., Keller, J. and Segl, M., Impact of climatic changes on the benthic foraminiferal fauna in the Ionian Sea during the last 330,000 years, *Paleoceanography*, 13, 1998, pp. 447–458; Cushman Foundation, Allen Press and Alve, E.: Benthic foraminiferal responses to estuarine pollution:

a review, *Journal of Foraminiferal Research*, 25, 1995, pp. 190–203; Cushman Foundation, Allen Press and Angel, D.L., Verghese, S., Lee, J.J., Saleh, A.M., Zuber, D., Lindell, D. and Symons, A.: Impact of a net cage fish farm on the distribution of benthic foraminifera in the northern Gulf of Eilat (Aqba, Red Sea), *Journal of Foraminiferal Research*, 30, 2000, 54–65; Cushman Foundation, Allen Press and Alve, E. and Olsgard, F.: Benthic foraminiferal colonization in experiments with copper-contaminated sediments, *Journal of Foraminiferal Research*, 29, 1999, 186–195; Cushman Foundation, Allen Press and Sloan, D.: Use of foraminiferal biostratigraphy in mitigating pollution and seismic problems, San Francisco, California, *Journal of Foraminiferal Research*, 25, 1995, 260–266; Linnean Society of London, Leiden, E.J. Brill and Athersuch, J., Horne, D.J. and Whittaker, J.E.: *Marine and Brackish Water Ostracods*. Synopsis of the British Fauna (New Series) No. 43, 1989, 343 pp.

CWS wishes to thank John Murray and Andy Gooday for all their helpful comments on the manuscript of Chapter 3. JWM and EA thank Professor K. L. Knudsen for providing some specimens of cold-water foraminifera illustrated in Figure 4.4. WRG would like to thank Simon Newman for the preparation of the plates. Figure 6.3A is from a photograph by Dr Richard Llewellyn Jones and reproduced by kind permission of the Natural History Museum, London. RWJ would like to thank Dr. Jeremy Young of the Natural History Museum, London, for providing extensive comments and photographs for chapter 9. BD and ALD give special thanks to Astri Dugan, recently retired from the Geology Dept., University of Oslo, for excellent sample preparations in these and many other projects. They also wish to thank the personnel of the Graphics and Photography Department of the Mathematics and Natural Sciences Faculty, University of Oslo, for their usual jovial, willing and competent help with illustrations. They are grateful for support and encouragement from marine geologist Fred Jansen (Netherlands Inst. for Sea Research), and marine archaeologist Dag Neverstad (Norwegian Maritime Museum) in the Angola Project and Inner Oslofjord archaeology, respectively. They gratefully acknowledge earlier financial support from Statoil for their ecological modelling work. They wish to thank Richard Furnas (Microcomputer Power, Ithaca, NY) and Warren L. Kovach (Kovach Computing Services, Pentraeth, Wales, UK) for help with developing the statistical approach presented in the Appendix. FMC thanks Andrew Lawrence, who redrew Figure 11.2, and Kathryn Sharp, who redrew Figures 11.3–11.7. Table 11.1 was adapted from a preparation procedure devised at Cheltenham by John Daniell.

EDITOR'S PREFACE

Micropalaeontology is an engaging discipline that is well suited to the study of environments and environmental change. Microfossils are fairly straightforward to collect from modern and fossil deposits, and their study enables undergraduate students to gain insight and valuable experience in general and specific field techniques and laboratory analysis, through sediment processing, microscopy and taxonomic identification. Post-data collection requires the use of various data-processing and statistical treatments (mainly computer-based), followed by mapping and evaluation. Therefore, the field of environmental micropalaeontology embeds many key skills expected of undergraduate study in geography and related environmental sciences, in a way that is exciting and rewarding to the student. It is this realisation that led to the initial idea for this book, and has encouraged the contributing authors to join me in fulfilling this end, for whereas environmental micropalaeontology is extremely valuable and appropriate at the undergraduate level, until now there hasn't been a supporting text. Hopefully, this book will cater for these growing needs. Also, the Quaternary time-frame used for this book should not suggest restricted use of the text, for the Quaternary Period embraces both the Pleistocene and Holocene Epochs, and therefore modern environments too. For example, at Bath Spa University College I employ environmental micropalaeontology in four apparently disparate modules: Quaternary Science, Global Environmental Issues, Climate and Environment, and Coastal Science and Management Issues. Such are the environmental applications of microfossils that it makes perfect sense to do so.

I am extremely grateful to many people who guided and encouraged me in the initial preparation for this book: the late Brian Funnell (to whom the book is dedicated) and Luciana O'Flaherty (formerly of Arnold Publishers); and those who helped to see the book through to fruition: Liz Gooster and Emma Heyworth-Dunn (both at Arnold), and my very good former and long-term colleagues at Bath Spa University College, Janice Ross, John Robb, Rick Curr, Tim Coles, Andy Skellern, Alan Marvell, Esther Edwards, Alex Koh, Fiona Strawbridge and Tina Jolly. Indebtedness, however, goes to my contributing authors: John Murray, Chris Smart, Elisabeth Alve, Roland Gehrels, Kevin Kennington, Ric Jordan, Jeremy Young, Barrie Dale, Amy Dale and Frank Chambers, without whom this book would not have been written. I would like to thank Paul Davies, who kindly reviewed the entire manuscript. I am also eternally grateful to my wife Sam and daughters Maya, Elinor and Rhiannon for supporting and tolerating my physical and mental absences while editing this book.

Simon Haslett
Newton Park, Bath
19th October 2001

BIG WORLD, SMALL FOSSILS: AN INTRODUCTION

Simon K. Haslett

Reconstructing changes in the environment – changes that have occurred over recent geological time – is fundamental in placing the modern environment in context; how has it evolved, what is the long-term trend, what is the frequency and magnitude of more rapid events, how should we view society's present concern for the environment, and how can we use the past to predict the future? Are present concerns justified, or has the Earth experienced environmental changes of greater magnitude in the past, and if so, why is society concerned? Issues such as environmental acidification, water pollution, air pollution, and the entourage of climate-change-related issues (i.e. variations in temperature and precipitation, increased storminess, sea-level change) all require benchmarks in the past, so that present conditions may be compared, differences measured and evaluated, and quantitative estimates of change made. Records of past environmental changes are to be found in a number of natural archives, such as ice sheets, coral reefs and sedimentary sequences. All of these accumulate through time, and information about the environment is incorporated: in ice as snow falls and traps atmospheric gases to become bubbles of 'fossil' air in the ice, in reefs as the chemical make-up of the coral skeleton, and in sediments as animal and plant fossils. It is this last archive that this book is concerned with – the use of fossils in the reconstruction of **palaeoenvironments**.

Palaeontology is the science of fossils, and all fossils of animals and plants relay some information to us about the environment in which they lived. They are often referred to as environmental **proxies**, because we cannot study the palaeoenvironment directly and so use fossils as indirect or proxy indicators of the palaeoenvironment. However, to arrive at a reliable palaeoenvironmental scenario, the fossils being employed should be abundant and well preserved in the sediment, as a single specimen may give extremely misleading information (some small algal fossils have been wind-transported from the Sahara Desert to Greenland!). Also, the modern ecology of the organism should be known in order to calibrate the fossil data. Furthermore, modern palaeoenvironmental studies often collect and use cores taken through sediment sequences (such as deep-sea sediments), and consequently the material available for study is often limited in volume. Therefore, ideal palaeoenvironmental proxies are those fossils whose modern ecology is understood, that are well preserved, and yield significant numbers in small-volume sediment samples. Large fossils (macrofossils) obviously do not match these criteria well, but microfossils do.

The study of microfossils is referred to as **micropalaeontology**, and microfossils are broadly defined as any fossil that requires a microscope for its study. Of course, under this definition, any fossil that is broken up into small pieces may be classed as a microfossil. However, for the purposes of this book, I have selected groups of organisms that may be classed as microfossils throughout their life-cycle, with the exception of pollen which, as sufferers of hay-fever will know all too well, is dispersed by flowering plants in vast quantities. Therefore, a number of fossil groups are excluded on these grounds, such as sponge spicules, plant phytoliths, ichthyoliths (fish bones), Coleoptera (beetles), and other insect groups. Also, as the temporal focus of this book is the Quaternary, pre-Quaternary microfossils such as conodonts are not included. Furthermore, this book is intended for undergraduate use, and therefore, in the interest of space and depth, only microfossils that are likely to be dealt with comprehensively at undergraduate level are included. This means that some very important 'true' microfossils that are employed mainly by professional researchers are omitted, such as testate amoeba.

The microfossil groups that are included in this book span the marine environment from the abyssal plains of the deep sea to the saltmarshes of the intertidal zone, the freshwater aquatic environments of rivers and lakes, and the terrestrial realm. Analyses of these microfossil groups can provide accurate Quaternary palaeoenvironmental reconstructions, and make significant contributions to studies of global environmental issues, such as climate change, sea-level change, acidification, water pollution, and a host of other applications in fields as diverse as oceanography and archaeology. In this book then, emphasis is placed on communicating the environmental applications of microfossils, so that little space is left for detailed descriptions of taxonomy, classification, anatomy and evolution. The reader is referred to the 'Further Reading' section at the end of this chapter for suggested books covering these topics.

The **Quaternary Period** (the most recent period of the **Cenozoic Era**) is the temporal focus here, because changes during this geological period have led up to and shaped the modern environment, and therefore are of great relevance in understanding global environmental issues. The start of the Quaternary has been dated to 1.81 million years ago; however, this is controversial because the Quaternary is often taken as being synonymous with the 'ice age', yet global ice volume had begun increasing markedly by 2.4 million years ago. Therefore, the end of the previous geological period, the **Pliocene Epoch** of the **Tertiary Period**, is often studied as part of the Quaternary. The Quaternary itself is subdivided into the **Pleistocene** and **Holocene Epochs**, the latter only beginning 10,000 years ago. The Pleistocene is characterised by cyclically alternating glacial and interglacial stages, but the Holocene is probably no more than a single ongoing interglacial. Suggestions for further reading on the Quaternary are given at the end of this chapter.

This book was conceived as a means to provide the undergraduate with a text suitable for use in a one-semester module. For this reason it is furnished with many topical case studies. It serves as a one-stop shop, so reducing pressure on library resources, especially interlibrary loan services, which frequently are not able to deliver requested articles before end-of-semester course-work submission dates and examinations. Also, several chapters include photographic plates of microfossil specimens, so that in a limited way the book may be used in the laboratory next to the microscope as an identification guide to environmentally important species. Chapter summaries, annotated guides to further reading, and keywords highlighted in **bold** are additional pedagogic features that the undergraduate will find beneficial. The book is structured so that each chapter deals with a different microfossil group. This represents the fact that active practitioners of environmental micropalaeontology are specialists in one or two microfossil groups only, and undoubtedly this is reflected in modules and courses that these and other lecturers currently deliver in higher-education institutions. It is conceivable, therefore, that students may only be interested in certain parts of this book – depending on the microfossil group(s) being studied in any given module. However, I did evaluate an alternative approach that would have examined different environments and the varied microfossil evidence pertaining to those environments. Although this approach was not adopted for editorial, academic and pedagogic reasons, it is still possible to use this book in this way, by cross-referencing between different chapters. The Appendix provides a very useful review of the statistical techniques that are applicable to environmental micropalaeontology.

FURTHER READING

Brasier, M.D., 1980. *Microfossils*. Unwin Hyman, London, 193 pp.
This a useful basic microfossil text covering all aspects of micropalaeontology, but is limited in depth.

Lipps, J.H. (ed.) 1993. *Fossil prokaryotes and protists*. Blackwell Scientific Publications, Oxford, 342 pp.
This is a well-illustrated textbook covering all aspects of single-celled microfossils. However, of course, the multicellular groups covered in our book are not dealt with therein.

Lowe, J.J. and Walker, M.J.C., 1997. *Reconstructing Quaternary environments* (2nd edn), Longman, Harlow, Essex, 446 pp.
This is a very comprehensive textbook on the Quaternary and reconstructing its environments using diverse methods.

Williams, M., Dunkerley, D., De Decker, P., Kershaw, P. and Chappell, J., 1998. *Quaternary environments* (2nd edn), Arnold, London, 329 pp.
This is similar to Lowe and Walker (1997) in its scope, but perhaps not of the same depth; however, it is very well written by a number of experts, and is a useful companion to the present book.

CHAPTER 2

INTRODUCTION TO BENTHIC FORAMINIFERA

John W. Murray

Foraminifera are single-celled organisms (protists), which have a long geological record (Cambrian to Recent) because they construct a shell (test) that can be preserved as a fossil. There are around 900 modern genera and 10,000 modern species. Foraminiferal tests are supplied to sediments in large numbers, so that small samples (such as those available from borehole cores) are usually adequate to provide a statistically acceptable number of individuals (≥250) for faunal studies. They have been used widely for biostratigraphical correlation and palaeoecology – not only in basic research but also in the search for petroleum. In recent years, they have become important in palaeoceanographic studies and the investigation of the effects of human impact on the environment.

2.1 WALL STRUCTURE AND CLASSIFICATION

The simplest tests have a wall of organic material which is not usually preserved in the fossil record. Tests that commonly fossilise are constructed either of tiny detrital sediment particles held together with cement (this being entirely organic or consisting of organic material plus secreted calcium carbonate – usually calcite), a structure that is given the term 'agglutinated' or 'arenaceous' – or the shell is entirely secreted by the foraminifer. In the latter case, the secreted mineral is normally calcite, but a few foraminifera produce aragonite. The calcareous forms are divided into two groups according to their wall structure.

Porcellaneous foraminifera were so named because their translucent walls sometimes resemble porcelain. The test wall comprises three layers: an inner layer of calcite crystallites arranged in a well-ordered sheet, a central layer of randomly arranged crystallites, and another sheet-like layer on the outside. There are no pores – this being an important feature that distinguishes porcellaneous forms from the other calcareous group, the hyaline foraminifera.

The **hyaline** foraminifera (hyaline means 'glassy') are so named because some have transparent shells. The calcite crystals are often arranged so that their long (c) axes are perpendicular to the wall surface. The test walls are also layered and contain pores with a transverse organic membrane that seals them. Thus the pores do not provide an open pathway from the inside to the outside of the test, although there may be diffusion of gases (e.g. oxygen) through them.

It should be emphasised, however, that the walls of both porcellaneous and hyaline foraminifera will appear to be opaque if the surface of the test is damaged through abrasion or slight dissolution, although they can be easily separated on the absence or presence of pores respectively.

Wall structure is the primary characteristic used to classify foraminifera into major groups. In modern and Holocene age (10,000 years and younger) benthic foraminifera, the most important groups are the Textulariina (agglutinated), the Miliolina (porcellaneous), and the Rotaliina (hyaline). The most commonly used classification is that of Loeblich and Tappan (1987). Although classification is never a primary objective in ecological studies, it is nevertheless essential to identify the foraminiferal genera and species correctly, because, otherwise, false interpretations will occur and comparison with other studies will be meaningless.

2.2 ECOLOGY

In foraminifera, the most conspicuous soft parts are the reticulopods or **pseudopodia**. These are protoplasmic extensions that form an actively streaming net (reticulum), which emerges from the aperture and spreads out around the test. The pseudopodia of foraminifera are unique in containing granules that migrate along their length. The term **'granuloreticulose'** is used to describe a pseudopodial net with granules. Pseudopodial nets are large in relation to the size of the test. Thus a 0.5-mm diameter test may have a pseudopodial net extending more than 1 cm. Pseudopodia provide the means for interaction with the surroundings, and are essential for movement, attachment, protection, constructing the test, gathering food, disposing of waste products, and respiration.

Benthic foraminifera live in a wide range of environments: from intertidal to abyssal; in or on the sediment (infaunal and epifaunal respectively); or on firm substrates such as other animals (shells), pebbles and plants, raised a little way above the sea-floor. In size, the shells of most of the fossilisable foraminifera range from a few tens of μm to a few mm, with most being less than 1 mm, but certain tropical forms collectively known as **'larger foraminifera'** reach sizes in excess of 1 cm in diameter. The global pattern of foraminiferal distribution is now reasonably well defined, but the understanding of exactly which environmental parameters are key in controlling distributions is far from complete. Most ecological studies are field based, meaning that samples are collected (often only during one time period) and compared with measured environmental parameters. For those forms living infaunally, the parameters need to be measured in the sediment rather than in the water above it. To do this, measurements can be made *in situ* (using an unmanned submersible and pushing a probe into the sediment), but more commonly cores are taken carefully to preserve the sediment/water interface, and measurements are made as soon as possible once the cores are on the ship. Another complication is that certain environmental parameters are closely linked (i.e. they covary). For instance, with increasing water depth there is often a decrease in water temperature, and, in the oceans, there is a decrease in the supply of food. In low-energy areas, muddy substrates are commonly rich in organic matter and low in oxygen. It then becomes difficult to unravel which environmental factor is the one that limits the foraminiferal population.

Although benthic foraminifera are sensitive to temperature as an environmental parameter, most shallow-water forms are adapted to tolerate the natural wide range experienced through the seasons (especially at mid latitudes). Nevertheless, it is clear that the major biogeographical distributions in lagoons, estuaries, deltas, and on the continental shelf are controlled by temperature. Other key factors are salinity, the nature of the substrate (which in part depends on the influence of tidally induced bottom currents and waves), the availability of oxygen (which is always limited with depth in the sediment, especially in muddy substrates, and related to the degree of water circulation), and the nature and abundance of food. Distributions of intertidal foraminifera appear to be controlled by salinity and by the duration of tidal submergence, or, inversely, by their tolerance to subaerial exposure. In oceanic areas, at water depths greater than 1000 m where the physical and chemical characteristics of the bottom waters are normally very stable, the amount of food input from surface-water primary production may be a key control on species distributions. In all environments, the number of individuals per unit area or volume of sediment depends partly on the amount of food available but also on biotic factors such as predation and competition.

Each species has its own individual requirements (lower and upper thresholds of tolerance) for a range of environmental parameters. This is collectively known as a niche. Ecologists distinguish between the **fundamental niche**, which is the ecospace where a species could potentially exist, and the **realised niche**, which is where the species really does exist and which is always a smaller part of the ecospace. At any one time, the factor or factors close to the threshold of tolerance for any given species are those that will limit its local distribution. It follows that for each species living in continually varying environments it is probable that different factors or a combination of factors may be limiting distributions both temporally and spatially. This in turn explains why, in such areas, there is a strong correlation between certain foraminifera and one particular factor, while in other areas there is not, and it also accounts for the lack of a consistent regional pattern of correlation between individual species and any single environmental factor.

In recent years, it has been recognised that the majority of foraminifera living on sediment substrates are present in the surface few millimetres, so in ecological studies it is essential not to lose this during sample collection. Other infaunal foraminifera live well below the sediment–water interface (down to at least 30 cm), but normally not in large numbers. Although in many environments there is depth stratification of taxa within the sediment, it is now clear that foraminifera are sufficiently mobile to track their niche optimum by moving vertically and/or laterally, e.g. in response to a moving redox boundary or in search of more food. Also, it is possible for foraminifera to live with their test in anoxic sediment and to extend their pseudopodia into the overlying oxygenated environment to feed and take up oxygen. Experiments have shown that where there is progressive oxygen depletion, causing the redox boundary to rise up through the sediment into the overlying water, infaunal foraminifera migrate upwards, and will even become epifaunal on structures projecting above the sediment surface in order to obtain oxygen (Alve and Bernhard, 1995).

The individuals of a species make up its **population**, and the populations of more than one species found together make up the **assemblage** or recurrent association. The number of species present in an assemblage is known as **species diversity**. The number of species found at a given locality depends partly on the size of the sample (i.e. the number of individuals present). The bigger the sample, the larger the number of individuals and the more species there will be. It is normal to express species diversity as a univariate statistical measure (a single figure) which takes these factors into account. Commonly used indices are the Fisher alpha index and the information function. The former can be calculated or read from a graph (see Murray, 1991) using the number of individuals and the number of species in the sample. The information function ($H(S)$) is calculated from the following formula:

$$H(S) = -\sum_{i=1}^{S} \cdot p_i \ln p_i$$

where p_i represents the proportion of a species (per cent divided by 100), and ln is the natural logarithm. Basically, the formal means add together the values of the proportion of each species multiplied by its log. The maximum value of H for any given number of species is attained when all species have equal abundance.

In general, species diversity is lowest in highly stressed environments (e.g. estuaries) and greatest in the least stressed environments (e.g. the deep sea). Likewise, the number of individuals for each species tends to be more equal in areas of high diversity, whereas high dominance of one species is commonly associated with low diversity. Attributes such as these are extremely useful in reconstructing past environments. Some general features of foraminiferal distributions are summarised in Table 2.1. However, it should be noted that the preparation of samples can affect the composition of the fauna. Certain small taxa such as *Stainforthia fusiformis* and *Epistominella* spp. are under-represented in samples prepared on sieves coarser than 63 μm (i.e. 125 μm or greater).

Table 2.1 Summary of the main features of modern foraminiferal faunas (based on Murray, 1991, which also gives details of recurrent associations from around the world) Diversity is given as the Fisher alpha index. Agglut. = agglutinated; Porcell. = porcellaneous

Environment	Diversity	Agglut.	Porcell.	Hyaline	Key agglutinated genera	Key calcareous genera
Brackish marsh	$\alpha < 1–5$	Low to high	Absent to rare	Low to high	Jadammina Miliammina Tiphotrocha Trochammina	
Marine marsh	$\alpha < 1–2$	Low to high	Up to 50%	Low to high	Miliammina Arenoparrella	Simple miliolids
Hypersaline marsh	$\alpha < 1–7$	Low to high	Up to 100%	Low to high	Miliammina Arenoparrella	Simple miliolids
Brackish lagoon	$\alpha < 1–5$	Low to high	Very low	Low to high	Miliammina	Ammonia Elphidium Haynesina
Marine lagoon	$\alpha < 1–12$		High			Ammonia miliolids
Hypersaline lagoon	$\alpha < 1–6$		High			Ammonia miliolids
Inner shelf 0–100 m	$\alpha\ 3–19$	Low to high	Occas. >20%	Low to high	Bigenerina Cribrostomoides Eggerelloides Eggerella Reophax Textularia	Bolivina Brizalina Bulimina Cassidulina Cibicides Elphidium Globocassidulina Nonionella Rosalina Stainforthia Uvigerina
Outer shelf 100–200 m	$\alpha\ 5–19$	Low to high	Low	Mainly high		As for inner shelf, plus: Hyalinea Trifarina
Upper slope (upper bathyal) 200–2000 m	$\alpha\ 1–>20$	Low	Very low	High		As for outer shelf, plus: Ehrenbergina Gyroidina Hoeglundina Nuttallides Oridorsalis Pullenia Reussella
Lower slope (lower bathyal) 2000–4000 m	$\alpha\ 1–22$	Low to high	Low	Low to high	Cyclammina Hyperammina Rhabdammina	Alabaminella Bulimina Cibicidoides Epistominella Hoeglundina Nuttallides
Abyssal >4000 m		Low to high	Absent	Low to high	Cyclammina Hormosina	Alabaminella Cibicidoides Epistominella Hoeglundina Nuttallides

Table 2.1 Continued

Environment	Diversity	Agglut.	Porcell.	Hyaline	Key agglutinated genera	Key calcareous genera
Dysoxic, various water depths and settings	Low	Mainly low	Absent	High		Bolivina Chilostomella Epistominella Globobulimina Loxostomum Nonionella Stainforthia Suggrunda

Foraminiferal assemblages and the environments in which they occur show a natural variability, which fluctuates around a mean on a time-scale ranging from daily to decadal. Environmental change requires that the mean itself alters through time. The causes of environmental change can be natural or induced by human activity. In order to reconstruct the timing and rates of environmental changes, it is necessary to date the sediments. Methods for doing this include stable-isotope stages, AMS radiocarbon dating, amino acids, ^{210}Pb and ^{137}Cs, for successively younger deposits (Williams *et al.*, 1998). However, it should not be assumed that these methods always provide a reliable date. For instance, the stable-isotope record is reliable for normal marine salinities, but corrections are required for brackish or hypersaline water. One problem is that reworking of older foraminifera into younger deposits can lead to false AMS dates and can affect amino-acid results. Macrofaunal bioturbation mixes sediment and blurs the record. Anoxic sediment which has not been reworked by bioturbation is the optimum environment for reliable dating by ^{210}Pb and ^{137}Cs. In order for useful dating results to be obtained, it is necessary to collect cores from an area of net sediment accumulation (i.e. as opposed to areas of erosion or non-deposition).

The term **proxy** is used for a chemical or biological attribute of the fauna which covaries with an environmental parameter. Proxies are potentially very useful for determining past environmental attributes using fossil forms. The chemical pathway is when an element is incorporated into a shell in proportion to an environmental parameter, and it operates over the whole range of that parameter (e.g. stable isotopes, nutrient proxies Cd and Ba, etc.). Such proxies are not dependent on the abundance of the shells; all that is required is enough shells to give analytical precision. By contrast, the biological pathway involves covariation between either the abundance of an organism – or the composition and structure of an assemblage of organisms – and a given environmental parameter. At present, benthic foraminifera are being used as proxies for various parameters (Table 2.2) and more refinements are under investigation. There are particular difficulties in using biological parameters as proxies for the flux of organic matter and availability of dissolved oxygen (Murray, 2001).

Table 2.2 Summary of foraminiferal proxies

Chemical parameter	Proxy for	Other uses
$\delta^{18}O$	Ice volume Water temperature	Isotope stratigraphy; high- resolution stratigraphy
$\delta^{13}C$	Age of water-masses Palaeoproductivity Sources of C (terrestrial vs. marine)	
Cd	Phosphate in water column	
Ba	Alkalinity	
Mg/Ca	Potentially for water temperature	
Sr/Ca	Under investigation	
$^{87}Sr/^{86}Sr$	Flux of Sr from rivers	Stratigraphy (correlation)
Biological parameter	**Proxy for**	**Other uses**
Marsh assemblages	Former sea-levels	Vertical land movements due to earthquakes
Taxa tolerant of low oxygen	Dysoxia	
Assemblages of taxa related to abundance of food	Flux of organic C and palaeoproductivity (but mainly deeper than 1000 m)	
Epistominella exigua and Alabaminella weddellensis	Seasonal input of food (phytodetritus) (deep sea)	

Palaeoecological studies are based on the assumption that 'the present is the key to the past'. For studies of Holocene, Quaternary and many Neogene deposits, this is a reasonable assumption. However, because of the increasing impact of pollution, care must be taken to select pristine modern analogues. There are several different kinds of palaeoecological study. In some cases it is desired to determine the environment of deposition. For this, as a starting point, fossil data can be compared with the summary information in Table 2.1 to determine the most likely environment. In other cases, the objective is to follow a progressive change, such as an increase in eutrophication, in essentially the same environment. For this, detailed comparisons with unpolluted modern analogues are essential.

2.3 METHODS OF STUDY

The choice of sampling strategy depends on the objective of the study. For ecological analysis, either the sediment surface or short cores should be collected carefully to avoid disturbing the sediment. Ideally, the samples should be of known area and thickness. They should be preserved in ethanol or buffered formalin. For studies of subrecent and fossil material, the samples should be of known volume or else should be dried and weighed before processing.

All samples need to be washed to remove the mud fraction. The sieve size used varies; most workers use either 63 or 125 μm, while some use 100, 150 or even 250 μm, but in order to retain small taxa it is desirable to use a 63-μm mesh. For ecological samples, the sediment retained on the sieve is stained with rose Bengal (to distinguish **live** individuals, in which the test contents are stained red, from **dead**, in which the empty tests remain unstained or take on a superficial slightly pink hue). Foraminifera can be extracted from the processed samples either while the sample is wet or after drying. Some researchers prefer the wet counting/picking method for intertidal samples as the high organic content of the sediment may cause clumping when dried. However, this can be overcome by spreading the sample thinly and drying at room temperature. Wet picking is also desirable if the aim is to study the whole foraminiferal assemblage (i.e. to include the fragile forms, which are destroyed by drying and normally do not fossilise, as well as the more robust ones that will readily fossilise). Examination of wet samples is very time-consuming, especially if there is a high terrigenous component in the sediment, and is only applicable to samples which contain living/stained foraminifera. Examination of dry samples is quicker because, if necessary, foraminifera can be concentrated using a heavy liquid such as trichloroethylene (although this must be done in a fume chamber for reasons of safety).

In most of the examples discussed below, the foraminifera were picked from dried samples. This is done by spreading the sample on a picking slide, examining it under a binocular microscope using reflected light, and picking out the foraminifera using a moistened fine paintbrush. The foraminifera are mounted on cardboard assemblage slides. Stained individuals representing living assemblages are treated separately from the dead assemblages. The living assemblage data can then be compared with measured environmental parameters. The dead assemblages give a time-averaged record of living assemblages; the dead may also have undergone post-mortem modification such as loss or gain of shells through transport or loss of shells through dissolution or destruction.

Because modern sediment samples used for live studies are not dried until after they have been processed, the foraminiferal results can be expressed only in relation to sample volume (e.g. 500 live individuals per 10 cm^3 of sediment). For fossil material, the samples are dried before processing, so that the foraminiferal results can be expressed as number per unit weight (e.g. 500 individuals per gram of sediment). Abundances of species can be either relative (per cent) or absolute (number per unit volume or unit weight of sediment). If the sediments are dated, it is possible to calculate the rate of accumulation of tests. The choice of methods for expressing the results depends on the objectives of the study. Numerical data can be analysed using multivariate techniques such as cluster analysis, principal component analysis, and non-metric multidimensional scaling (see Sen Gupta, 1999, chapter 5) or univariate measures such as the diversity indices described above. The most reliable interpretations are made using numerical data rather than just presence/absence (see Appendix).

2.4 SUMMARY

Foraminifera have a single cell (which is never preserved), housed in a shell that is commonly preserved because generally the wall is made either of material cemented together (agglutinated) or of secreted calcium carbonate (hyaline or porcellaneous). In Quaternary studies, a major use of benthic foraminifera is in determining former environments of deposition and reconstructing environmental or climatic change. The key to interpreting the past is a good understanding of modern ecology. Most Quaternary taxa still have living representatives, and this is true also of a large part of the Neogene fauna. Therefore, direct comparisons can made between the fossil and modern forms. Interpretations are more reliable if studies are based on numerical data.

FURTHER READING

Loeblich, A.R. and Tappan, H., 1987. *Foraminiferal genera and their classification.* Van Nostrand Reinhold, New York, 970 pp., 847 pls.
The most comprehensive taxonomic reference book.
Murray, J.W., 1991. *Ecology and palaeoecology of benthic foraminifera.* Longman, Harlow, Essex, 397 pp.
Sen Gupta, B.K. (ed.), 1999. *Modern foraminifera.* Kluwer Academic Publishers, Dordrecht, 371 pp.
Williams, M., Dunkerley, D., De Deckker, P. and Chappell, J., 1998. *Quaternary environments* (2nd edn), Arnold, London, 329 pp.

ENVIRONMENTAL APPLICATIONS OF DEEP-SEA BENTHIC FORAMINIFERA

Christopher W. Smart

The deep-ocean floor covers around 60 per cent of the Earth's surface. Topographically, the deep sea starts at the edge of the continental shelf, but in terms of hydrography it is usually considered to be the region below the permanent thermocline, around 1000 m, where conditions become more uniform and only change gradually (e.g. Douglas and Woodruff, 1981; Gage and Tyler, 1991). The ocean floor is a relatively stable environment in terms of its physical and chemical characteristics (e.g. temperature and salinity). Major differences, however, exist between continental margins (slope and rise) and abyssal regions, particularly with respect to the amount of organic matter arriving on the sea-floor, with higher fluxes occurring in marginal settings compared with the abyssal deep-sea. On a regional scale, the deep-sea environment exhibits considerable heterogeneity due to variations in topography, organic-matter inputs and bottom-current activity. The boundary between the benthic faunas of the continental margins and those that occupy the deep sea is not distinct. The benthic foraminifera occurring below middle bathyal depths (>1000 m) are considered to be deep-sea species, and it is these that are discussed in this chapter.

Benthic foraminifera are a highly successful, diverse and widely distributed group of marine protists. They dominate modern ocean-floor meiobenthic and macrobenthic communities (e.g. Gooday *et al.*, 1992; Gooday, 1999) and are the most abundant benthic deep-sea organisms preserved in the fossil record. Living species of deep-sea benthic foraminifera often contain numerous, delicate, soft-bodied agglutinated forms which are very rarely, if ever, preserved in the fossil record. However, fossil faunas usually consist of calcareous and more robust agglutinated taxa (Fig. 3.1).

The study of deep-sea benthic foraminifera has contributed significantly to our understanding of past oceanic and climatic conditions. They provide valuable information on the physical and chemical characteristics of the ocean-floor environment. The use of benthic foraminifera as palaeoceanographic and palaeoclimatic proxies essentially involves two approaches: (1) the study of the patterns of distribution, abundance, species composition and structure of the fossil assemblage, and (2) the analysis of the elemental and isotopic composition of their calcareous shells (tests). This chapter is primarily concerned with the faunal assemblages of benthic foraminifera, i.e. the former approach. However, studies involving the integration of stable isotopes, geochemistry, other microfossils, and sedimentology, etc., with the benthic foraminiferal faunal records (multiproxy approach) ultimately offer the most powerful approach in our understanding of past oceanic and climatic conditions.

Many modern deep-sea benthic foraminiferal species have their origins in the Miocene and some in the Palaeogene. This allows for direct comparisons and interpretations to be made between modern faunal distribution patterns and those in the Quaternary and Neogene. However, in older sediments that contain few extant forms, the causes of patterns of distribution become increasingly difficult to interpret. Studies involving analyses of morphotypes, i.e. certain morphologies that are associated with particular epifaunal or infaunal microhabitats, offer a useful alternative approach (Corliss, 1985; Koutsoukos and Hart, 1990; Thomas, 1990; Kaiho, 1991). Much of our knowledge of Cenozoic palaeoceanography has stemmed from the work of the Ocean Drilling Program (ODP) and its predecessor the Deep-Sea Drilling Project (DSDP), which have recovered many high-quality, undisturbed, and, in many cases, complete sediment cores from the world's ocean basins. In Quaternary studies, deep-sea sediment cores are typically recovered using piston and gravity coring methods.

3.1 METHODS COMMONLY USED IN THE STUDY OF DEEP-SEA BENTHIC FORAMINIFERA

Although many of the standard preparation methods are adequately covered in textbooks such as Brasier (1980) and Murray (1991), some have important palaeoenvironmental implications and therefore need to be discussed.

3.1.1 Size-fractions

The choice of size-fractions has important ecological and palaeoecological implications. Some researchers use the >63 μm fraction, while others use the >125 μm, >150 μm and even >250 μm fractions. Due to the fact that certain benthic foraminifera are restricted to particular size ranges, it has been argued that in order to include the large number of smaller-sized and ecologically significant species in the analysis, only the >63 μm (fine) size-fraction should be used (Schröder et al., 1987; Sen Gupta et al., 1987). For example, small taxa (typically 63–125 μm), such as Epistominella exigua and Bolivina spp. and

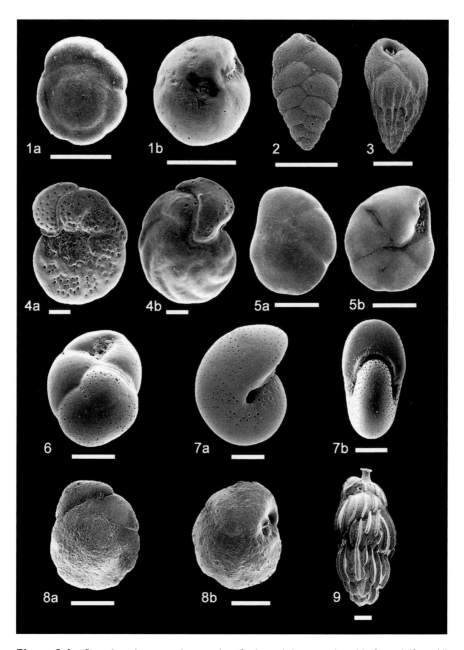

Figure 3.1 Scanning electron micrographs of selected deep-sea benthic foraminifera. All from Recent, Porcupine Abyssal Plain, Northeast Atlantic Ocean, 4800 m water depth unless otherwise stated. All scale bars = 100 µm. 1a, *Alabaminella weddellensis* (Earland), spiral view. 1b, *Alabaminella weddellensis* (Earland), umbilical view. 2, *Bolivina* sp., side view (Miocene, Somali Basin, northwest Indian Ocean, 1600 m water depth). 3, *Bulimina* sp., side view (Miocene, Walvis Ridge, southeast Atlantic Ocean, 3000 m water depth). 4a, *Cibicides wuellerstorfi* (Schwager), spiral view. 4b, *Cibicides wuellerstorfi* (Schwager), umbilical view.

elongate forms (e.g. *Stilostomella* spp., *Pleurostomella* spp., *Fursenkoina* spp.) are under-represented or even absent in studies that employ size-fractions of >125 μm (e.g. Thomas, 1986; Gooday, 1993; Smart and Murray, 1994). Some workers (e.g. Pawlowski, 1991; Moodley *et al.*, 1998) argue that the 32–63 μm size-fraction is very important because a large number of benthic foraminifera occur within this size range. The sieve size used is a compromise between the information gained and the time required for analysis. In other words, the finer the fraction used, the more information obtained, but it takes longer to acquire it. Clearly, in order to achieve reliable comparisons between benthic foraminiferal faunas of different workers, only the same size-fraction should be considered.

3.1.2   Assemblage Counts

Deep-sea benthic foraminiferal assemblages are usually represented by highly diverse, often low-dominance faunas (with no one species dominating the fauna) that are outnumbered by planktonic organisms by several orders of magnitude. It is therefore important to study (and therefore pick) large numbers of benthic foraminiferal specimens, in order to obtain a statistically valid representation of total species richness, and to include those species which are commonly represented by only one or two specimens. Most authors are in agreement that in order to achieve statistically reliable assemblage data, a minimum of 250 specimens (ideally >300) are required per sample. Live and dead specimens are treated separately. Live individuals are usually determined using rose Bengal or fluorescent stains (e.g. Bernhard and Bowser, 1996). Dead specimens are not stained and do not fluoresce.

3.1.3   Abundances

Relative abundances refer to the proportion of species (or any other taxonomic group such as species groups, genera, etc.) in the assemblage, and are usually expressed as a percentage. Absolute abundances refer to the number of individuals in a unit area of sea-floor or volume of sediment. In fossil samples, only the weight of sediment can be used for estimates of the absolute abundance, and benthic foraminifera are usually recorded in terms of the number of specimens per gram of dry sediment. Relative and absolute abundance data provide different information (see e.g. Murray, 1973,

---

5a, *Epistominella exigua* (Brady), spiral view. 5b, *Epistominella exigua* (Brady), umbilical view. 6, *Globocassidulina subglobosa* (Brady), side view. 7a, *Melonis barleeanum* (Williamson), side view. 7b, *Melonis barleeanum* (Williamson), apertural view. 8a, *Nuttallides umbonifera* (Cushman), spiral view (Recent, Cape Verde Abyssal Plain, North Atlantic Ocean, 4500 m water depth). 8b, *Nuttallides umbonifera* (Cushman), umbilical view (Recent, Cape Verde Abyssal Plain, North Atlantic Ocean, 4500 m water depth). 9, *Uvigerina peregrina* Cushman, side view (Recent, northeast Atlantic Ocean, 1000 m water depth).

pp. 1–6) and are both useful methods, particularly when used together, in studies of modern and fossil benthic foraminifera.

3.1.4   Benthic Foraminiferal Accumulation Rates

The accumulation rate of benthic foraminifera expressed as number of specimens/cm^2 per ka is a useful method for estimating the flux of organic carbon to the sea-floor, and surface productivity (and palaeoproductivity) (Herguera and Berger, 1991; Herguera, 1992, 1994; Berger *et al.*, 1994). The benthic foraminiferal accumulation rate (BFAR) can be calculated as follows:

$$BFAR\ (number\ of\ specimens/cm^2\ per\ ka) = BF \times LSR \times DBD$$

where BF is the number of benthic foraminifera per gram of dry sediment, LSR is the linear sedimentation rate (cm/ka), and DBD is the dry bulk density (g/cm^3) of the sediment.

Herguera and Berger (1991) suggested that for each 1 mg of organic carbon arriving on the sea-floor, one benthic foraminiferal test (>150 μm) is deposited. However, as pointed out by Loubere and Fariduddin (1999b), the use of BFARs as a productivity indicator is dependent on the premise that a given benthic foraminiferal test, regardless of species, relates to a set amount of organic matter arriving on the sea-floor. Although such an interpretation requires further work, BFARs nevertheless provide a very useful tool, particularly in palaeoceanographic studies. It has been shown (Naidu and Malmgren, 1995), however, that in oxygen minimum zones (e.g. Oman Margin, Arabian Sea), the BFARs do not reflect productivity variations. In addition, BFAR values are affected by dissolution of calcareous tests. In the Arctic Ocean, Wollenburg and Kuhnt (2000) found that the highest BFAR values occurred in areas under permanent ice where the organic carbon flux was lowest, whereas seasonally ice-free areas subject to carbonate dissolution yielded low values. While BFARs appear to provide a good estimate of organic-matter flux and (palaeo)productivity in many areas, species and assemblages of benthic foraminifera serve as useful proxies of the type of organic-matter inputs (e.g. Caralp, 1989; Gooday, 1993, 1994).

3.1.5   Diversity

Diversity is a measure of the number of species in a sample and the distribution of specimens among species. Deep-sea benthic foraminifera are usually characterised by high diversity (e.g. Douglas and Woodruff, 1981; Gooday *et al.*, 1998; Gooday, 1999; Culver and Buzas, 2000). For example, modern core-top (0–1 cm) samples from well-oxygenated bathyal and abyssal areas typically have >100 live species (e.g. Gooday *et al.*, 1998; Gooday, 1999). Fossil samples have fewer species, primarily because many of the abundant and diverse modern, delicate, soft-bodied agglutinated species have a low fossilisation potential.

There are many measures of diversity (e.g. see Gage and Tyler, 1991; Murray, 1991). Common indices of diversity used in the study of fossil deep-sea benthic foraminifera include the following:

1. Fisher's alpha index (Fisher *et al.*, 1943) (a measure of species richness);
2. the information function, $H(S)$, based on information theory, which uses the Shannon–Weiner Function (Shannon and Weaver, 1963) (a measure of heterogeneity);
3. equitability (E') (Buzas and Gibson, 1969) (a measure of evenness);
4. Hurlbert's Index, $ES(n)$, which is concerned with the expected number of species in a rarefied sample of n individuals, usually 100 (Hurlbert, 1971).

Since benthic foraminifera have long fossil records, they provide valuable information on changes in benthic species diversity through geological time. This is important because modern diversity patterns are affected by both prevailing environmental factors acting on the living organisms over ecological time-scales and the historical development of the habitats over geological time-scales.

3.1.6 Planktonic/Benthic Ratios

The planktonic/benthic (P:B) ratio refers to the number of planktonic foraminifera relative to the number of benthic foraminifera in a sample, and provides a useful tool in palaeoceanographic reconstructions. Typically, in deep-sea sediment samples unaffected by dissolution, planktonic foraminifera outnumber benthic forms by at least 99 to 1. The P:B ratio increases with increasing water depth and increasing distance from the coast. Traditionally workers have used the P:B ratio to estimate water depth, although it is known that food supply (productivity) and dissolution strongly affect the signal (Berger and Diester-Haas, 1988; Van der Zwaan *et al.*, 1990, 1999).

3.1.7 Dissolution Indices

The dissolution of calcareous tests occurs in waters undersaturated with respect to calcium carbonate. Dissolution strongly influences the distribution and composition of calcareous fauna and flora in the oceans. It is therefore important to understand the influence of dissolution in the reconstruction of past environments. Planktonic foraminifera are more susceptible to dissolution than benthic foraminifera. It is known that certain benthic foraminifera (and planktonic foraminifera) are more susceptible to dissolution than others (Corliss and Honjo, 1981). Although the P:B ratio reflects a number of environmental factors (see above), it may be useful as a proxy of levels of dissolution (e.g. Thunell, 1976).

3.1.8 Multivariate Analyses

Various multivariate statistical techniques are commonly used in micropalaeontology. A review of some multivariate methods employed in ecological and palaeoecological research is given in the Appendix , and is also dealt with adequately elsewhere (Malmgren and Haq, 1982; Dillion and Goldstein, 1984; Davis, 1986; Shi, 1993; see also Loubere and Qian, 1997 for some limitations to techniques). For a review of the methods of data analysis, including multivariate techniques, as they apply to foraminiferal ecology, the reader is referred to Parker and Arnold (1999). Multivariate analyses provide a useful objective method for simplifying and arranging large data sets (e.g. species abundances) into groups/factors for palaeoenvironmental interpretation. Common techniques include cluster analysis, principal components analysis (including VARIMAX rotation), factor analysis, correspondence analysis, discriminant function analysis, regression analysis and multidimensional scaling. The technique chosen depends on the specific aims of the project. For example, for the differentiation of (palaeo)ecologically significant deep-sea benthic foraminiferal associations, principal component analysis is often used.

3.1.9 Transfer Functions

Transfer functions are mathematical functions relating species abundances to environmental parameters (e.g. temperature, organic-matter flux, etc.). Models which relate modern species distribution patterns to a given environmental parameter are developed, which can then be applied to the fossil record. The application of transfer functions for estimating sea-surface temperatures using planktonic foraminiferal data is a widely used and powerful technique (e.g. Imbrie and Kipp, 1971; McIntyre *et al.*, 1976; Thunell *et al.*, 1994; see Chapter 7 and Appendix). In benthic foraminiferal studies, transfer functions have been developed to calculate surface-water productivity and organic-matter fluxes using deep-sea benthic foraminiferal data (Herguera and Berger, 1991; Loubere, 1991, 1994; Kuhnt *et al.*, 1999; Wollenburg and Kuhnt, 2000). Clearly, this offers a useful technique in future investigations.

3.2 GEOCHEMISTRY OF DEEP-SEA BENTHIC FORAMINIFERAL SHELLS

The chemical composition (stable isotopes and trace elements) of calcareous foraminiferal shells (tests) is used extensively by palaeoceanographers to understand past oceanic and climatic conditions. Although the central theme of this chapter is the environmental application of benthic foraminiferal faunas and assemblages, the geochemistry of foraminiferal shells is so important and widely used in palaeoceanography that it is considered appropriate to mention it here. The reader is referred to a number of selected references which provide good background information on the techniques and some specific references which deal with the application of foraminiferal shell geochemistry to palaeoceanography and palaeoclimatology with particular reference to the Quaternary.

3.2.1   Stable Isotopes

The oxygen and carbon stable-isotope composition preserved in the shells of calcareous benthic and planktonic foraminifera has important applications in palaeoceanography. Palaeoceanographers use variations in the oxygen stable-isotope composition ($^{18}O/^{16}O$ ratio), which are expressed as deviations from a standard (denoted $\delta^{18}O$) of foraminifera to estimate past ice volumes and ocean temperatures. In addition, oxygen isotopes are widely used in stratigraphy for dating and correlation. Carbon stable isotopes ($^{13}C/^{12}C$ ratio expressed as deviations from a standard, denoted $\delta^{13}C$) are used to estimate carbon fluxes and to reconstruct water-mass circulation patterns. Stable isotopes have also proved useful in understanding biological aspects of foraminifera, including life processes. Microhabitat preferences of benthic foraminifera have proved useful for understanding stable-isotope ratios. Good general background information on foraminiferal stable isotopes is given in Murray (1991) and Rohling and Cooke (1999).

There are numerous publications on the application of benthic stable isotopes in Quaternary palaeoceanography and it is beyond the scope of this chapter to provide a review of these. Some useful examples include: Shackleton and Opdyke (1973), Curry and Lohmann (1982), Duplessy *et al.* (1984, 1988), Oppo and Fairbanks (1987), Zahn *et al.* (1987), Curry *et al.* (1988), Oppo *et al.* (1990), Zahn and Mix (1991), Charles and Fairbanks (1992), De Menocal *et al.* (1992), Ku and Luo (1992), Mackensen *et al.* (1993a, 1994), Sarnthein *et al.* (1994), Charles *et al.* (1996), Raymo *et al.* (1997), Lund and Mix (1998), McCorkle *et al.* (1998), Vidal *et al.* (1998), Matsumoto and Lynch-Stieglitz (1999), Flower *et al.* (2000), among many others.

3.2.2   Trace Elements

Studies of the trace-element composition of calcareous benthic (and planktonic) foraminifera shells provide a useful method for investigating past oceanic conditions. For a good review of trace elements in foraminiferal calcite, the reader is referred to Lea (1999). Table 3.1 provides a summary of selected benthic foraminiferal trace elements and their uses in Quaternary palaeoceanography. Benthic foraminiferal Cd/Ca ratios are particularly useful in palaeoceanographic studies as they provide a good proxy for nutrients (phosphate) and deep ocean circulation, since phosphate becomes depleted as water-masses age.

Table 3.1 Summary of selected trace elements found in benthic foraminifera, and their uses

Trace element	Proxy/uses	Reference
Cadmium (Cd)	Sea-water nutrients, phosphate, oceanic circulation changes	Boyle and Keigwin (1982, 1985/1986) Boyle (1988) Oppo and Rosenthal (1994) Rosenthal et al. (1997a)
Barium (Ba)	Sea-water nutrients, alkalinity, oceanic circulation changes	Lea and Boyle (1989, 1990a, b) Lea (1993)
Magnesium (Mg)	Temperature	Rosenthal et al. (1997b)
Strontium (Sr)	Sea-water chemistry, pressure	Rosenthal et al. (1997b)
Fluorine (F)	Temperature, salinity	Rosenthal et al. (1997b)
Zinc (Zn)	Sea-water chemistry, oceanic circulation changes	Marchitto et al. (2000)
Strontium isotopes ($^{87}Sr/^{86}Sr$)	Changes in Sr supply from rivers, stratigraphy	Raymo et al. (1988)
Boron isotopes ($^{10}B/^{11}B$)	pH	Sanyal et al. (1995)

3.3 FACTORS CONTROLLING THE DISTRIBUTION OF MODERN DEEP-SEA BENTHIC FORAMINIFERA

If we are to interpret the fossil record of deep-sea benthic foraminifera, it is clearly important to understand the environmental factors that control their distribution and abundance in the modern ocean. These factors are complex and have been debated for some time (see Schnitker, 1994; Gooday, 1994; Murray, 1995 for reviews). In early studies, benthic foraminiferal assemblages were related to water depth (e.g. Phleger, 1960). When it became apparent that bathymetric ranges were not consistent, attention turned to the physicochemical properties of water masses (e.g. temperature, dissolved oxygen, salinity, dissolved carbonate) as controlling factors (e.g. Streeter, 1973; Lohmann, 1978; Schnitker, 1974, 1980) and comparisons between modern and fossil assemblages were made to infer the existence of similar water-masses in the past (e.g. Schnitker, 1980; Weston and Murray, 1984; Murray, 1988). However, over the years, as more data were collected, these water-mass associations were found to be globally inconsistent. Furthermore, as yet, no model based on ecological theory has been developed that would explain how relatively small differences in the physicochemical properties of water-masses would control the distribution of benthic species.

More recently, it appears that the abundance and distribution of deep-sea benthic foraminifera is controlled largely by two inversely related parameters, the flux of organic matter (food) to the sea-floor, and the oxygen concentrations of the bottom water and sediment pore-water (e.g. Gooday, 1994; Jorissen et al., 1995; Gooday and Rathburn, 1999). It is believed, however, that food input is the most important variable, and oxygenation only becomes a significant factor when concentrations fall to low values (<1.0 mL/L) (Sen Gupta and Machain-Castillo, 1993; Gooday, 1994; Jorissen *et al.*, 1995; Gooday *et al.*, 2000). According to Levin and Gage (1998) on their studies of bathyal macrofauna (e.g. polychaetes, crustaceans, molluscs), oxygen controls species richness (i.e. presence or absence of species), whereas food controls the abundance of species. Van der Zwaan *et al.* (1999) came to the same conclusions in the case of benthic foraminifera. Food and oxygen levels also control the microhabitat preferences of species. Specific benthic foraminiferal assemblages are often related to particular types of organic-matter input or with low-oxygen settings.

Mackensen *et al.* (1995) and Schmiedl *et al.* (1997) have suggested that the most important environmental factors controlling the broad-scale distribution patterns of live (rose-Bengal-stained) benthic foraminiferal assemblages in the South Atlantic Ocean are:

1. lateral advection and bottom-water ventilation (which affect factors such as oxygen concentrations and temperature);
2. surface primary productivity and organic-matter inputs to the sea-floor;
3. bottom-water carbonate corrosivity (which will affect mainly calcareous taxa); and
4. the energy of the benthic boundary layer.

Loubere (1991, 1994, 1996) and Fariduddin and Loubere (1997) have proposed that benthic foraminiferal assemblages, defined by factor analysis, are strongly related to organic-matter fluxes to the sea-floor associated with surface-water productivity. Schnitker (1994) argues that deep-sea benthic foraminifera are good indicators of productivity in areas where productivity is high, although in areas of low or very uniform productivity the composition and distribution of benthic foraminifera are more clearly related to the distribution of bottom-water masses. It appears that a combination of various parameters of a region controls the regional composition of the benthic foraminiferal fauna. These parameters include organic-matter (food) fluxes (associated with surface-water productivity), oxygen concentrations of the bottom water and sediment pore-water, oceanographic differences (i.e. water-mass properties), and sedimentological characteristics (i.e. sediment type and grain size).

In a discussion on the niche of benthic foraminifera, critical thresholds and proxies, Murray (2001) argues that, in order to explain the patterns of distribution of benthic foraminifera, a consideration of a broad range of environmental factors is necessary. He also draws attention to the fact that it is often too simplistic to explain distribution patterns in terms of just a few factors, such as flux of organic matter and oxygen, particularly in shallow-water environments. Furthermore, within a local area, different factors may reach critical thresholds singly or in combination at different times and in different places. However, Murray (2001) suggests that in deep-sea environments, where there is less spatial and temporal variability, it is occasionally possible to isolate a single environmental factor controlling the distribution of benthic foraminifera. The most significant appear to be organic matter and oxygen.

Since organic matter (food) and oxygen appear to be important factors controlling the distribution and abundance of deep-sea benthic foraminifera, the following account discusses these variables in more detail. For further information on the biological and ecological aspects of deep-sea benthic foraminifera the reader is referred to the following reviews: Gooday *et al.* (1992), Gooday (1994), Gooday and Rathburn (1999), Loubere and Fariduddin (1999b), and Van der Zwaan *et al.* (1999).

3.3.1 Relationship of Benthic Foraminifera to Organic-matter Fluxes and Oxygen Concentrations

To a large extent, deep-sea communities are fuelled by organic matter derived from phytoplankton primary production in the euphotic zone. As this material settles out through the water column, it undergoes progressive degradation so that the flux arriving at water depths of >1000 m represents only a few per cent of the euphotic zone production (Loubere and Fariduddin, 1999b). In the modern ocean, the supply of organic matter (potential food) to the sea-floor appears to be a major factor controlling the abundance and distribution of deep-sea benthic foraminifera (Lutze and Coulbourn, 1984; Gooday, 1988, 1993; Herguera and Berger, 1991; Loubere, 1991, 1994, 1996; Fariduddin and Loubere, 1997; Gooday and Rathburn, 1999; Loubere and Fariduddin, 1999a, b; Van der Zwaan *et al.*, 1999). Like that of other deep-sea organisms (Rowe, 1983; Sibuet *et al.*, 1989), the abundance and biomass of benthic foraminifera is related to organic-carbon fluxes to the ocean floor (Altenbach, 1988, 1992; Altenbach and Sarnthein, 1989; Altenbach *et al.*, 1999). A closely related factor influencing deep-sea benthic foraminiferal faunas is oxygen availability, particularly in continental-margin settings (Sen Gupta and Machain-Castillo, 1993). On a broad scale, ocean-basin variations in bottom-water oxygen concentrations are caused by global thermohaline circulation patterns. Local or regional oxygen depletion is related, in part, to rates of biological meta-bolism, which are associated with organic-matter (food) fluxes. It appears, however, that food input is the most important variable controlling benthic foraminiferal faunas and that oxygenation only becomes a significant factor where high organic-matter fluxes lead to low-oxygen conditions (e.g. Sen Gupta and Machain-Castillo, 1993; Gooday, 1994; Jorissen *et al.*, 1995). It should be noted that the term 'food' refers to labile particulate organic matter which can be eaten by heterotrophic organisms, as opposed to nutrients (dissolved inorganic elements, e.g. N and P) which cannot. Particular species of benthic foraminifera have different food and oxygen requirements. Therefore, distinct assemblages are related to areas with different inputs of organic matter.

Organic-matter and oxygen concentrations control the faunal composition and microhabitat preferences of benthic foraminifera. In eutrophic areas (bathyal continental margins), particularly in oxygen minimum zones where oxygen is severely depleted due to high organic-matter inputs associated with upwelling, both food and low oxygen availability exert strong influences on the composition and microhabitat preferences of benthic foraminifera. In more oligotrophic (abyssal) regions food supply is more significant (Jorissen *et al.*, 1995; De Stigter, 1996) since oxygen is generally available.

3.3.2 Microhabitats

It has been shown that deep-sea benthic foraminifera occupy both epifaunal and infaunal microhabitats (see Jorissen, 1999a for a review). Epifaunal species live on the sediment surface, while some, e.g. *Cibicides wuellerstorfi* live on elevated structures above the substrate (Lutze and Thiel, 1989). Infaunal species are predominantly free-living, mobile and occupy different levels within the sediment down to depths of 10–15 cm below the sediment–water interface in well-oxygenated areas (e.g. Corliss, 1985, 1991). As argued by Buzas *et al.* (1993), the term 'epifaunal' should only be used for species that occupy elevated habitats (e.g. on mollusc shells, grains of sand, etc.) and species described as living in the top 1 cm of the sediment should be referred to as shallow infaunal.

The microhabitat preference of benthic foraminifera is important in stable isotope studies, as significant differences exist between the $\delta^{13}C$ values of epifaunal (e.g. *Cibicidoides* spp.) and infaunal taxa (e.g. *Uvigerina* spp.), which have important implications for palaeoceanographic reconstructions (see e.g. Berger and Wefer, 1988; McCorkle *et al.*, 1990, 1997; Mackensen *et al.*, 2000). In sediment pore-waters, the proportion of ^{12}C increases (i.e. $\delta^{13}C$ values become lower) due to the oxidation of isotopically light (^{12}C-enriched) organic matter. Therefore, a trend towards progressively lower $\delta^{13}C$ values exists from epifaunal through to shallow infaunal through to deep infaunal benthic foraminifera. These trends have been used to infer $\delta^{13}C$ pore-water gradients (associated with organic-matter inputs and oxygen concentrations) in the geological record (e.g. McCorkle *et al.*, 1990).

A number of workers have used terms such as epifaunal, epibenthic, elevated epibenthic, shallow infaunal, intermediate infaunal, deep infaunal and preferentially infaunal in a rather rigorous sense to differentiate vertical distribution patterns of benthic foraminifera. However, it has been demonstrated that these categories are oversimplistic, as benthic foraminifera show large spatial and temporal variations in their microhabitat and a number of taxa are known to migrate within the sediment (e.g. Barmawidjaja *et al.*, 1992; Linke and Lutze, 1993; Alve and Bernhard, 1995; Kitazato and Ohga, 1995; Ohga and Kitazato, 1997). In addition, the microhabitat preferences listed above are commonly based, not on actual observations of rose-Bengal-stained living individuals, but inferred from presumed relationships between test morphology and microhabitat preferences.

3.3.2.1 Relationships between Test Morphology and Microhabitats

A relationship exists between test morphology of calcareous deep-sea benthic foraminifera and their microhabitat preferences (Corliss, 1985, 1991; Corliss and Chen, 1988; Corliss and Fois, 1990; Rosoff and Corliss, 1992). Although less is known about agglutinated foraminifera, they also occupy different microhabitats (Jones and Charnock, 1985; Gooday, 1986, 1994). According to Corliss (1985, 1991) and Corliss and Chen (1988), epifaunal calcareous foraminifera are characterised by rounded, planoconvex or biconvex trochospiral tests, with large pores absent or restricted to one side of the test, or milioline forms. Infaunal calcareous species are rounded planispiral, flattened ovoid, tapered and cylindrical triserial, spherical, or flattened tapered biserial, and usually have pores distributed evenly over the test. There are, however, exceptions to these generalisations and the relationship between test morphology and microhabitat preferences is not as straightforward as once believed. Nevertheless, despite this, the morphotype approach offers a useful tool in palaeoenvironmental studies if the limitations are appreciated (see Jorissen, 1999a). For example, as suggested by Corliss and Chen (1988), higher relative abundances of infaunal species compared with epifaunal species may be used as an indicator of increased organic-matter fluxes to the sea-floor. Kaiho (1991, 1994, 1999) uses a benthic foraminiferal oxygen index (BFOI) based on morphotypes as a means for estimating dissolved oxygen concentrations in ancient sediments. However, Kaiho's approach is problematic, because he claims to be able to detect oxygen levels well above the critical threshold (c. 0.4–0.5 mL/L) where oxygen limitation becomes important.

3.3.2.2 Relationship of Microhabitats to Organic-matter and Oxygen Concentrations

The vertical distribution of benthic foraminifera within deep-sea sediments is believed to be controlled primarily by a combination of organic-matter (food) availability and oxygen concentrations (Shirayama, 1984; Corliss and Emerson, 1990; Jorissen et al., 1995), although other factors, such as competitive interactions with other organisms, may play a role (Gooday, 1986). A model explaining the microhabitat distribution in terms of these parameters was developed by Jorissen et al. (1995), which is known as the TROX (TRophic and OXygen) model (Fig. 3.2). In oligotrophic settings, most of the metabolisable organic matter is consumed at the sediment surface, and the underlying sediment contains little organic matter and is well oxygenated to a great depth. These food-limited areas are dominated by epifaunal species (intolerant of low oxygen), while infaunal species are uncommon. In mesotrophic regions, the depth of the sediment profile occupied by foraminifera deepens as organic matter is supplied to deeper sediment layers by bioturbation. In eutrophic settings, the flux of organic matter and the consumption of oxygen are high, which leads to oxygen depletion close to the sediment

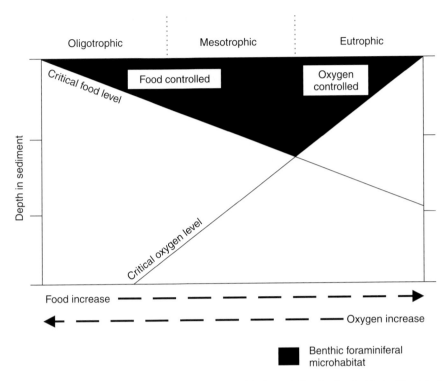

Figure 3.2 The TROX model explaining benthic foraminiferal microhabitats. The black area represents the living depth, which is related to organic matter (food) and oxygen concentration. (Modified after Jorissen *et al.* (1995).)

surface, and to the dominance of low-oxygen tolerant infaunal taxa. In such situations, the redox boundary shallows and the infaunal species move closer to the sediment surface. Van der Zwaan *et al.* (1999) proposed a modification of the TROX model, which they called TROX-2, where they attempted to identify the processes which affect the vertical distribution and abundance of benthic foraminifera. In this model, competition and redox gradients are regulated in turn by food supply.

It has been suggested that increases in the relative proportion of infaunal species of benthic foraminifera correlate with increases in organic-matter fluxes (e.g. Corliss and Chen, 1988; Jorissen *et al.*, 1995). However, as pointed out by a number of authors (Van der Zwaan *et al.*, 1990, 1999; Den Dulk *et al.* 1998), the proportion of infaunal taxa is related more to the storage of organic matter in the sediment rather than to the flux of organic matter. Furthermore, these authors (see also Naidu and Malmgren, 1995) showed that the abundance (BFARs) of epifaunal taxa is a reliable proxy of organic-matter fluxes (e.g. Herguera and Berger, 1991), except where high organic-matter fluxes lead to low-oxygen conditions. When oxygen levels start to become a limiting factor, epifaunal species are affected first and their abundances decrease, while the infaunal species are most tolerant and prevail for longer periods.

3.3.3 Faunal Associations

Major environmental and faunal differences exist between bathyal continental margin (slope and rise) settings and the abyssal central oceanic areas. Most noticeably, there is a significant difference in organic-matter fluxes between the two regions. In continental-margin settings, organic-matter inputs resulting from surface-water productivity and terrigenous inputs are considerably higher and are often associated with oxygen depletion in the surface sediments as a result of the oxidation of organic matter. In these settings, benthic foraminiferal faunas are dominated by infaunal, high-productivity/low-oxygen tolerant species. In abyssal central oceanic areas, and more oligotrophic continental margins, the flux of organic matter is not sufficient to lead to oxygen depletion in the sediments. These regions are food limited and variations in the population densities of benthic foraminifera are associated with seasonal food fluxes (phytodetritus) rather than oxygen. In these settings, the benthic foraminifera are dominated by epifaunal or shallow infaunal species which are most likely to be intolerant of low-oxygen conditions. In oligotrophic areas where seasonal food fluxes are not a significant factor, epifaunal, low-productivity taxa dominate. In these settings, the benthic foraminiferal assemblages may be related more to water-mass properties such as carbonate corrosivity and bottom-water ventilation (see Mackensen *et al.*, 1995).

3.3.3.1 Assemblages Associated with High Organic-matter Inputs/Low-oxygen Conditions

Low-oxygen conditions (dysoxic, O_2 0.1–1.0 mL/L) occur in shelf and slope areas where a mid-water oxygen minimum zone (OMZ) impinges on the sea-floor, in areas of high surface-water productivity (upwelling zones), in basins with sluggish circulation and in fjords. Reviews of the foraminifera found in these settings are given by Sen Gupta and Machain-Castillo (1993) and Bernhard and Sen Gupta (1999). Low-oxygen foraminiferal assemblages (LOFAs) tend to be characterised by high standing crops and low-diversity faunas (e.g. Phleger and Soutar, 1973). The high standing crops may be due to the exclusion of macrofaunal predators. Low-oxygen assemblages are typically dominated by calcareous species with infaunal microhabitat preferences, and contain notably low proportions of soft-shelled monothalamous taxa (Gooday *et al.*, 2000). Although no particular species occurs exclusively in low-oxygen environments (Sen Gupta and Machain-Castillo, 1993; Bernhard and Sen Gupta, 1999), there are certain calcareous taxa which are typically indicative of oxygen depletion, and these include *Bolivina*, *Bulimina*, *Cassidulina*, *Chilostomella*, *Epistominella*, *Globobulimina*, *Fursenkoina*, *Nonionella* and *Uvigerina*. Since these taxa also occur in oxygenated conditions, they only tend to be indicative of oxygen depletion when they are abundant. However, studies by Jorissen *et al.* (1998) and Gooday *et al.* (2001) have shown that abundant low-oxygen faunas can occur in areas overlain by fully oxygenated bottom water. Particular agglutinated species are also known to occur in oxygen-depleted environments (e.g. *Reophax, Spiroplectammina, Trochammina*), although they are not usually abundant (Phleger and Soutar, 1973; Bernhard, 1992; Bernhard *et al.*, 1997), at least in relative terms (Gooday *et al.*, 2000).

Benthic foraminifera that can withstand low-oxygen conditions have particular physiological, ultrastructural and morphological adaptations (Sen Gupta and Machain-Castillo, 1993; Bernhard and Sen Gupta, 1999). Experimental studies have shown that certain foraminifera can survive anoxic conditions (i.e. no oxygen) for considerable periods (Bernhard, 1993; Bernhard and Alve, 1996; Moodley *et al.*, 1997, 1998), although the physiological mechanism responsible for this is currently unknown. High sulphide concentrations, which are sometimes associated with low-oxygen conditions, may be an important limiting factor (Bernhard, 1993; Moodley *et al.*, 1998).

Bottom-water oxygenation and organic-matter (food) supply are closely interrelated variables, and their effects on benthic foraminiferal faunas are difficult to distinguish. For metazoan macrofauna, and probably for foraminifera too, oxygen availability becomes limiting only when concentrations fall to values well below 1.0 mL/L (probably <0.5 mL/L) (Levin and Gage, 1998). Above this level, oxygen appears to have no influence on the abundance and the presence or absence of particular species. As suggested by Jorissen (1999a), the percentage of infaunal low-oxygen tolerant species may be the most useful benthic foraminiferal proxy of oxygen concentrations.

Distinctive species assemblages also occur in areas of high surface-water productivity along continental margins where bottom-water oxygen concentrations are not severely depleted. However, oxygen may be depleted within the sediment, even if the overlying bottom water contains oxygen (see e.g. Jorissen *et al.* (1998); Gooday *et al.*, 2000). These are commonly referred to as 'high-productivity taxa' and, in particular, include *Melonis barleeanum* and *Uvigerina peregrina* (Lutze and Coulbourn, 1984; Lutze *et al.*, 1986; Mackensen *et al.*, 1993b; Schmiedl *et al.*, 1997). Similarly, taxa related to high organic-matter fluxes have also been shown to occur in open-ocean, well-oxygenated settings by a number of authors (Loubere, 1991, 1994, 1998; Fariduddin and Loubere, 1997; Loubere and Fariduddin, 1999a, b) (see below). Sen Gupta *et al.* (1981) recorded high relative abundances of a number of species (namely *Bolivina lowmani*, *Bolivina subaenariensis* and *Globocassidulina subglobosa*) on the upper continental slope (185 m) of northern Florida, which they related to high organic-matter fluxes associated with upwelling.

Uvigerina is an infaunal taxon associated with high organic-matter fluxes irrespective of oxygen concentrations (e.g. Miller and Lohmann, 1982; Van der Zwaan, 1982; Lutze and Coulbourn, 1984; Pedersen *et al.*, 1988; Rathburn and Corliss, 1994). It is known to occur in low-oxygen conditions (Sen Gupta and Machain-Castillo, 1993; Bernhard *et al.*, 1997), but it appears that high, continuous fluxes of organic matter to the sea-floor constitute the most important factor controlling its distribution. Similarly, *Melonis* spp. are also abundant in areas of high organic-matter fluxes where the sediments are oxygenated (e.g. Lutze *et al.*, 1986; Caralp, 1984, 1988). Caralp (1989) suggested that *Melonis barleeanum* is related to more degraded organic matter than a co-occurring species *Bulimina exilis*.

3.3.3.2   *Assemblages Associated with Well-oxygenated Conditions*

The supply of food also plays a major role in the distribution and abundance of benthic foraminiferal faunas in open-ocean (abyssal) environments where oxygen depletion of the bottom water is not a factor. The strong relationship between benthic foraminifera and organic-carbon flux to the sea-floor associated with surface-water productivity has clearly been demonstrated by the work of Loubere and co-workers (see Loubere and Fariduddin, 1999b for a review). Using multivariate analyses (e.g. principal component analysis, discriminant factor analysis and multiple regressions), modern benthic foraminiferal faunas have been related to organic-carbon fluxes and surface-water productivity in depth transects (2300–3600 m) in the Pacific Ocean (Loubere, 1991, 1994), Atlantic Ocean (Fariduddin and Loubere, 1997), Indian Ocean (Loubere, 1998) and global ocean (Loubere and Fariduddin, 1999a). These studies involved using samples from particular areas within limited water depths where surface ocean productivity was the only significant environmental variable, and thus eliminated the effects of other environmental controls on the benthic foraminiferal assemblages. In his later papers, Loubere (1998, 1999) focused on the effects of seasonality in food supply by comparing Indian Ocean (seasonal) and Pacific (non-seasonal) faunas. Loubere and co-workers have identified certain taxa associated with high productivity and low productivity areas which appear to be regionally and globally consistent ($r^2 = 0.97$, eastern Pacific, Loubere, 1994; $r^2 = 0.89$, global ocean, Loubere and Fariduddin, 1999a). Taxa associated with areas of high productivity include: *Uvigerina* spp., *Globobulimina* sp., *Bulimina alazanensis*, *B. mexicana*, *Melonis barleeanum*, *M. pompilioides* and *Sphaeroidina bulloides*. Low-productivity taxa include: *Nuttallides umbonifera*, *Bulimina translucens* and *Globocassidulina subglobosa*. These higher and lower productivity groupings of Loubere and co-workers have been found to be consistent with other studies (e.g. Mackensen *et al.*, 1993b; Schmiedl *et al.*, 1997).

It is well documented that in a number of oceanic settings there are seasonal (and occasionally unpredictable) inputs of food in the form of phytodetritus. With the exception of Loubere (1998, 1999), who discusses the importance of seasonality, the discussion presented above mainly considers the supply of organic matter (food) to the deep ocean floor as a relatively continuous phenomenon, i.e. mean annual flux to the sea-floor. Phytodetritus is phytoplankton detritus and other organic material resulting from surface production, and represents an important way in which organic matter is delivered from the euphotic zone to the ocean floor (e.g. Gooday and Turley, 1990; Gage and Tyler, 1991). The presence of phytodetritus, resulting from seasonal spring blooms, was first reported from the Porcupine Seabight of the NE Atlantic (1370–4100 m) (Billett *et al.*, 1983; Rice *et al.*, 1986), and is now known to occur in other areas, including the central oceanic NE Atlantic (Thiel *et al.*, 1990; Rice *et al.*, 1994), the NW Atlantic (Hecker, 1990), South Atlantic (Mackensen *et al.*, 1993b), the eastern North Pacific (Smith *et al.*, 1994), the equatorial Pacific (Smith, 1994; Smith *et al.*, 1996) and the bathyal western Pacific (Kitazato and Ohga, 1995; Ohga and Kitazato, 1997; Kitazato *et al.*, 2000), among other areas. Phytodetritus originates in the euphotic zone, primarily during the spring bloom, and settles rapidly over a period of a few weeks through the water column to form a light fluffy deposit with a patchy distribution on the ocean floor.

In the abyssal NE Atlantic (>4500 m) it has been shown that phytodetritus, which contains a relatively high proportion of labile organic matter, is exploited by certain calcareous opportunistic benthic foraminifera, notably *Epistominella exigua* and *Alabaminella weddellensis* (Gooday, 1988, 1993, 1994, 1996; Gooday and Turley, 1990). In the bathyal Porcupine Seabight (at a depth of 1340 m) other species respond to the presence of phytodetritus, notably *Eponides pusillus* (Gooday and Lambshead, 1989; Gooday, 1993). These 'phytodetritus species', as they have been called, react dramatically to the presence of phytodetritus arriving on the deep-sea floor by quickly colonizing and feeding on the detritus and subsequently reproducing rapidly and building up large populations (e.g. Gooday, 1988, 1993, 1996). *Epistominella exigua*, the best-known phytodetritus species, is abundant and widely distributed (almost cosmopolitan) in the modern ocean.

Other benthic foraminifera known to respond to phytodetritus in the NE Atlantic include the calcareous species *Fursenkoina* sp. (=*Stainforthia* sp.) and *Globocassidulina subglobosa* and the allogromiid *Tinogullmia riemanni* (e.g. Gooday, 1988, 1993, 1996; Gooday and Alve, 2001). In the bathyal western Pacific (Sagami Bay, Japan), benthic foraminifera associated with phytodetritus include the calcareous species *Bolivina pacifica* and *Stainforthia apertura* and the agglutinated species *Textularia kattegatensis* (Kitazato *et al.*, 2000). The calcareous 'phytodetritus species' are usually small (<150 μm) and have thin, smooth, hyaline walls reflecting an epifaunal (trochospiral) or shallow infaunal microhabitat. All 'phytodetritus species' are most likely to be opportunists that are able to respond rapidly to fluctuating (and unpredictable) food fluxes. It has been proposed that the distribution and abundance of these species may be related to phytodetritus in the modern ocean (Gooday, 1988, 1993, 1996; Mackensen *et al.*, 1993b; Schmiedl, 1995; Smart and Gooday, 1997; Loubere and Fariduddin, 1999a). The proposal that *E. exigua* may be used as a proxy of pulsed organic-matter fluxes in the fossil record (Smart *et al.*, 1994) has also been investigated by other workers (e.g. Mackensen, 1992; Thomas *et al.*, 1995; Thomas and Gooday, 1996).

3.4 APPLICATION OF DEEP-SEA BENTHIC FORAMINIFERA IN QUATERNARY PALAEOCEANOGRAPHY: CASE STUDIES

There have been numerous studies published on the application of deep-sea benthic foraminifera in palaeoenvironmental analyses, particularly in palaeoceanographic reconstructions. Our understanding of Cenozoic deep-water palaeoceanography has come from studies based on the analysis of deep-sea benthic foraminiferal faunal and geochemical data, and sedimentological records. As mentioned above, this chapter is concerned primarily with the uses of the benthic fauna (i.e. faunal composition, abundance and distribution) rather than shell geochemistry (i.e. stable isotopes, trace elements) in palaeoenvironmental studies with particular reference to the Quaternary record, i.e. the last *c.* 2 Ma of Earth's history.

Table 3.2 provides a summary list of selected published references on Quaternary deep-sea benthic foraminiferal faunas, which the reader can investigate further. In addition, a number of recently published specific case studies have been chosen that illustrate the ways in which benthic foraminiferal faunas have contributed to our understanding of Quaternary palaeoceanography and palaeoclimatology. These case studies are dealt with in detail and are indicated by asterisks (*) in Table 3.2. They have been selected from different ocean basins and they emphasise the importance that organic-matter fluxes resulting from surface production and oxygen concentrations have for the distribution and abundance of deep-sea benthic foraminifera. For a good review of Cenozoic deep-sea circulation based on benthic foraminifera, the reader is referred to Thomas (1992), and, for reviews and background information on Quaternary deep-water palaeoceanography, the reader is referred to Corliss *et al.* (1986) and Boyle (1990). Excellent examples of the application of deep-sea benthic foraminifera in palaeoceanography can be found in many issues of journals dealing with micropalaeontology and marine geoscience, e.g. *Journal of Foraminiferal Research*, *Marine Geology*, *Marine Micropaleontology* and *Paleoceanography*, among others.

3.4.1 Atlantic Ocean

The circulation pattern of the modern ocean is characterised by a thermohaline-driven conveyor-belt system initiated by the formation of North Atlantic Deep Water (NADW) in the northern North Atlantic Ocean. This circulation pattern affects global climate by transporting heat from the low-latitude North Atlantic to high northern and southern latitudes, and by influencing CO_2 partitioning between deep water and the atmosphere. Deep water also forms in the Southern Ocean, to give rise to Antarctic Bottom Water (AABW).

Table 3.2 Selected references on the application of deep-sea benthic foraminifera faunas in Quaternary palaeoceanography

Author(s)	Area of study	Size-fraction	Main aspects of study
Streeter (1973)	North Atlantic Ocean	>149 µm	Bottom-water-mass reconstruction – last 150 ka
Schnitker (1974)	West Atlantic Ocean	>125 µm	Bottom-water-mass reconstruction – last 120 ka
Corliss (1979)	Southeast Indian Ocean	>150 µm	AABW history – last 500 ka
Schnitker (1979)	West Atlantic Ocean	>125 µm	Bottom-water-mass reconstruction – last 24 ka
Streeter and Shackleton (1979)	North Atlantic Ocean	>150 µm	Bottom-water-mass reconstruction – last 150 ka
Schnitker (1980)	Global	Various	Review of Quaternary benthic foraminifera and bottom-water masses
Mullineaux and Lohmann (1981)	Eastern Mediterranean	>150 µm	Low-oxygen conditions, bottom-water circulation and sapropels – last 80 ka
Corliss (1982)	Southeast Indian Ocean	>150 µm	Bottom-water-mass circulation changes – 128–440 ka
Schnitker (1982)	North Atlantic Ocean	>125 µm	Bottom-water-mass circulation and climate change – last 24 ka
Sen Gupta et al. (1982)	Caribbean Sea, Atlantic Ocean	>63 µm	Stratigraphy and palaeoceanography – last 127 ka
Corliss (1983a)	Southeast Indian Ocean	>150 µm	Circulation of ACC – last 450 ka
Corliss (1983b)	Southwest Indian Ocean	>150 µm	Bottom-water-mass reconstruction – Holocene
Ross and Kennett (1983)	Eastern Mediterranean	>150 µm	Late Quaternary water-mass reconstruction
Weston and Murray (1984)	Northeastern Atlantic Ocean	>125 µm	Bottom-water masses – Recent and Neogene
Gaby and Sen Gupta (1985)	Eastern Caribbean Sea	>63 µm	Late Quaternary water-mass reconstruction
Corliss et al. (1986)	Global	Various	Review of Late Quaternary deep-ocean circulation
Murray (1986)	Northeastern Atlantic Ocean	>125 µm	Bottom-water masses – Late Neogene to Pleistocene
Murray et al. (1986)	Northeastern Atlantic Ocean	>125 µm	Bottom-water masses – Miocene to Recent
Caralp (1987)	Northeastern Atlantic Ocean	>250 µm	Bottom-water mass circulation changes – last 30 ka
Braatz and Corliss (1987)	Southeast Indian Ocean	>150 µm	Calcium carbonate undersaturation of bottom waters – last 3.2 Ma
Oggioni and Zandini (1987)	Eastern Mediterranean	>63 µm	Low-oxygen episodes – 5–250 ka; sapropels
Alavi (1988)	Sea of Marmara, Mediterranean	>63 µm	Organic carbon and oxygen variations – late Holocene
Caralp (1988)	Northeastern Atlantic and Western Mediterranean	>250 µm	Palaeoceanographic evolution: Atlantic side of Gibraltar Strait compared with Mediterranean side – last 18 ka

Table 3.2 Continued

Author(s)	Area of study	Size-fraction	Main aspects of study
Gupta and Srinivasan (1990)	Northern Indian Ocean	>150 μm	Palaeoclimatic and palaeoceano-graphic history – Pliocene – Pleistocene
Nolet and Corliss (1990)	Eastern Mediterranean	>150 μm	Environmental conditions and sapropels – 116–125 ka
Herguera and Berger (1991)	Western equatorial Pacific Ocean	>150 μm	BFARs and palaeoproductivity – glacial to post-glacial
Herguera (1992)	Western equatorial Pacific Ocean	>150 μm	BFARs and palaeoproductivity – glacial to post-glacial
Clark et al. (1994)	Southwest Pacific Ocean	>63 μm	Bottom-water masses – Holocene
Wells et al. (1994)	Southeast Indian Ocean	>150 μm	Palaeoceanographic and palaeo-climatic reconstruction – Late Quaternary
Hermelin and Shimmield (1995)	Northwest Arabian Sea, Indian Ocean	>125 μm	Productivity changes – last 150 ka
Stuck (1995)	Northeastern Norwegian Sea, North Atlantic Ocean	>125 μm	Last Glacial to Holocene record of benthic foraminiferal migra-tion into deep-sea environment
*Thomas et al. (1995)	Northeastern Atlantic Ocean	>63 μm	Productivity changes – last 45 ka
Almogi-Labin et al. (1996)	Red Sea	>149 μm	Bottom-water-mass changes – last 380 ka
Jian and Wang (1997)	South China Sea, Pacific Ocean	>150 μm	Bottom-water -mass reconstruction – last 200 ka
Nees (1997)	Tasman Sea, western Pacific Ocean	>150 μm	Palaeoproductivity and palae-oceanographic changes – last 200 ka
*Nees et al. (1997)	Northern North Atlantic Ocean	>125 μm	Reconstruction of sea-ice retreat during the last glacial–inter-glacial transition (last 20 ka)
*Schmiedl and Mackensen (1997)	Eastern South Atlantic Ocean	>125 μm	Productivity and deep-water circulation changes – last 450 ka
*Schmiedl et al. (1998)	Ionian Sea, eastern Mediterranean	>125 μm	Palaeoceanographic and palaeoclimatic reconstruction – last 330 ka
*Den Dulk et al. (1998)	Northern Arabian Sea, Indian Ocean	>150–595 μm	Productivity and oxygenation variations – last 120 ka; OMZ
Gupta (1999)	Eastern Indian Ocean	>149 μm	Palaeoceanographic changes – latest Pliocene to Holocene
Jian et al. (1999)	South China Sea, Pacific Ocean	>150 μm	Productivity and water-mass property changes – last 40 kyr
Jorissen (1999b)	Mediterranean	Various	Review of benthic foraminifera across Last Quaternary sapropels
Loubere (1999)	Eastern equatorial Pacific Ocean	>63 μm	Palaeoproductivity and palae-oceanography – last 30 ka

Table 3.2 Continued

Author(s)	Area of study	Size-fraction	Main aspects of study
Nees et al. (1999)	Southeastern Indian Ocean	>150 μm	Diatoms, benthic foraminifera and palaeoceanography – last 200 ka
Herguera (2000)	Eastern equatorial Pacific Ocean	>150 μm	BFARs and palaeoproductivity – Last Glacial
*Ohkushi et al. (2000)	Northeastern Pacific Ocean	>75 μm	Productivity changes – last 298 ka

AABW – Antarctic Bottom Water
ACC – Antarctic Circumpolar Current
OMZ – Oxygen Minimum Zone
BFAR – Benthic Foraminiferal Accumulation Rate
ka – 1000 years
* – further details given

During the glacial–interglacial cycles of the Quaternary, the oceanic environment underwent major changes. Evidence from changes in faunal patterns and the stable-isotope and trace-element composition of benthic foraminifera, particularly in the Atlantic Ocean, indicates that the formation of NADW was greatly reduced or absent during glacial periods, which contributed to cooling of high-latitude areas. These glacial intervals were also associated with higher surface-water productivity. Interglacials were characterised by high fluxes of NADW production and reduced surface-water productivity.

3.4.1.1 Productivity Changes during the Last 45,000 Years in the Northeastern Atlantic (50°N and 58°N)

Thomas et al. (1995) recognised major changes in the composition, abundance and diversity of deep-sea benthic foraminifera (>63 μm) in the northeastern Atlantic during the last 45,000 years, which they suggested were the result of variations in surface productivity rather than deep-water circulation. They studied the abundances (relative and absolute), BFARs and diversity of benthic foraminifera from two cores recovered as part of the Biogeochemical Ocean Flux Studies (BOFS): Site 5K (50°41.3'N, 21°51.9'W, depth 3547 m) and Site 14K (58°37.2'N, 19°26.2'W, depth 1756 m) at a time resolution corresponding to several hundreds to a thousand years. Major changes occurred in the abundances of *Epistominella exigua* and *Alabaminella weddellensis* ('phytodetritus species'), particularly at the deepest Site 5K, which they used as palaeoproductivity indicators.

During the last glacial maximum (LGM), the relative abundances and accumulation rates of 'phytodetritus species' were very low and benthic foraminiferal diversity was relatively high. This suggested that during this time the flux of phytodetritus to the sea-floor was low, and hence productivity was lower, probably because of extensive ice-cover. Similarly, the benthic foraminifera indicate major decreases in productivity during Heinrich events (periods of ice-rafting/meltwater formation) and a slight reduction during the Younger Dryas.

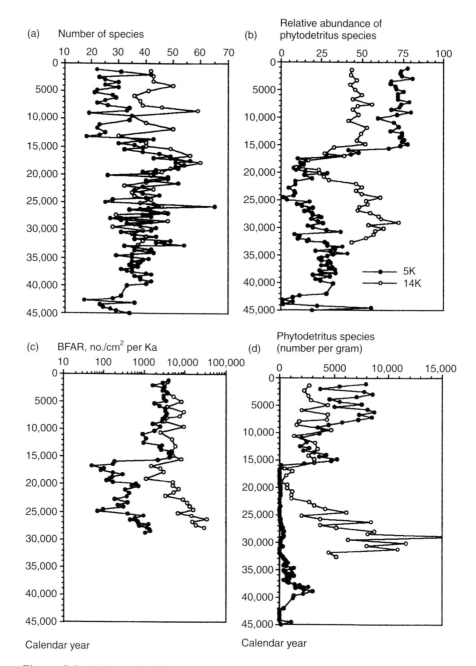

Figure 3.3 Variations in (a) species richness, (b) relative abundance (per cent) of 'phytodetritus species' (*Epistominella exigua* and *Alabaminella weddellensis*), (c) BFARs, and (d) absolute abundance of 'phytodetritus species' at Sites 5K and 14K, northeastern Atlantic Ocean during the last 45,000 years. Note that the BFAR plot has a logarithmic scale. (Reprinted from: *Paleoceanography*, v. 10, 1995, pp. 545–562, Thomas, E., Booth, L.,

During deglaciation, a major increase occurred in the absolute and relative abundance of 'phytodetritus species' and BFARs (Fig. 3.3). This increase was associated with a coeval decrease in the abundance of *Neogloquadrina pachy-derma* (sinistral) (a planktonic foraminiferal species indicative of cold/glacial conditions) together with an increase in the abundance of *Globigerina bul-loides* (a planktonic foraminiferal species indicative of warmer conditions and upwelling) (Fig. 3.4). Thomas *et al.* (1995) suggested that the increase in abundance of 'phytodetritus species' was associated with an increase in the supply of phytodetritus to the deep ocean floor (and thus probably of sur-face productivity) as the polar front retreated northwards across the coring sites.

Increases in the relative abundance of 'high-productivity/low-oxygen' taxa (*Pullenia*, *Cassidulina*, bolivinids, buliminids and uvigerinids) occurred during glacials (Oxygen Isotope Stages 2, 3 and 4) at both sites. These increases were negatively correlated with BFARs, and Thomas *et al.* (1995) believe that this indicates that productivity did not increase during glacials. During deglaciation, a slight decrease in the relative abundance of 'high-pro-ductivity/low-oxygen' taxa (calculated on a 'phytodetritus species'-free basis) occurred only at the deeper Site 5K. They proposed that decreased production of North Atlantic Deep Water during the LGM affected benthic foraminifera at Site 5K, but not at Site 14K.

3.4.1.2   Sea-ice Retreat during the Last 20,000 Years in the Norwegian–Greenland Sea, Northern North Atlantic (53°N to 76°N)

Nees *et al.* (1997) recorded a significant benthic foraminiferal (>125 μm) response to sea-ice retreat in the Norwegian–Greenland Sea during the last glacial–interglacial transition (Oxygen Isotope Stage 2/1, Termination I). They studied BFARs and abundances of *Cibicides wuellerstorfi* and *Melonis* spp. from four sites (M 23414: 53°32.2'N, 20°17.4'W, depth 2196 m; M 23068: 67°50.0'N, 1°30.3'E, depth 2231 m; M 23256: 73°10.3'N, 10°56.6'E, depth 2061 m; PS 1906: 76°50.1'N, 2°9.1'W, depth 2939 m) along a north–south transect at a time resolution corresponding to ±200 years.

During Termination I, a major increase occurred in BFARs and the accu-mulation rates of *Cibicides wuellerstorfi* and *Melonis* spp. The abundance peak appeared first in the southern Site (M 23414, Rockall Plateau at 12,400 years BP) and progressively moved northward, with peaks at 11,499 years BP at Site M 23068, 10,050 years BP at Site M 23256, and 8,900 years BP at the northern site in the southern Fram Strait (PS 1906). The mean velocity of movement

Maslin, M. and Shackleton, N.J.: Northeastern Atlantic benthic foraminifera during the last 45,000 years: changes in productivity seen from the bottom up, fig. 9, copyright (1995) American Geophysical Union, with permission from American Geophysical Union.)

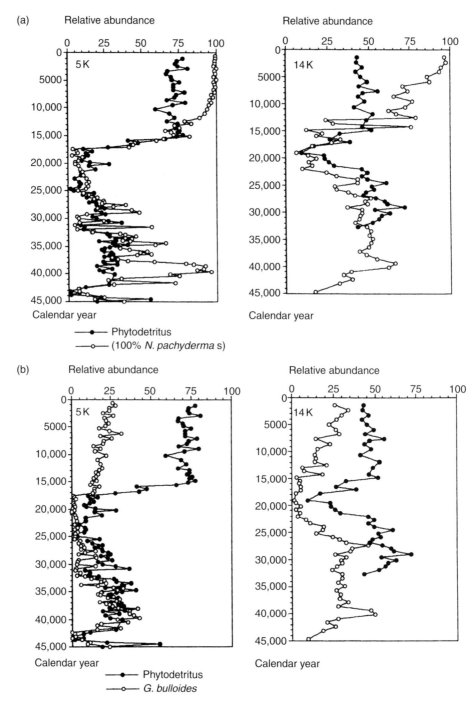

Figure 3.4 Comparison of the relative abundances (per cent) of 'phytodetritus species' (*Epistominella exigua* and *Alabaminella weddellensis*) and the planktonic foraminiferal species:

was calculated as 0.77 km/yr. The amplitude of peaks in BFARs also showed an increase from south (184 specimens/cm² per ka) to north (5863 per cm² per ka) (Fig. 3.5). Nees *et al.* (1997) interpreted the observed patterns in terms of a high-productivity area at the edge of the sea-ice cover which caused heightened organic-carbon fluxes (recorded as high BFARs), which moved progressively northward during gradual deglaciation (sea-ice retreat).

3.4.1.3 Productivity and Deep-water Circulation Changes During the Last 450,000 Years in the Eastern South Atlantic (23°S and 24°S)

Schmiedl and Mackensen (1997) used benthic foraminiferal faunas (>125 μm) and benthic stable isotopes from two areas in the eastern South Atlantic in order to investigate changes in productivity and bottom-water circulation patterns during the last 450,000 years. They studied two cores which were both located within present-day North Atlantic Deep Water (NADW) but under surface waters of distinctly different productivities. One core (GeoB1214: 24°41.4′S, 7°14.4′E, depth 3210 m) was located in the central Walvis Ridge area where surface productivity is low and the other (GeoB1710: 23°25.9′S, 11°41.9′E, depth 2987 m) was located in the southwest African continental margin area, where coastal upwelling causes moderately high organic-matter fluxes. Abundances of benthic foraminifera, including BFARs, were studied, together with epibenthic (*Cibicides wuellerstorfi*) oxygen and carbon stable isotopes at a time resolution corresponding to 6600–9100 yr for core GeoB1214 and 3700–4900 yr for core GeoB1710. The late Quaternary benthic foraminiferal data were compared with modern faunal distributions and $\delta^{13}C$ values from the South Atlantic (Mackensen *et al.*, 1990, 1993a, b, 1995; Schmiedl *et al.*, 1997) and were analysed using Q- and R-mode principal component analysis.

Schmiedl and Mackensen (1997) used the abundances of *Cibicides wuellerstorfi* and *Bulimina alazanensis* as proxies for NADW, and *Cassidulina laevigata*, *Melonis barleeanum* and *M. zaandami* as indicators of high surface-water productivity. They proposed that during the last 450,000 years the flux of NADW was restricted to interglacial periods, with enhanced flow occurring in Oxygen Isotope Stages 1, 9 and 11 (Fig. 3.6, NADW > 50 per cent). However, $\delta^{13}C$ records from ODP Site 607 (North Atlantic Ocean) suggest

(a) *Neogloboquadrina pachyderma* (sinistral) and (b) *Globigerina bulloides* at Sites 5K and 14K, in the northeastern Atlantic Ocean. Note that the percentage of sinistral forms of *N. pachyderma* out of the total *N. pachyderma* is plotted. (Reprinted from *Paleoceanography*, v. 10, 1995, pp. 545–562, Thomas, E., Booth, L., Maslin, M. and Shackleton, N.J.: Northeastern Atlantic benthic foraminifera during the last 45,000 years: changes in productivity seen from the bottom up, fig. 10, copyright (1995) American Geophysical Union, with permission from the American Geophysical Union.)

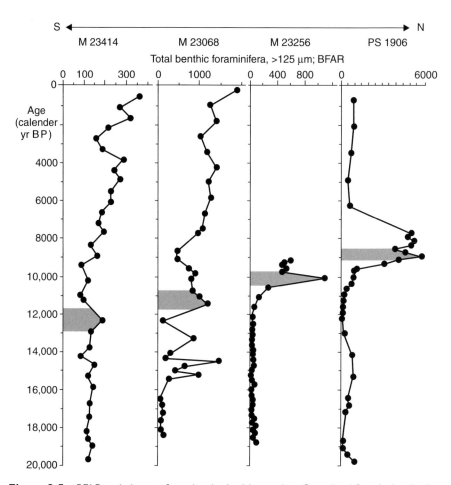

Figure 3.5 BFAR variations at four sites in the Norwegian–Greenland Sea during the last 20 ka. Maximum abundance peaks are represented by shaded areas in each core. Note how the peak appears first at the southern site and then migrates progressively northward. (Reprinted from *Geology*, v. 25, 1997, pp. 659–662, Nees, S., Altenbach, A.V., Kassens, H. and Thiede, J.: High-resolution record of foraminiferal response to late Quaternary sea-ice retreat in the Norwegian–Greenland Sea, fig. 2, copyright (1997) Geological Society of America, with permission from Geological Society of America.)

that an increase in NADW (>50 per cent) occurred during glacial Oxygen Isotope Stage 8 (Fig. 3.6).

Schmiedl and Mackensen (1997) also suggested that higher productivity and fluxes of organic matter occurred during glacial periods (Fig. 3.7). At the Walvis Ridge site (GeoB1214), organic-matter fluxes showed only slight changes during the last 450,000 years. Here the faunas were dominated by *Epistominella exigua*. During OI Stages 8, 10 and 12, increases in the abundance of moderately high-productivity faunas (*Uvigerina peregrina* and

Cassidulina laevigata) occurred in response to enhanced coastal upwelling and associated organic-matter fluxes. At the Southwest African continental margin site (GeoB1710), high organic-matter fluxes also occurred during glacial isotope stages 2, 3 and 6.0–6.5. During these glacials, the *Epistominella exigua* fauna was replaced by high-productivity faunas (*Cassidulina laevigata*, *Melonis barleeanum* and *M. zaandami*). In addition, these times were associated with particularly depleted epibenthic $\delta^{13}C$ values, which, as the authors note, occur in the modern South Atlantic Ocean in areas of enhanced seasonal phytodetritus production. Schmiedl and Mackensen (1997) also recognised that the long-term glacial–interglacial cycles are superimposed by high-frequency fluctuations in organic-matter fluxes with a periodicity of 23,000 years (i.e. precession orbital cycles).

3.4.2 Pacific Ocean

3.4.2.1 Productivity Changes During the Last 298,000 Years in the Northwestern Pacific 32°N

Ohkushi *et al.* (2000) recognised large fluctuations in the composition, abundance and diversity of deep-sea benthic foraminifera (>75 μm) in the northwestern Pacific during the last 298,000 years, which they related to changes in palaeoproductivity associated with the glacial–interglacial reorganisation of the Kuroshio gyre system and the southward advance of the Subarctic front within the area. They investigated the abundances (relative and absolute), BFARs and diversity (Hurlbert's index) of benthic foraminifera from one gravity core (NGC102: 32°19.84'N, 157°51.00'E, depth 2612 m) at a time resolution of approximately 6000 years. The relative abundance data were analysed using Q-mode VARIMAX factor analysis. The benthic foraminiferal faunal records were compared with published records of planktonic foraminiferal fragmentation, organic carbon (C_{org}) and biogenic opal contents, together with calculated sediment mass-accumulation rates (MAR) and planktonic foraminiferal oxygen stable isotopes.

The largest fluctuations occurred in the three most abundant species: *Epistominella exigua*, *Alabaminella weddellensis* and *Uvigerina peregrina*, which Ohkushi *et al.* (2000) suggested were the result of variations in productivity and seasonal productivity fluxes (i.e. phytodetritus production). The relative abundance peaks of *E. exigua* and *A. weddellensis* ('phytodetritus species') showed strong temporal differences (Fig. 3.8). In addition, the overall species diversity was found to be negatively correlated with percentages of *E. exigua* as expected, but not with *A. weddellensis* (Fig. 3.8). These differences were attributed to the different environmental preferences exhibited by the two species – perhaps their preference for different types of phytodetritus.

Peaks in the accumulation rate of *E. exigua* occurred during glacial stages 4, the middle of 8 and 6, and during interglacial stages 5 and 7 (Fig. 3.9). Peaks in the accumulation rate of *A. weddellensis* occurred in the early part of

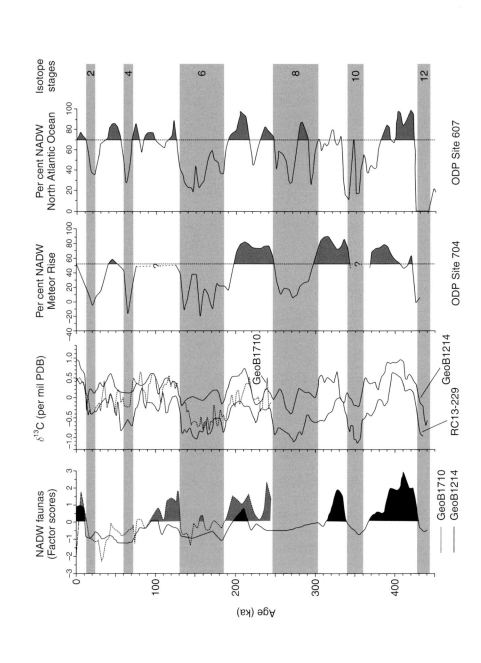

Figure 3.6 Variations in the NADW faunas (*Cibicides wuellerstorfi* and *Bulimina alazanensis*) and benthic δ^{13}C records at two sites (GeoB1214 and GeoB1710) in the eastern South Atlantic during the last 450,000 years. The δ^{13}C record of sites GeoB1214 and GeoB1710 is based on *C. wuellerstorfi* (Bickert and Wefer, 1996; Bickert, unpubl. data) and the δ^{13}C record of Site RC13–229 (Cape Basin) is based on *Cibicidoides* spp. (Oppo et al., 1990). Also shown are estimates of per cent NADW fluctuations derived from δ^{13}C records at the Meteor Rise (ODP Site 704; Hodell, 1993) and North Atlantic Ocean (upper NADW, ODP Site 607; Raymo et al., 1990). Significant increases are represented by shaded areas. Glacial stages are represented by shaded bands. (Reprinted from *Palaeogeography, Palaeoclimatology, Palaeoecology*, v. 130, 1997, pp. 43–80, Schmiedl, G. and Mackensen, A.: Late Quaternary paleoproductivity and deep water circulation in the eastern South Atlantic Ocean: evidence from benthic foraminifera, fig. 9, copyright (1997) Elsevier Science, with permission from Elsevier Science. Also reproduced from *Paleoceanography*, v. 5, 1990, pp. 43–54, Oppo, D.W., Fairbanks, R.G., Gordon, A.L. and Shackleton, N.J.: Late Pleistocene Southern Ocean δ^{13}C variability, fig. 1, copyright (1990) American Geophysical Union, with permission from the American Geophysical Union; *Earth and Planetary Science Letters*, v. 5, 1990, pp. 353–368, Raymo, M.E., Ruddiman, W.F., Shackleton, N.J. and Oppo, D.W. Evolution of Atlantic–Pacific δ^{13}C gradients over the last 2.5 m.y., fig. 2, copyright (1990) Elsevier Science, with permission from Elsevier Science; *Paleoceanography*, v. 8, 1993, pp. 47–67, Hodell, D.A.: Late Pleistocene paleoceanography of the South Atlantic sector of the Southern Ocean: Ocean Drilling Program Hole 704A, fig. 10, copyright (1993) American Geophysical Union, with permission from the American Geophysical Union; *The South Atlantic: present and past circulation*, Wefer, G., Berger, W.H., Siedler, G. and Webb, D.J. (eds), 1996, pp. 599–620, Bickert, T. and Wefer, G.: Late Quaternary deep water circulation in the South Atlantic: reconstruction from carbonate dissolution and benthic stable isotopes, fig. 7, copyright (1996) Springer-Verlag, with permission from Springer-Verlag.)

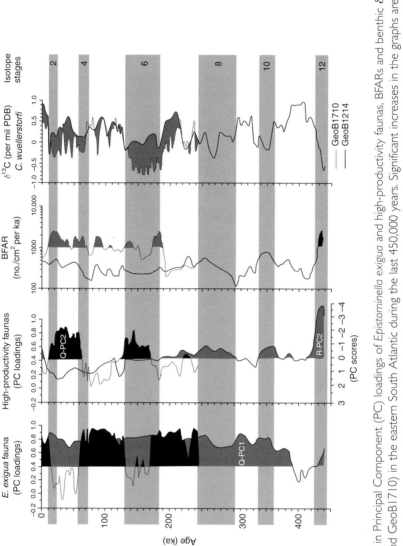

Figure 3.7 Variations in Principal Component (PC) loadings of *Epistominella exigua* and high-productivity faunas, BFARs and benthic δ¹³C values at two sites (GeoB1214 and GeoB1710) in the eastern South Atlantic during the last 450,000 years. Significant increases in the graphs are represented by shaded areas. Glacial stages are represented by shaded bands. (Reprinted from *Palaeogeography, Palaeoclimatology, Palaeoecology*, v. 130, 1997, pp. 43–80, Schmiedl, G. and Mackensen, A.: Late Quaternary paleoproductivity and deep water circulation in the eastern South Atlantic Ocean: evidence from benthic foraminifera, fig. 10, copyright (2000) Elsevier Science, with permission from Elsevier Science. Also reproduced from: *The South Atlantic: present and past circulation*, Wefer, G., Berger, W.H., Siedler, G. and Webb, D.J. (eds), 1996, pp. 599–620, Bickert, T. and Wefer, G.: Late Quaternary deep water circulation in the South Atlantic: reconstruction from carbonate dissolution and benthic stable isotopes, fig. 7, copyright (1996) Springer-

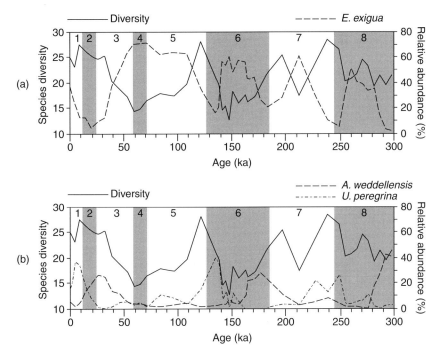

Figure 3.8 Comparison of (a) species diversity and relative abundance of *Epistominella exigua*, and (b) species diversity and relative abundance of *Alabaminella weddellensis* and *Uvigerina peregrina* at Site NGC102 in the northwestern Pacific Ocean during the last 298,000 years. Glacial stages are represented by shaded bands. (Reprinted from *Marine Micropaleontology*, v. 38, 2000, pp. 119–147, Ohkushi, K., Thomas, E. and Kawahata, H.: Abyssal benthic foraminifera from the northwestern Pacific (Shatsky Rise) during the last 298 ka, fig. 6, copyright (2000) Elsevier Science, with permission from Elsevier Science.)

glacial stages 8 and 6 and the late part of interglacial stage 3 (Fig. 3.9). Ohkushi *et al.* (2000) interpreted the abundance peaks of these 'phytodetritus species' as reflecting increased delivery of fresh, pulsed organic matter (phytodetritus) to the deep ocean floor in response to enhanced seasonal productivity caused by the southward movement of the Subarctic front over the Shatsky Rise during glacial periods and the associated changes in frontal systems and surface circulation.

In contrast, peaks in the accumulation rate of *U. peregrina* occurred with a periodicity of 100 ka at the end of glacial periods (end of stages 2, 6 and 8) and at the beginning of the present interglacial stage 1 and stage 7 (Fig. 3.9). Peaks in the accumulation rate of *U. peregrina* were found to correspond to increases in the MAR of organic carbon (Fig. 3.9). The authors proposed that during the end of glacial stages the area was characterised by year-round, high fluxes of organic matter, as opposed to pulsed, seasonal events, which are reflected in the abundance increases in *U. peregrina*. They suggested that

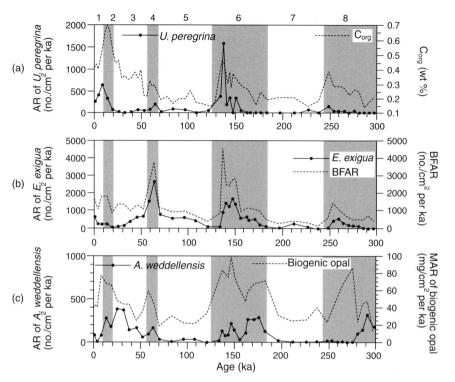

Figure 3.9 Variations in (a) the accumulation rate (AR) of *Uvigerina peregrina* and the organic carbon (C$_{org}$) content; (b) AR of *Epistominella exigua* and the mass-accumulation rate (MAR) of organic carbon; and (c) AR of *Alabaminella weddellensis* and the MAR of biogenic opal in core NGC102, northwestern Pacific Ocean during the last 298,000 years. Glacial stages are represented by shaded bands. (Reprinted from *Marine Micropaleontology*, v. 38, 2000, pp. 119–147, Ohkushi, K., Thomas, E. and Kawahata, H.: Abyssal benthic foraminifera from the northwestern Pacific (Shatsky Rise) during the last 298 ka, fig. 9, copyright (2000) Elsevier Science, with permission from Elsevier Science.)

these high-productivity events were the result of a greatly increased supply of dust (carrying nutrients) from the dry Asian continent at the end of glacial periods.

3.4.3 Indian Ocean

3.4.3.1 Productivity and Oxygenation Changes during the Last 120,000 Years in the Northwest Arabian Sea (Base of OMZ; 23°N, 1002 m)

The modern Arabian Sea, in the northwestern Indian Ocean, is characterised by an annually reversing wind system, associated with the Asian monsoon, which causes seasonal changes in oceanic upwelling and biological productivity. During the summer (southwest) monsoon, increased upwelling of nutrient-rich waters from depth results in high surface-water productivity and elevated fluxes of organic matter to the sea-floor in the northwestern Arabian Sea. During the winter (northeast) monsoon, productivity is generally reduced except in the northern Arabian Sea, where the surface waters are cooled, causing convective overturn and associated enhanced productivity.

High surface-water productivity together with moderate thermocline ventilation causes an intense Oxygen Minimum Zone (OMZ) between 150 and 1200 m where oxygen concentrations fall well below 0.5 mL/L. Where the OMZ impinges on the sea-floor, the bottom waters are severely oxygen depleted. The benthic foraminifera which occupy such environments are characterised by low diversities and high dominances (see above).

Den Dulk *et al.* (1998) studied the benthic foraminifera (150–595 μm) from the base of the OMZ in the Pakistan Margin, northern Arabian Sea, during the last 120,000 years. They analysed the relative abundances, species diversity, equitability and dominance patterns of benthic foraminifera from one piston core (NIOP455: 23°33′N, 65°57′E, depth 1002 m), at a time resolution of around 2700 years. The benthic foraminiferal data were compared with proxies for palaeoproductivity (C_{org} and abundances of *Globigerina bulloides* — a planktonic foraminiferal species indicative of intensity of upwelling) and bottom-water oxygenation (the preservation of pteropods – small pelagic gastropods with aragonitic shells, and sediment geochemistry of redox-sensitive elements – Mo/Al, V/Al and Mn/Al), together with MARs and planktonic foraminiferal oxygen stable isotopes.

The relative abundance data were subjected to principal component analysis, which produced two factors, although only the first factor yielded ecologically

Table 3.3 Summary of Assemblages 1 and 2 identified by Den Dulk *et al.* (1998)

Assemblage 1 ('High' O_2, 'low' C_{org})	Assemblage 2 ('Low' O_2, 'high' C_{org})
Bulimina striata	*Bulimina exilis*
Gavelinopsis lobatula	*Rotaliatinopsis semiinvoluta*
Chilostomella oolina	*Bolivina alata*
Monothalamous spp.	*Bolivina pygmaea*
Sphaeroidina bulloides	*Globobulimina* spp.
Cibicides ungerianus	*Bulimina* sp.
Hyalinea balthica	
Hoeglundia elegans	
Bulimina alazanensis	
Melonis barleeanum	
Quinqueloculina spp.	
Globocassidulina subglobosa	
Cassidulina carinata	
High diversity	Low diversity
Low equitability	Low equitability
Infaunal and epifaunal species	Infaunal species

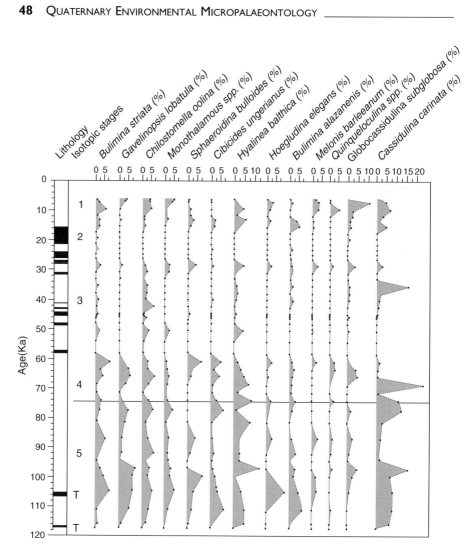

Figure 3.10 Variations in the relative abundances of Assemblage 1 ('high' O$_2$, 'low' C$_{org}$) from Site NIOP455, northwest Arabian Sea during the last 120,000 years. In the lithology column, white = homogeneous hemipelagic mud; black = laminated hemipelagic mud; T = turbidites. (Reprinted from *Marine Micropaleontology*, v. 35, 1998, pp. 43–66, Den Dulk, M., Reichart, G.J., Memon, G.M., Roelofs, E.M.P., Zachariasse, W.J., Van der Zwaan, G.J.: Benthic foraminiferal response to variations in surface water productivity and oxygenation in the northern Arabian Sea, fig. 3, copyright (1998) Elsevier Science, with permission from Elsevier Science.)

useful and meaningful species groups. This first factor was split into two distinct species assemblages: Assemblage 1, which revealed high positive loadings (>0.6), and Assemblage 2 which had high negative loadings (less than –0.36). Den Dulk *et al.* (1998) interpreted Assemblage 1 as indicating relatively oxygenated bottom-water conditions and low organic-matter flux, and Assemblage 2 as indicative of low bottom-water oxygen conditions and high

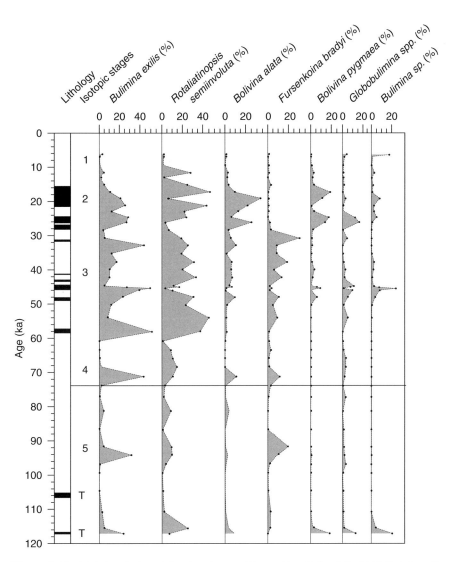

Figure 3.11 Variations in the relative abundances of Assemblage 2 ('low' O₂, 'high' C_org) from Site NIOP455, northwest Arabian Sea during the last 120,000 years. The relative abundance of *Fursenkoina bradyi* is also shown because it was found to be quantitatively important. In the lithology column, white = homogeneous hemipelagic mud; black = laminated hemipelagic mud; T = turbidites. (Reprinted from *Marine Micropaleontology*, v. 35, 1998, pp. 43–66, Den Dulk, M., Reichart, G.J., Memon, G.M., Roelofs, E.M.P., Zachariasse, W.J., Van der Zwaan, G.J.: Benthic foraminiferal response to variations in surface water productivity and oxygenation in the northern Arabian Sea, fig. 4, copyright (1998) Elsevier Science, with permission from Elsevier Science.)

fluxes of organic matter. The compositions of these assemblages are given in Table 3.3. Fluctuations in the percentage of Assemblage 1 are shown in Fig. 3.10, and variations in the percentage of Assemblage 2 are shown in Fig. 3.11.

Den Dulk *et al.* (1998) interpreted the fluctuations in the benthic foraminiferal faunas and other palaeoproductivity proxies during the last 120,000 years as reflecting variations in summer monsoon productivity and intensity of the OMZ. These records were found to vary with periodicities of 23,000 years (i.e. precession cycles).

Assemblage 1 (high oxygen, low C_{org}) dominated during isotope stages 1, 4 and 5 (Fig. 3.10) and was associated with reduced summer monsoon productivity and a weaker OMZ. Increased dominance of Assemblage 2 (low oxygen, high C_{org}) occurred during precession-driven increases in summer productivity and associated intensification of the OMZ. In addition, Assemblage 2 was a - significant component during glacial stages 2 and 3 (Fig. 3.11). Den Dulk *et al.* (1998) speculated that during these colder, glacial times an increase in the intensity of the winter (northeast) monsoon occurred, which resulted in enhanced winter productivity that, together with the precessional forcing of productivity, caused lowered bottom-water oxygen concentrations (see also section 4.2.1).

3.4.4   Mediterranean Sea

3.4.4.1   Palaeoenvironmental Changes During the Last 330,000 Years in the Ionian Sea (37°N)

The Mediterranean Sea is a semi-enclosed marginal basin, where the regional climate and topography control the hydrography and sedimentation processes. During the Quaternary, the climate of the region was characterised by humid interglacials (low $\delta^{18}O$ values) and arid glacials (high $\delta^{18}O$ values). In the eastern Mediterranean, the Neogene and Quaternary deep-sea record is notable for the presence of numerous cyclic layers of dark-coloured, organic-rich deposits known as sapropels. The formation of sapropels has been debated for some time. Most sapropels are associated with interglacials, although some formed during glacials. Bottom-water anoxia due to stagnation and/or increased organic-carbon flux associated with elevated surface-water productivity has been suggested as a possible explanation. It has been proposed that the formation of sapropels is related to changes in humidity associated with precessional orbital forcing of the African Monsoon. During increased northern hemisphere insolation, heightened seasonality and freshwater input occurred, which caused increased productivity and flux of organic carbon to the sea-floor, together with reduced deep-water formation (see Rohling, 1994 for a review).

Schmiedl *et al.* (1998) found large fluctuations in the composition, abundance and diversity of benthic foraminifera (>125 μm) in the Ionian Sea of the eastern Mediterranean during the last 330,000 years. The relative abundances and species diversity (*H(S)*) of benthic foraminifera were investigated from one piston core (M25/4-KL13: 37°33.2'N, 17°49.2'E; depth 2533 m) at a time resolution of approximately 4400 years (Fig. 3.12). The data were compared with planktonic (*Globigerina bulloides* and *Globigerinoides ruber*) foraminiferal oxygen and carbon-isotope records. Ten sapropel layers (S1–S10) were recorded from the core, which correlated with decreases in $\delta^{18}O$ values. The benthic foraminiferal abundances were examined using Q- and R-mode (VARIMAX-rotated) principal component analysis, which produced three and four main assemblages respectively (Figs 3.13 and 3.14).

Large fluctuations in both the benthic foraminiferal faunas and planktonic foraminiferal isotope records during the last 330,000 years were interpreted by Schmiedl *et al.* (1998) as representing significant environmental and climatic changes in the Ionian Sea. In particular, they suggested that the faunal records were related to changes in rainfall and river runoff associated with the Asian Monsoon System, and to long-term variations in the North Atlantic climate system (westerlies and anticyclones), which affected humidity and wind stress within the studied area. The most extreme environmental conditions were associated with the sapropel layers.

Interglacials (with sapropel layers S1–S5 and S7–S10) were associated with very low benthic foraminiferal numbers and diversities (Fig. 3.12), which Schmiedl *et al.* (1998) suggested were related to oligotrophic conditions. Following most sapropels, there was complete faunal recovery within a short time interval. Peaks in abundance of low-oxygen tolerant, deep infaunal taxa (*Globobulimina affinis* and *Chilostomella ovoidea*) were found to occur directly below or above the sapropel layers (Fig. 3.12), which were suggested to indicate lowered oxygen conditions associated with reduced bottom-water circulation. Schmiedl *et al.* (1998) related the abundance peaks in *G. affinis* and *C. ovoidea* to maxima in northern hemisphere summer insolation. A time lag of 1.8 ka below and 8 ka above sapropel formation was found between insolation maxima and faunal abundance peaks. In other words, faunal abundance peaks followed insolation maxima, which were found to be consistent with time-lag estimates of *c.* 3 ka between insolation maxima and onset of sapropel formation (Hilgen *et al.*, 1993).

Using the faunal and isotope data, Schmiedl *et al.* (1998) suggested that during the formation of interglacial sapropels S1–S5 and S7–S10, bottom-water circulation was strongly reduced. Most sapropels were found to be associated with low planktonic foraminiferal $\delta^{18}O$ and $\delta^{13}C$ values, which the authors related to increased freshwater fluxes resulting in lowered sea-surface salinities from the eastern Mediterranean borderlands following northern summer insolation maxima, and to reduced deep-water formation. Glacial sapropel S6 was also associated with a maximum in summer insolation, but it differed from all the other interglacial sapropels. Just below and above sapropel S6, the benthic foraminiferal numbers were high and the species were dominated by shallow and deep infaunal forms, e.g. *Bulimina exilis*, *Fursenkoina acuta*, *Nonionella turgida* and *Gyroidinoides neosoldanii*, which Schmiedl *et al.* (1998) interpreted as reflecting an abundant supply of organic matter to the sea-floor. Furthermore, the high planktonic $\delta^{13}C$ values during sapropel S6 suggest increased organic-matter fluxes. It was proposed that sapropel S6 was formed during glacial boundary conditions, i.e. low sea-level, a large Eurasian ice-sheet, and different wind stresses.

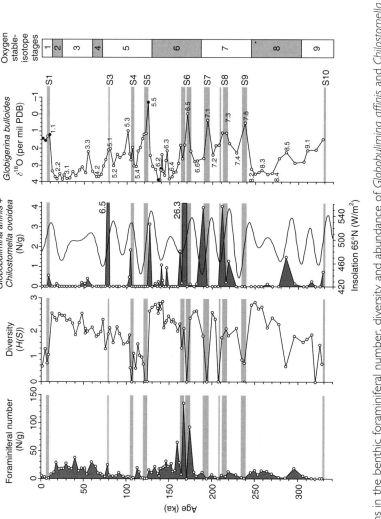

Figure 3.12 Variations in the benthic foraminiferal number, diversity and abundance of *Globobulimina affinis* and *Chilostomella ovoidea* (high-productivity species) at Site M25/4-KL13 in the Ionian Sea, eastern Mediterranean during the last 330,000 years. Also shown is the northern hemisphere summer insolation at 65° N, the planktonic foraminiferal (*Globigerina bulloides* – open circles, *Globigerinoides ruber* adjusted to *G. bulloides* – solid circles) δ¹⁸O record, stable oxygen-isotope stages (glacial stages shaded), and positions of the sapropel layers S1–S10. Significant increases are represented by shaded areas. (Reprinted from *Paleoceanography*, v. 13, 1998, pp. 447–458, Schmiedl, G., Hemleben, C., Keller, J. and Segl, M.: Impact of climatic changes on the benthic foraminiferal fauna in the Ionian Sea during the last 330,000 years, fig. 4, copyright (1998) American Geophysical Union, with permission from American Geophysical Union.)

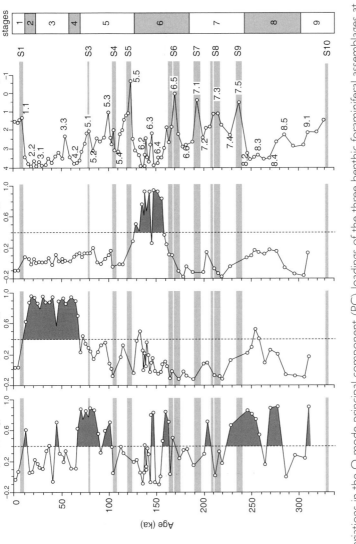

Figure 3.13 Variations in the Q-mode principal component (PC) loadings of the three benthic foraminiferal assemblages at Site M25/4-KL13 in the Ionian Sea, eastern Mediterranean during the last 330,000 years. Also shown is the planktonic foraminiferal (*Globigerina bulloides* – open circles, *Globigerinoides ruber* adjusted to *G. bulloides* – solid circles) $\delta^{18}O$ record, stable oxygen-isotope stages (glacial stages shaded), and positions of the sapropel layers S1–S10. (Reprinted from *Paleoceanography*, v. 13, 1998, pp. 447–458, Schmiedl, G., Hemleben, C., Keller, J. and Segl, M.: Impact of climatic changes on the benthic foraminiferal fauna in the Ionian Sea during the last 330,000 years, fig. 5, copyright (1998) American Geophysical Union, with permission from American Geophysical Union.)

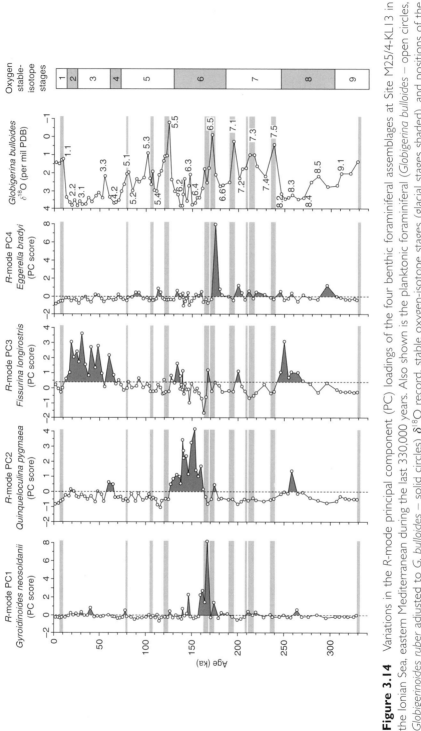

Figure 3.14 Variations in the *R*-mode principal component (PC) loadings of the four benthic foraminiferal assemblages at Site M25/4-KL13 in the Ionian Sea, eastern Mediterranean during the last 330,000 years. Also shown is the planktonic foraminiferal (*Globigerina bulloides* – open circles, *Globigerinoides ruber* adjusted to *G. bulloides* – solid circles) $\delta^{18}O$ record, stable oxygen-isotope stages (glacial stages shaded), and positions of the sapropel layers S1–S10. Significant increases are represented by shaded areas. (Reprinted from *Paleoceanography*, v. 13, 1998, pp. 447–458, Schmiedl, G., Hemleben, C., Keller, J. and Segl, M.: Impact of climatic changes on the benthic foraminiferal fauna in the Ionian Sea during the last 330,000 years, fig. 6, copyright (1998) American Geophysical Union, with permission from American Geophysical Union.)

Peaks in *G. affinis* and *C. ovoidea* were also recorded during cold isotope stages 8.5, 6.3 and 3.3, and are not related to sapropels, which Schmiedl *et al.* (1998) suggested were times of slightly increased freshwater fluxes and reduced bottom-water circulation.

Glacials were characterised by high benthic foraminiferal numbers and high-diversity faunas, with a dominance of shallow infaunal species indicative of mesotrophic conditions. It was suggested that during glacial periods the southward shift and intensification of the westerlies would have caused increased vertical mixing and associated enhanced surface nutrients, which would have led to increased fluxes of organic matter to the sea-floor.

3.5 SUMMARY

At the present state of understanding, it appears that the abundance and distribution of modern deep-sea benthic foraminifera is controlled largely by two inversely related parameters, the flux of organic matter (food) to the sea-floor and the oxygen concentrations of the bottom-water and sediment pore-water. Food input appears to be the most important variable controlling faunal abundance and possibly the presence or absence of certain taxa, whereas oxygenation only becomes a significant factor when concentrations fall to low values of <1.0 mL/L (probably <0.5 mL/L). Above this level, oxygen appears to have no influence on faunal abundance and the presence or absence of particular species. However, the situation is almost certainly more complex and it is unlikely that food and oxygen are the only controlling variables. A number of authors (e.g. Schnitker, 1994; Mackensen *et al.*, 1995; Schmiedl *et al.* 1997) argue that, in more oligotrophic oceanic settings, benthic foraminiferal faunas are influenced more by the physicochemical properties (e.g. temperature, salinity, dissolved oxygen content, carbonate corrosivity) and current flow of the ambient bottom-water. As Schnitker (1994) points out, benthic foraminifera are good indicators of productivity in areas where productivity is high, but in areas where productivity is low or very uniform the composition of the benthic fauna is clearly related to bottom-water mass structure. Most of the available data suggest that it is a combination of various parameters of a region that controls the composition of the benthic foraminiferal fauna. These parameters include organic-matter (food) fluxes (derived from surface primary productivity), oxygen concentrations of the bottom-water and sediment pore-water, oceanographic differences (i.e. water-mass properties), and sedimentological characteristics (i.e. sediment type and grain size).

In modern deep-sea benthic foraminiferal assemblages, it can be possible to distinguish distinct faunas associated with:

1. areas of high organic-matter fluxes and low bottom-water oxygen concentrations (infaunal, low-oxygen tolerant species);
2. areas of high organic-matter fluxes where bottom-water oxygen concentrations are not severely depleted (infaunal species);

3. well-oxygenated eutrophic regions where seasonal fluxes of phytodetritus occur (epifaunal and shallow infaunal species); and
4. well-oxygenated oligotrophic regions (low-productivity, epifaunal species).

However, it is not always possible to make these distinctions because
'low-oxygen' faunas also occur in areas where the bottom-water is well oxygenated (e.g. Sen Gupta *et al.*, 1981; Jorissen *et al.*, 1998; Gooday *et al.*, 2001).
Nevertheless, accepting the limitations, it may be possible to apply these
generalisations to fossil faunas.

It is almost certain that different benthic foraminiferal species have different food requirements, and it is highly unlikely that they are all competing
for the same food. It is likely that through resource partitioning, species are
selecting the type and size of food particles. It is also probable that each
species may be controlled in its abundance and distribution by different factors in different areas according to the local critical limits (see Murray, 2001).

For a species to act as a proxy for a given factor, it should show a linear
correlation between increased abundance and increase/decrease of the
factor. Since such correlations have not yet been established and neither have
transfer functions been determined (except for organic-matter fluxes and
sea-levels), all palaeoceanographic interpretations are inevitably very sub-
jective. Instead, they depend entirely on the assumptions being made. For
example, an abundance of *E. exigua* and/or *A. weddellensis* is taken as an
indication of the availability of phytodetritus, and an abundance of
Uvigerina is taken as an indication of high organic-matter input. Although
these generalisations may be true, it is also possible that some high abundances of these taxa are due to causes that have not yet been considered.
Also, in many modern studies an attempt is made to correlate the
foraminiferal data with only one environmental variable, so the outcome
may be predetermined. This is partly because in deep-sea studies very few
workers measure any environmental parameters. Usually, if there are any
environmental data they are derived from published records of the nearest
observation area (which is commonly a great distance away).

If certain species are always taken to indicate a particular environment then
there is only one possible environmental interpretation. In the same way, a
few decades ago, the correlation between benthic foraminifera and bottom-
water-masses seemed the best explanation, because there was a vague idea of
the distribution of water masses and there were few data on surface-water
productivity. As more data have been collected, this water-mass correlation
has been found to be globally inconsistent and more emphasis is placed on
the influence of food and oxygen availability. Furthermore, no model based
on ecological principles has yet been developed that would explain how relatively small physical and chemical differences in water-masses would control
the distribution of benthic foraminiferal species.

In palaeoceanographic studies some authors use similar data to derive quite
different interpretations. For example, in the Quaternary case studies presented
in this review, Nees *et al.* (1997) use the BFARs of *Cibicides wuellerstorfi*

and *Melonis* spp. as evidence of high productivity as the ice margin retreated. Conversely, Schmiedl and Mackensen (1997) used the abundances of *C. wueller-storfi* and *Bulimina alazanensis* as proxies for NADW. This poses the question: is *C. wuellerstorfi* both an indicator of high productivity (and how high is 'high'?) and of bottom water? In fact, other authors (Altenbach, 1988; Altenbach and Sarnthein, 1989; Altenbach *et al.*, 1999) have claimed that *C. wuellerstorfi* is a low productivity indicator. In addition, Lutze and Thiel (1989) showed that *C. wuellerstorfi* is associated with vigorous bottom currents because of its epifaunal suspension-feeding mode of life. These different interpretations are not surprising considering the comments of Murray (2001). However, in order to interpret fully the significance of faunal data it is necessary to evaluate the alternative possibilities and avoid relative (non-quantitative) information.

It is clear that the study of deep-sea benthic foraminifera has contributed significantly to our understanding of past oceanic and climatic conditions. However, a sound understanding of the environmental factors controlling modern deep-sea benthic foraminiferal species is essential if we are fully to understand and interpret their distribution and abundance in the fossil record. Although deep-sea palaeoceanography is very much in its infancy it does have exciting possibilities for the future.

FURTHER READING

The following books provide useful background reading and further information on various aspects of foraminiferal biology, taxonomy, ecology and palaeoecology.

Gage, J.D. and Tyler, P.A., 1991. *Deep-sea biology: a natural history of organisms at the deep-sea floor*. Cambridge University Press, Cambridge, 504 pp.
This is a useful account of the biology of the deep-sea environment, including infor-mation on the historical aspects of its study, the physical environment, methods of study and sampling, the organisms inhabiting the deep sea, and their patterns in space and time.

Jones, R.W., 1994. *The Challenger foraminifera*. The Natural History Museum and Oxford University Press, Oxford, 151 pp.
This beautifully illustrated book presents a revised taxonomy of the famous and clas-sic monograph by Henry Bowman Brady on his *Report of the foraminifera dredged by HMS Challenger during the years 1873–1876*, which was originally published in 1884. Brady's original plates are reproduced in colour, together with some additional fig-ures. This book is particularly useful for identification and taxonomic purposes of deep-sea (and other) benthic foraminifera. For a review of the book, the reader is referred to Smart, C.W., 1996. *Marine Micropaleontology*, **29**, 63–64.

Lee, J.J. and Anderson, O.R., 1991. *Biology of foraminifera*. Academic Press, London, 368 pp.
This is a useful book, with contributions from a number of recognised experts on various aspects of the biology of foraminifera, including their cell biology, life-cycle, form and function, mechanisms of test construction, motility, symbiosis, ecology and distribution.

Loeblich, A.R. and Tappan, H., 1987. *Foraminiferal genera and their classification*. 2 volumes, Van Nostrand Reinhold, New York, 970 pp., 847 pls.

This is the most comprehensive classification of foraminifera ever produced and, although it has been the subject of discussion, it is the current standard. A revised classification scheme was published by the authors in 1992: Loeblich, A.R. and Tappan, H., 1992. Present status of foraminiferal classification. In Y. Takayanagi and T. Saito (eds), *Studies in benthic foraminifera. Proceedings of the Fourth International Symposium on Benthic Foraminifera, Sendai, 1990 (Benthos '90)*. Tokai University Press, Japan, pp. 93–102.

Murray, J.W., 1991. *Ecology and palaeoecology of benthic foraminifera*. Longman, Harlow, Essex, 397 pp.

This is an excellent book, which synthesises and summarises a large volume of data on the ecology and distribution of benthic foraminifera in all environments and areas, as a basis for the interpretation of palaeoecology.

Sen Gupta, B.K. (ed.), 1999. *Modern foraminifera*. Kluwer Academic Publishers, Dordecht, The Netherlands, 371 pp.

This is an excellent, up-to-date book with contributions from a number of recognised experts on various aspects of modern foraminifera. From the deep-sea point of view, particularly useful detailed information is given on quantitative methods of data analysis (Parker and Arnold), microhabitats (Jorissen), organic-matter fluxes (Loubere and Fariduddin), low-oxygen faunas (Bernhard and Sen Gupta), stable isotopes (Rohling and Cooke), and trace elements (Lea).

Van Morkhoven, F.P.C.M., Berggren, W.A. and Edwards, A.S., 1986. Cenozoic cosmopolitan deep-water benthic foraminifera. *Bulletin des Centres de Recherches Exploration–Production Elf-Aquitaine, Memoir*, 11, Elf-Aquitaine Production, Pau, France, 421 pp.

This atlas provides useful taxonomic, biostratigraphical and palaeobathymetric information on mainly bathyal, cosmopolitan Cenozoic benthic foraminifera that are considered to be stratigraphically useful. The book is well illustrated with many SEM micrographs and line drawings.

CHAPTER

4

BENTHIC FORAMINIFERA AS INDICATORS OF ENVIRONMENTAL CHANGE: MARGINAL-MARINE, SHELF AND UPPER-SLOPE ENVIRONMENTS

John W. Murray and Elisabeth Alve

This chapter deals with the subtidal parts of estuaries and lagoons, continental shelf, and upper continental slope environments. In all these areas the water shows thermohaline circulation. This means that the waters may be layered based on their density (with the densest water forming the basal layer). Temperature and salinity determine the density – the higher the salinity, the greater the density; the higher the temperature, the lower the density. Therefore, the surface water will be warmer and/or less saline than the bottom water.

The distinction between estuaries and lagoons is primarily geomorphological; an estuary has an open mouth and receives river water, mainly from the inner part, whereas a lagoon is separated from the open sea by barrier islands and may have river water entering from the landward side. The continental shelf is the open region between the shore and the shelf break. The latter is a change in gradient which marks the beginning of the upper continental slope. The width of the continental shelf and the depth of the shelf break vary geographically. The latter may be as shallow as 120 m, but is commonly around 180–200 m, and is sometimes as deep as 500 m, depending on the tectonic setting. Shelf seas generally have normal salinity (around 35 psu, where psu are practical salinity units) but some are brackish (eastern North America) while others are hypersaline (Arabian Gulf). Over some temperate shelves the water is seasonally stratified, with a warm surface layer (up to 40 m thick) formed during the summer, separated by an abrupt temperature change (thermocline) from cooler bottom waters. During the autumn, storms destroy the thermocline and the warm water is mixed right the way down to the sea-floor, so that the bottom-water temperature increases slightly. In areas where tidal energy is sufficient to cause vertical mixing, the seasonal layer does not develop. The boundary between the seasonal layer and the adjacent vertically mixed water is termed a tidal front and is normally a region of higher nutrient content in the water, and therefore of higher primary productivity. The bottom water in contact with the upper continental slope is relatively cool and is of normal salinity (around 35 psu).

The distribution of foraminifera in modern environments is discussed by Murray (1991), Sen Gupta (1999), Martin (2000) and Scott *et al.* (2001), and recent advances in biology by Lee and Hallock (2000).

4. I TAPHONOMY

4.1.1 Transport

The sea-floor may be disturbed by waves and currents and the greatest disturbance is normally in shallow water. Most currents are driven by tidal flow and because of this they are commonly bidirectional, flowing in opposite directions according to the state of the tide. The effects of tides are small in deeper water, but in the shallow areas of the shelf and in estuaries friction causes the tidal wave to slow down and increase in height. Some estuaries have an extreme tidal range (macrotidal, >4 m), and some a very low tidal range (microtidal, 0–2 m) and in between are mesotidal examples (2–4 m). Benthic foraminifera smaller than 1 mm in size may be entrained by currents and transported as bed load. On the open shelf, the combined effects of tidal currents and waves may throw into suspension small tests (<200 μm in diameter or <300 μm in length). Transport has two consequences: the source area loses tests; the depositional area receives tests. If transport carries tests outside the area of origin, the depositional region will have a mixed assemblage comprising the indigenous and transported (exotic) components (Murray, 1991). It is therefore important to recognise the effects of transport when interpreting fossil assemblages. The criteria for recognising transport are: broken and abraded tests for bed load; thin-walled, size-sorted tests for suspended load. Also, shallow-water assemblages found in deeper water sequences may be the result of downslope mass transport.

Small, thin-walled tests are readily transported in suspension from a source area where the sea-floor is stirred by waves during storms and/or by tidal currents. Away from the source, transported tests are deposited in low-energy areas and commonly occur in estuarine mud deposits. Their abundance shows a general relationship with tidal range (Wang and Murray, 1983). For assemblages in mud substrates, microtidal estuaries contain <22 per cent exotic forms; macrotidal deposits generally have >50 per cent and the values for mesotidal deposits are intermediate (but both show a big range, 0–85 per cent). The exotic forms are derived from the adjacent continental shelf and are readily recognised because they do not live in the depositional area and they are size-sorted. Thus the macrotidal Severn estuary in western Britain has forms such as *Bolivina pseudoplicata, Bolivinellina pseudopunctata, Gavelinopsis praegeri, Cassidulina obtusa, Fissurina marginata, Trifarina angulosa, Lamarckina haliotidea* and *Miliolinella subrotunda* derived from the Celtic Sea shelf (Murray and Hawkins, 1976). In the adjacent Somerset Levels, Holocene deposits from the past 9000 years record environments from subtidal to marsh. Five forms transported in from the shelf are common in the estuarine subtidal deposits: *Buccella frigida, Bolivina pseudoplicata, Bolivinellina pseudopunctata, Fissurina marginata* and *Miliolinella subrotunda*. Thus the process of the onshore transport of foraminiferal tests in suspension has been operating throughout the past 9000 years (Murray and Hawkins, 1976). If these forms were not recognised as being transported, then the environmental interpretation of the estuarine subtidal deposits would be that they represent an inner-shelf environment.

In the early to middle Miocene of the Czech Republic, there are changes in the foraminiferal assemblages which have been related to variations in sea-level (Holcová, 1999). Unsorted and well-preserved tests are considered to be indigenous, but other types of preservation indicate transport: small, thin-walled, well-sorted tests indicate transport by suspension load; similar tests but showing abrasion are interpreted as having been transported in suspension and then abraded during later reworking; corroded, robust, size-sorted tests larger than 200 μm are considered to have been transported as bed load. In some parts of the shelf succession, fewer than 10 per cent of the tests are considered to be indigenous; the majority are considered to have been transported in as suspended load during a sea-level rise of around 100 m, with some reworking back to the shelf during times of lowered sea-level.

Off the west coast of Norway in the Storegga area there have been enormous submarine slides caused by slope failure triggered by earthquakes and/or release of gas from the decomposition of gas hydrates (Jansen *et al.*, 1987). The slides are well documented through seismic profiles and detailed bathymetric studies. The most recent events happened between 8 and 5 ka BP. Evidence of the extent to which material moved downslope can be seen in the incorporated foraminiferal assemblages. By comparison with the modern faunas it can be shown that shelf assemblages derived from 200–600 m water depth form around 30 per cent of the preserved assemblages in turbidites and debris flows from nearly 3000 m water depth.

4.1.2 Destruction of Tests

Apart from transport, the other main taphonomic change is loss of tests through destruction. Preservation of foraminiferal tests depends on microhabitat, biogeochemical conditions in the sediments, and test composition. Agglutinated tests with an organic cement can be broken down through bacterial degradation of the organic material leading to fragmentation of the tests. Tests may also be destroyed through metazoan predation. Fiddler crabs, for instance, have been suspected to be partly responsible for breaking foraminiferal tests in half (Goldstein and Watkins, 1999). Calcareous tests may be dissolved if the pH of the sediment pore-water or the bottom water becomes acidic. In the extreme case of total loss of all calcareous tests, a residual assemblage may be left, composed entirely of non-calcareous agglutinated forms, as in some shelf areas around Antarctica. However, even if all calcareous tests are lost through dissolution, the remaining agglutinated assemblages may still provide useful environmental information. In shallow-water sediments, evidence of dissolution loss of calcareous tests is sometimes provided by the presence of their organic linings (especially of *Ammonia* where the lining is thick). In Long Island Sound, a lagoon in eastern USA, carbonate dissolution is active and tests are considered to last for 86 ± 13 days following death (Green *et al.*, 1993). Around the Kattegat and Oslofjord, Scandinavia, partial dissolution takes place in the shallow subtidal zone (Murray and Alve, 1999). The taphonomically active zone for foraminifera appears to be the upper 10 cm in most depositional settings, from shallow water to the deep sea (Walker and Goldstein, 1999).

4.2 NATURAL ENVIRONMENTAL CHANGE

Most natural environmental change is ultimately related to climate and therefore to the oceans. Over the last decade or so there has been public awareness and concern about changing climate on a short time-scale. On the other hand, the geological record shows that global climate has varied considerably throughout the history of the Earth. The concern about the recent climatic development stems partly from the fear that the climate may cool to give another ice age or that there may be global warming through the accumulation of so-called greenhouse gases. The impact of climate change is greatest in nearshore and marginal-marine environments, which are already stressed, e.g. in some silled fjords. These areas are also most affected by local or global changes of sea-level.

4.2.1 Quaternary

The Arabian Sea experiences a summer monsoon, when nutrient-rich waters upwell in the northwest, causing very high primary productivity. The combination of high surface-water productivity and moderate rates of thermohaline circulation leads to the formation of an oxygen minimum zone (OMZ) between 150 and 1200 m with low oxygen values (<2 μM). Since the intensity of the OMZ varies from season to season, then it may also have varied over a longer time-span, especially involving climate change. A 14.6-m core on the Pakistan margin (1002 m) provides a 120,000 year record (oxygen-isotope stages 1–5) of OMZ development (Fig. 4.1) (Den Dulk *et al.*, 1998). The benthic foraminiferal record comprises two distinct assemblages: 1. *Bulimina striata, Gavelinopsis lobatula, Chilostomella oolina, Sphaeroidina bulloides, Cibicides ungerianus, Hyalinea balthica* and *Hoeglundina elegans*; 2. *Bulimina exilis, Rotaliatinopsis semiinvoluta, Bolivina alata, Bolivina*

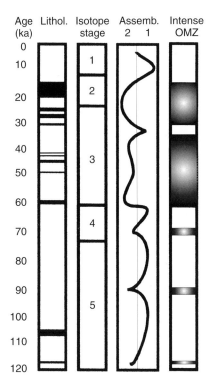

Figure 4.1   Core NIOP455 from the Oxygen Minimum Zone, 1002 m water depth, Pakistan margin. Dominance of assemblage 2 indicates intense development of the OMZ, whereas dominance of assemblage 1 indicates greater availability of oxygen. (After Den Dulk *et al.* (1998).)

pygmaea and *Globobulimina* spp. Assemblage 1 is interpreted as requiring oxygenated bottom waters and it occurs during periods of low C_{org} flux (i.e. low surface productivity), while assemblage 2 comprises species that are tolerant of low oxygen and flourish under high C_{org} flux (thus indicating monsoon conditions) and is associated with laminated sediments. From multivariate factor analysis, it is seen that assemblage 1 dominates in isotope stages 1, 4 and 5 (positive score), but there were three brief intervals with a negative score in isotopic stages 4 and 5 (i.e. when assemblage 2 dominates). These are at approximately 23 ka intervals, and were interpreted to represent periods of lower oxygen. This may indicate that the benthic foraminifera respond to variations in the Earth's orbit with changes in surface productivity being caused by changes in the dates of the precession of the equinoxes. Assemblage 2 dominates throughout isotope stages 2 and 3, which may mean that there was higher winter surface production during this period, superimposed on the precession-driven changes in the summer surface-water productivity (see also section 3.4.3.1).

The North Atlantic region plays a key role in understanding global as well as regional climates and oceanographic conditions. The climate and oceanography of NW Europe and its seaboard are strongly influenced by sources of water entering the North Sea. At present, the North Sea receives most of its water from southerly sources, especially from the North Atlantic Current, a continuation of the Gulf Stream. However, during the Quaternary glacial events, the North Atlantic Current was displaced south towards Iberia, and most of the water entering the North Sea had a northerly, arctic origin. A vibrocore of Middle Pleistocene silty clays from Inner Silver Pit (southern North Sea) yielded foraminiferal assemblages dominated throughout by *Elphidium excavatum* (Fig. 4.2) (Kristensen *et al.*, 1998). The important features for the rest of the fauna are summarised in Table 4.1. *Haynesina orbiculare* is regarded as an indicator of brackish water and is present near the base of zone 52–1. *Cassidulina reniforme* is a marine arctic species. *Ammonia beccarii, Bulimina marginata, Cassidulina laevigata, Melonis barleeanus* and *Sigmoilopsis schlumbergeri* are indicators of warmer temperate water. Thus, the succession shows four ecologically defined zones with a change from brackish, ice-proximal, glaciomarine (52–1) into a transitional state where arctic forms are associated with a few warm-water taxa (52–2), then the warmer water forms become more important, indicating temperate interglacial conditions (52–3) and finally a return to glacial conditions (52–4). This is a complete glacial–interglacial–glacial cycle.

The Kattegat, between Denmark and Sweden, is linked to the North Sea via the Skagerrak and to the brackish Baltic Sea to the south via shallow channels (Fig. 4.3). In the Kattegat at present, there is a surface outflow, to the north, of brackish water derived from the Baltic. Marine waters from the Jutland

Table 4.1   Inner Silver Pit core (all zones contain *Elphidium excavatum* as the dominant taxon). The diversity data are given as the Fisher alpha index. See Fig. 4.1 for illustrations. Data from Kristensen *et al.* (1998)

Data	Zone	Interpretation
Diversity α 2 *Cassidulina reniforme*	52–4	Glaciomarine shelf
Diversity α 4 *Ammonia beccarii* *Bulimina marginata* *Cassidulina laevigata* *Sigmoilopsis schlumbergeri* *Melonis barleeanus*	52–3	Interglacial marine shelf
Diversity α 3 *Cassidulina reniforme* *Cassidulina laevigata* *Stainforthia fusiformis*	52–2	Subarctic transitional shelf
Diversity α 1–2 *Cassidulina reniforme* *Haynesina orbiculare*	52–1	Ice-proximal glaciomarine Brackish at base

Current, which carries water to the northeast along the west coast of Denmark, enter along the bottom from the north. A borehole drilled on the northern tip of Denmark (Skagen) gave a record of changes over the past 10,000 years. Water depths were calculated taking into account the eustatic rise and the compensating isostatic uplift. The foraminiferal evidence (Table 4.2) shows that there was a Preboreal cold spell indicated by the arctic–subarctic markers *Cassidulina reniforme* and *Nonionellina labradorica* (Fig. 4.2) (Conradsen

(a)

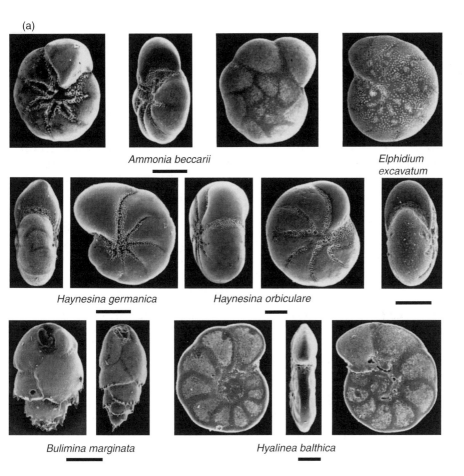

Ammonia beccarii

Elphidium excavatum

Haynesina germanica   *Haynesina orbiculare*

Bulimina marginata   *Hyalinea balthica*

Figure 4.2 Scanning electron micrographs of modern and Quaternary benthic foraminifera from NW Europe: (a) brackish – *Haynesina germanica* and *H. orbiculare*; marginal-marine to temperate shelf – *Elphidium excavatum* and *Ammonia beccarii*; the remainder are temperate shelf taxa. (b) temperate and cold-water shelf taxa. *L. goesi* = *Liebusella goesi*, *B. ps.* = *Bolivinellina pseudopunctata*, *S. biformis* = *Spiroplectammina biformis*. Scale bars = 100 μm. NB. The *Ammonia beccarii* group includes variants that are separately named *A. batavus* and *A. tepida*.

(b)

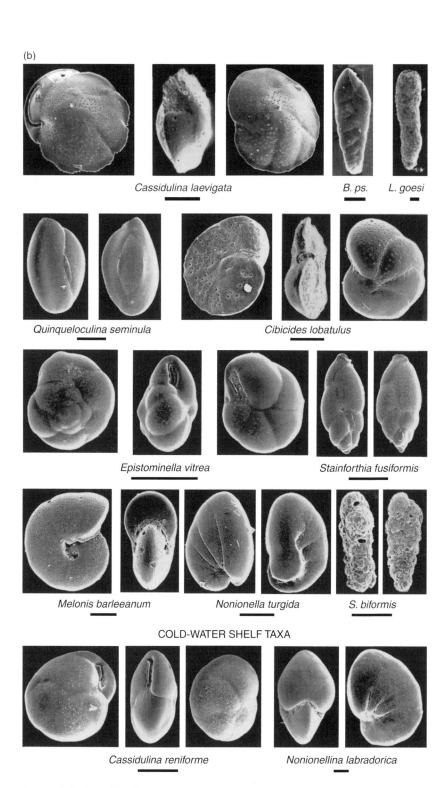

Cassidulina laevigata   B. ps.   L. goesi

Quinqueloculina seminula   *Cibicides lobatulus*

Epistominella vitrea   *Stainforthia fusiformis*

Melonis barleeanum   *Nonionella turgida*   S. biformis

COLD-WATER SHELF TAXA

Cassidulina reniforme   *Nonionellina labradorica*

Figure 4.2 Continued

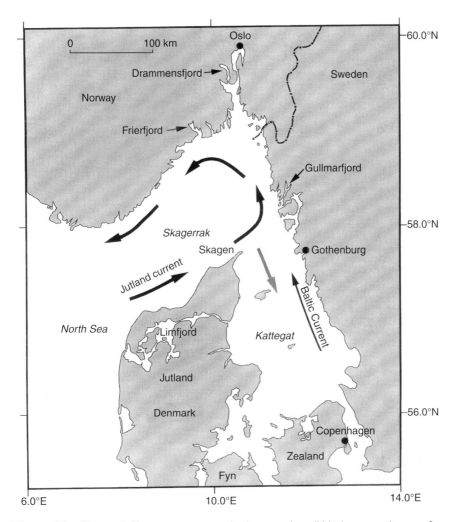

Figure 4.3 Skagerrak–Kattegat oceanography. Large and small black arrows show surface currents; grey arrow shows bottom current. From at least 9.6 to around 5 ka, a 100-m deep basin occupied the Skagen area. The map also shows the position of the Skagen borehole, Drammensfjord, Gullmarfjord and Limfjord, mentioned elsewhere in the text.

and Heier-Nielsen, 1995). Only after this time was there an influence of warm Atlantic water strong enough to create interglacial conditions. Then the Skagerrak–Kattegat formed a large fjord-like system and the position of Skagen was a region with >100 m depth of water. After 8.6 ka the rate of sedimentation increased and the foraminiferal assemblages include reworked miliolids of unknown source. The water circulation at this time is uncertain. At around 8 ka the English Channel formed a connection with the North Sea due to the rise in sea-level. The foraminiferal fauna became the

Table 4.2 The Skagen borehole in northern Denmark. See Fig. 4.2 for illustrations (after Conradsen and Heier-Nielsen, 1995)

Age (ka)	Palaeo-water depth (m)	Foraminifera	Interpretation
7.6–1.0	110–30	*Elphidium excavatum*	Modern conditions
8.6–7.6	110	*Bulimina marginata* + reworked miliolids	Uncertain
9.1–8.6	110	*Bulimina marginata*	Continued warming
9.6–9.1	105–110	*Bulimina marginata*	Gradual warming – influx of Atlantic Water
>9.6	105	*Elphidium excavatum* dominant *Cassidulina reniforme* *Nonionellina labradorica*	Preboreal arctic–subarctic water masses

same as that found in the area at present. Therefore, it is assumed that the modern circulation pattern was established at that time.

The post-glacial rise of sea-level from around –125 m to its present level was a major trangressive event that has had a profound effect on both continental shelf deposits and those of estuaries. As sea-level rose, rivers adjusted their profiles by silting up their estuaries. Thus, the effect of the transgression was for the shoreline to move up former river valleys, and the valley's lower reaches became infilled with sediment. Most of the sediments accumulated in the intertidal zone, but some were subtidal. In the Somerset Levels of southern England discussed above (section 4.1.1), the borehole at New Passage penetrated 20 m of Holocene fill. This records the progressive effects of the transgression during the past 9000 years. The subtidal deposits were recognised on the basis of having a sparse indigenous fauna and >70 per cent exotic forms transported in as suspension load (Murray and Hawkins, 1976).

The effects of a transgression on a continental shelf is to increase the water depth. The degree of tidal mixing controls whether or not stratification takes place. Layering develops during the summer where $\log_{10}(H/U^3)$ exceeds a value of 2.0 (H = water depth and U = the maximum tidal stream velocity at the surface). The surface layer becomes warm, while the water beneath the thermocline remains cool throughout the year. Along the

Table 4.3 Summary of environmental changes in the Holocene deposits of the Somerset Levels, England (after Murray and Hawkins, 1976)

Lithology	Dominant foraminifera	Environment	Depth in borehole (m)
Muds, silts	*Jadammina macrescens* *Trochammina inflata*	High tidal flat	0–5
Muddy sand	*Ammonia beccarii* *Elphidium williamsoni* *Haynesina germanica*	High tidal flat to low tidal flat	5–12
Sand	Mainly transported	Subtidal	12–20

frontal boundary, nutrient renewal takes place intermittently due to mixing by wind and tide, and this creates conditions suitable for rapid phytoplankton growth. Consequently, there is increased organic flux to the sea-floor in frontal areas. These processes of cooler water and enhanced organic flux have an effect on the benthos beneath and marginal to stratified areas. In the Celtic Sea between Eire and SW England, a borehole from 118 m water depth beneath the modern stratified area yielded a 5.6 m core giving a record of the past 12 ka (Fig. 4.4) (Austin and Scourse, 1997). There are changes in lithology from mud at the base to coarse shelly sand and gravel and back to mud. Foraminifera are absent from the basal mud unit. In the coarse sands there is an upward change in the composition of the foraminiferal fauna, with *Cibicides lobatulus* giving way to *Quinqueloculina seminula* and then to *Bulimina marginata* in the upper mud unit. These changes suggest an increase in water depth and a decrease in energy, but by themselves do not provide evidence of thermal stratification. However, the oxygen stable-isotope records for *Ammonia batavus* (infaunal) and *Quinqueloculina seminula* (epifaunal) both show a marked positive trend between 9 and 5 ka, indicating a temperature decrease of 4–5 °C. This suggests that prior to 9 ka the waters were vertically mixed but at that date water depth had reached the critical limit for seasonal stratification to develop, and progressive development continued

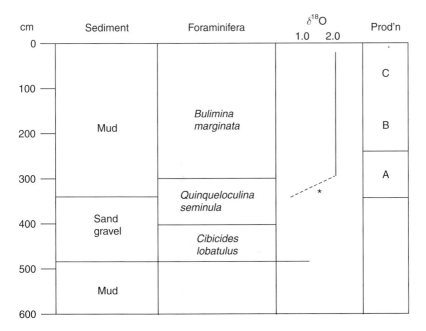

Figure 4.4 Data and interpretation of a core from the Celtic Sea, showing the onset of thermal stratification (*) and the three stages of production development. The oxygen stable-isotope curve is generalised. (Based on Austin and Scourse (1997).)

until 5 ka, when essentially the modern pattern was established. The carbon stable-isotope records from the same two species show a negative trend that indicates decomposition of organic matter beneath the frontal zone. However, the productivity pattern does not exactly coincide with the onset of the stratification. It may be that by the end of productivity stage A, nutrient levels were limiting productivity or else the frontal zone had moved away from the core site. It is important to recognise that in this example the temperature decrease of the bottom water was caused by a eustatic rise in sea-level due to global warming and not due to climatic cooling.

The occurrence of foraminifera in lake deposits is sometimes due to marine transgression (if the lake is close to the sea), but need not be so. Since the Miocene/Pliocene, freshwater and hypersaline lakes have occupied the Dead Sea rift in Israel. Until recently, the occurrence there of foraminifera in fossil lake deposits has been interpreted as evidence of a marine transgression. However, about a decade ago *Ammonia tepida* was found living on microbial mats in a shallow saline pool (salinity 40–55 psu) in a former quarry 2 km west of the Dead Sea (Almogi-Labin *et al.*, 1992, 1995). The high salinity is due to the total dissolved solids of the source waters and high evaporation associated with the dry hot climate; the ionic composition differs from seawater in several respects. Dead tests of *Ammonia tepida* also occur down to 18 cm below the sediment surface. In both living and preserved dead populations the deformed tests are much more abundant than is normal for marine environments (an average of 17 per cent compared with 1–2 per cent respectively) and must reflect the stressed conditions. The most likely method of transporting live foraminifera to colonise the pool is through mud (derived from coastal areas) on the feet of birds, as the Dead Sea rift area is a main route for seasonal bird migration. It seems probable that records of *Ammonia tepida* in the fossil lake deposits there were also due to migrating birds rather than to marine transgressions. The revised interpretation has profound consequences for reconstructing the tectonic history and palaeogeography of the area.

The water circulation of lagoons is controlled to a large extent by the size of their connection with the sea and their tidal range. In northern Denmark, a large microtidal lagoon (the Limfjord, generally <10 m deep) stretches from the North Sea to the Kattegat (Fig. 4.3). Because the tides are small there is little diurnal variation in salinity. Throughout the Holocene, the connection with the North Sea has repeatedly been closed through sand deposition, and has been periodically reopened through breaching by storms. The foraminiferal record of the past 2000 years shows that major salinity changes in the central part of the lagoon have been caused by the alternating restriction and opening of the North Sea connection (Fig. 4.5) (Kristensen *et al.*, 1995). The higher-salinity periods are characterised by large numbers of tests (indicating higher production) and higher-diversity assemblages (*Ammonia beccarii* and *Elphidium excavatum* generally with *Haynesina germanica* and sometimes with *Elphidium albiumbilicatum* and *Aubignyna perlucida*). The lower-salinity periods have much lower numbers of tests (indicating lower production), low diversity and an absence of more marine taxa such as *Aubignyna perlucida*. The higher salinity

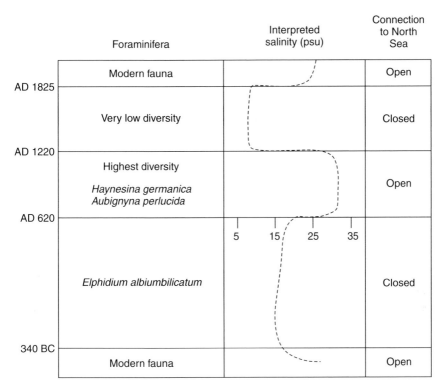

Figure 4.5 Record of salinity changes and inferred changes in the connection with the North Sea for the Limfjord, Denmark. *Elphidium excavatum* and *Ammonia beccarii* are important components of the fauna at all levels, so only the additional taxa are listed. (Based on Kristensen *et al.* (1995).)

episodes represent a good connection with the North Sea. There is corroboration of this from historical records. For instance, the Danish Viking King Knud the Great sailed into the Limfjord from the west after an expedition to Britain in AD 1027. A salinity measurement made in AD 1811, at a time when the North Sea connection was closed, was 8 psu. In AD 1825 a major storm surge that was 7 m above present sea-level reopened the North Sea connection and now it is maintained artificially.

4.2.2   The Last Few Hundred Years

4.2.2.1   Benthic Foraminifera as Indicators of the Healthiness of Modern Coral Reefs

Reefs formed of hermatypic corals grow in warm, shallow, tropical to subtropical waters that are low in nutrients (oligotrophic). Hermatypic corals contain algal endosymbionts in the ectoderm; these recycle coral waste products and aid the process of calcification of the corallite. Because corals grow incrementally, it is possible to determine from growth lines that some slow-growing forms live for hundreds of years, although others, such as stagshorn coral, grow more rapidly.

Benthic foraminifera are commonly abundant in modern reefs, and one group, the larger foraminifera, have algal endosymbionts similar to those of corals. The life-cycle of larger foraminifera is normally completed in one to two years. Thus, they are potentially able to respond to environmental stresses much faster than corals. Reefs are at risk from a range of human influences, including nutrient and other chemical pollution, sediment deposition, ultraviolet B (UVB) radiation, destruction through certain types of fishing and also from shipping (Hallock, 2000).

In the most extreme nutrient-depleted environments of Pacific atolls, larger foraminifera dominate the sand-grade sediments, but when the availability of nutrients increases, they become less abundant, whereas bioeroded fragments of corals, calcareous algae, and mollusc bioclasts increase, as do the numbers of smaller benthic foraminifera. Under these conditions, there is not normally a measurable increase in nutrients in the water column, because the benthic and planktonic communities utilise all available nutrients.

Organisms with endosymbionts take their colour from the latter, so when these are lost or damaged the colour intensity decreases. In corals, this is known as bleaching, and it is a process that has increased in frequency since the 1980s. Recently, bleaching has been reported in the larger foraminifer *Amphistegina*. It results from the deterioration and digestion of the diatom endosymbionts and associated changes in the cytoplasm of the host. The foraminiferal shells become much more susceptible to breakage, as illustrated by 40 per cent breakage subsequent to bleaching, as compared with 5 per cent pre-bleaching breakage found in specimens from the Florida Keys, USA. The process is not confined to Florida but is occurring worldwide and affects photosynthesis and shell-protein synthesis, and causes reproductive damage and loss of disease defence mechanisms. Consequently, recorded decreases in the severity of bleaching with increasing water depth are consistent with UVB radiation (rather than local impacts) being the cause. Although UVB is rapidly absorbed in waters containing refractory organic matter such as terrestrially derived tannins, the oligotrophic waters in which reefs form lack these substances, so that UVB penetrates tens of metres.

4.2.2.2   *Periodic Anoxia Associated with the Oxygen Minimum Zone (OMZ)*

On the US Californian borderland shelf, there are deep basins. In the Santa Barbara Basin, the sill is at 475 m and the maximum depth is 627 m. Water is supplied to the basin from the OMZ, and, since circulation is restricted, the bottom waters are either devoid of dissolved oxygen (anoxic) or have

Table 4.4 Tolerance of taxa to decreasing levels of oxygen in the Santa Barbara Basin, California, USA (from Bernhard et al., 1997)

Oxygen	Taxa
Decreasing downwards	*Uvigerina juncea*
	Suggrunda eckisi
	Bolivina argentea and *Loxostomum pseudobeyrichi*
	Trochammina pacifica, Buliminella tenuata and *Bolivina seminuda*
	Chilostomella ovoidea and *Spiroplectammina earlandi*
	Nonionella stella

only a small amount (dysoxic) during the spring renewal. In the centre of the basins the sediments are laminated and foraminifera were common during February, June and October 1988 during dysoxic conditions, but had died out by July 1989 when the waters were anoxic (Bernhard and Reimers, 1991; Bernhard, et al., 1997). Although some foraminifera can survive brief intervals of anoxia (weeks to a few months), they die out during prolonged anoxia. The tolerance of some taxa to decreasing levels of oxygen is given in Table 4.4.

4.2.2.3   North Atlantic Oscillation (NAO)

The Gullmarfjord on the west coast of Sweden is more than 110 m deep and has a sill at 42 m water depth through which it connects to the Skagerrak, the most northeasterly arm of the North Sea (Fig. 4.3). The waters are stratified, with a low-salinity surface layer (0–60 m, 20–25 psu) and a deeper normal-marine layer (>60 m, 34–35 psu). Whereas the surface layer is renewed biweekly to monthly, the deep layer is renewed annually during winter or early spring. The bottom waters show a weak decline in oxygen from the 1970s onwards, and during the 1970s there was also a period of reduced salinity. Sediment cores from the deep basin were dated by ^{210}Pb. The foraminiferal record commenced with a normal inner-shelf fauna present until 1976 (Nordberg et al., 2000). Then faunal changes were initiated, particularly the increase in *Stainforthia fusiformis* (an opportunistic species which can withstand severe dysoxia), culminating in its total dominance at the time the core was collected (Table 4.5). These changes coincide with the historical records of declining oxygen availability and periodic dysoxia in the deep basin. Gullmarfjord has not been affected by any major source of pollution since the late 1960s, so the changes are interpreted to be natural, i.e. climatic/hydrographic. One possibility is a link to the NAO, which is a temporal fluctuation of the zonal wind strength across the Atlantic Ocean. During the 1970s there was a positive NAO winter change characterised by mild wet winters and westerly winds. These counteracted the upwelling of the Skagerrak deep water. Upwelling is an essential requirement for renewal of the oxygen-depleted fjord deep water with new oxygenated water. Another

Table 4.5 Summary of data and interpretation of core G110 from the Gullmar Fjord deep basin. M = miliolids; A = agglutinated; H = hyaline test wall (from Nordberg *et al.*, 2000)

Depth (cm)	Date	Foraminifera	Interpretation
0–2	1996–1994	*Stainforthia fusiformis* H dominant *Nonionella turgida* H	Low-oxygen conditions
2–13	1994–1982	*Stainforthia fusiformis* H increasing Few agglutinated forms	Declining oxygen
13–19	1982–1976	Decrease of: *Textularia earlandi* A Increase of: *Stainforthia fusiformis* H *Quinqueloculina stalkeri* M *Spiroplectammina biformis* A *Bolivinellina pseudopunctata* H *Epistominella vitrea* H	Transition, start of declining oxygen
19–42	Top = 1976, Base = 1950	*Liebusella goesi* A *Textularia earlandi* A *Hyalinea balthica* H *Cassidulina laevigata* H *Nonionellina labradorica* H *Bulimina marginata* H (*Stainforthia fusiformis* H rare)	Typical Skagerrak inner-shelf fauna of normal salinity and adequate oxygen

possibility is that the marine waters entering the fjord may be slightly polluted. At present, neither theory can be proven.

4.3 HUMAN-INDUCED ENVIRONMENTAL CHANGE

Recently, the application of benthic foraminifera has entered a new arena, as it emerges that they can be used as an excellent tool in environmental monitoring of present-day contaminated and polluted areas (Martin, 2000; Scott *et al.*, 2001). Many different biological techniques are in use today (e.g. various diversity indices, log-normal distributions, K-dominance curves, abundance–biomass comparisons, and different multivariate statistical methods) in order to assess pollution effects on marine communities and to separate these effects from natural environmental variability. However, since coastal environments (and particularly estuaries) are extremely heterogeneous, no two areas have exactly the same environmental characteristics, and hence their benthic communities cannot be expected to be the same either. Consequently, comparison with pre-pollution surveys forms the best basis for drawing conclusions about possible impacts on the biota. It follows that since long-term biological and hydrographical time-series (extending back to pre-impacted times) are virtually non-existent for most areas, retrospective studies of the sedimentary record and its fossil content are the only means to reconstruct past biological and environmental changes and thereby distinguish natural from human-induced effects. In other words, the fossil record is the only source that can provide us with information about ecosystem variability at the same site over historical time-scales. However, the reliability and usefulness of this method requires that the study area has a reasonably high sediment accumulation rate (preferably more than about 1–2 mm/year) and that taphonomic processes and vertical sediment mixing through bioturbation are not too extensive. Fortunately, marginal-marine environments best fulfill these requirements, and these areas are also the main sites of pollution.

As the majority of benthic foraminiferal pollution impact studies have focused on different aspects of organic-matter and metal enrichments, the case studies presented here are chosen from these topics. It must be noted, however, that, so far, most foraminiferal impact studies lack quantitative considerations of the faunal responses to the concentrations and bioavailability of the different kinds of pollutants to which they are/have been exposed (i.e. dose–response relationships). Future studies should be more interdisciplinary and involve biological (especially physiological) and geochemical as well as geological approaches.

4.3.1 Organic Pollution and Eutrophication

Coastal areas around estuaries and fjords have traditionally been places for human settlement, with the development of cities and industry. Consequently, these environments have experienced the deleterious effects of human activity, particularly the increased sedimentation of organic matter and nutrients deriving from sewage outlets, agriculture, paper and pulp mills, and the industrial processes such as food production, fertiliser manufacturing, etc. Over the last couple of decades, aquaculture has also become a common industry in coastal areas selected because they were not previously affected by contaminants.

The most characteristic features of benthic foraminiferal assemblages in environments that originally were nutrient limited and have been enriched subsequently in organic matter include:

1. a dead, azoic (devoid of eukaryotes) zone in the immediate vicinity of a point source, if the organic load is so high that anoxic sediment pore-waters have developed right up to the sediment surface;
2. increased abundance in the 'hypertrophic zone', relative to the natural background level at, or close to, the outfall if the organic load causes increased nutrient supply without creating permanent anoxia in the sediment surface pore-waters;
3. if anoxic pore-water conditions develop up to the sediment surface only occasionally, this will reduce the species diversity and cause a change in the community structure, with an increase in opportunistic species in between the anoxic events (Alve, 1995a).

These patterns have been recorded in spatial studies of sediment-surface samples and they can be preserved in the fossil record. The latter is demonstrated in Fig. 4.6, which gives a schematic representation of variations in the main faunal parameters through a sediment core which penetrates a succession from pre-impacted deposits (lower part) through a dead, azoic zone deposited during prolonged anoxia, to a recolonisation phase (in the upper part).

Between 1957 and 1962, the benthic foraminiferal assemblages around the Los Angeles County **sewer outfall** at 50–60 m water depth at Whites Point, USA, were severely affected (Stott *et al.*, 1996). The zone adjacent to the discharge diffusers was characterised by an azoic zone with no living foraminifera, and the numbers of most species were unusually low (compared with adjacent areas) as far away as several kilometres. About 30 years

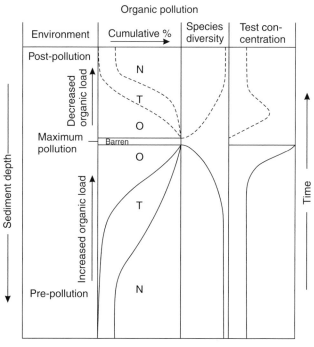

O = Opportunistic populations; T = Transitional populations
N = 'Natural' populations

Figure 4.6   Simplistic diagram showing how benthic foraminiferal analyses of sediment cores penetrating through organically contaminated deposits may provide comparative information for evaluating: (1) how severely the original, 'natural' assemblages have been affected, and (2) the positive effects of environmental improvements. Solid lines = pre-pollution patterns, dashed lines = post-pollution patterns. (Modified from Alve, 1995a, fig. 2. Reproduced with permission from the *Journal of Foraminiferal Research.*)

later, in 1995, following significant decline in the volume of contaminants, the environment around the outfall was re-inhabited by benthic foraminifera in numbers similar to those found in other non-affected parts of the Southern California shelf. The benthic foraminiferal populations had nearly returned to 'normal' shelf abundances, although the environmental conditions adjacent to the outfall still precluded the growth of the most sensitive species such as *Nonionella* spp.

In the northern Adriatic Sea, an arm of the Mediterranean, the fossil record of benthic foraminifera documents geographical shifts in the major outlet channels of the River Po as a result of human activity during the middle part of the nineteenth century. The gradual disappearance of epiphytic taxa (those living attached to plants, e.g. *Cibicides lobatulus*, *Gavelinopsis praegeri*, *Buccella granulata*) and the concurrent increase in mud-dwellers (e.g. *Bulimina marginata*, *Epistominella vitrea*) showed that the vegetation cover gradually disappeared as the sedimentation rate increased closer to the new position of the outlet. From about 1900 AD onwards, the increasing relative abundance of *Nonionella turgida* suggests a continuously increasing nutrient enrichment due to **eutrophication** caused by agriculture and waste-water disposal (Barmawidjaja *et al.*, 1995).

The upper reaches of Saguenay fjord, Quebec, Canada, received fluctuating – but generally increasing – effluent discharges rich in organic matter (primarily from the **pulp and paper industries**) from the early twentieth century until the mid 1960s. Since that time, a large area of contaminated sediments has been buried by a layer of clay due to a catastrophic landslide in 1971, and governmental regulations were imposed on local industrial polluters in the early 1970s. Sediment core investigations from the area affected by the landslide burial showed a highly resolvable unbioturbated record of organic-matter fluxes corresponding with the scale and character of the industrial activity over that time period (Schafer *et al.*, 1991). The organic-waste deposition showed a first-order inverse relationship to benthic foraminiferal assemblage diversity in the cores. The presence of only allochthonous thecamoebians (shelled freshwater protists derived from rivers) and reworked planktonic foraminiferal tests (introduced from the ocean) in many core intervals, which otherwise were barren of benthic foraminifera, suggests that indigenous estuarine species, such as the agglutinated *Spiroplectammina biformis*, were excluded from those areas at certain times as a result of pollution stress, rather than having been removed from the sediment record through taphonomic effects. Faunal analyses of surface sediments collected in 1982 and 1988 suggested that a successive recolonisation of the upper reaches and of some formerly barren benthic environments near the head of the fjord occurred during this six-year period. Several agglutinated species expanded their distribution and *Cassidulina reniforme* (calcareous) showed a roughly threefold increase, while the relative abundance of the previously dominant *S. biformis* declined (Fig. 4.7). The recolonisation probably occurred as a response to decreased organic flux and consequent improvement in the benthic environments brought about by the

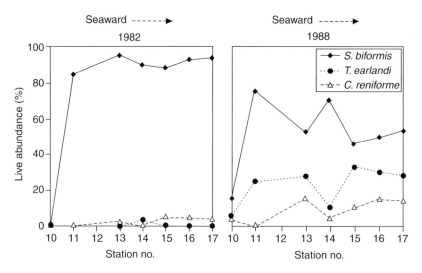

Figure 4.7   Distribution of the relative abundance of the most abundant live (stained) benthic foraminifera recorded from east to west (seaward direction) in the upper reaches of the Saguenay Fiord, Quebec, Canada, in 1982 and 1988. (Data from Schafer *et al.* (1991).)

combined effects of the landslide and the reduction in pollution due to enforcement of the governmental regulations.

Dysoxic and even anoxic bottom-water conditions dominate in a large number of Norwegian silled fjords along the Skagerrak coast. For some of them, the development of oxygen depletion is clearly due to human activity, whereas for others the development is due to natural environmental change as described above. Analyses of the historical distribution of benthic foraminifera can help to clarify the reason. Some fjord basins with an estuarine circulation pattern and very small tidal range are especially vulnerable to organic pollution because the bottom-water renewal is restricted by the presence of one or more sills, which, additionally, make the basins function as sediment sinks (and pollution traps). In such basins, the oxygen consumption, both in the water-masses and in the surface sediments, increases with the increasing supply of reactive organic matter. The organic matter may be introduced directly from outlets or terrestrial runoff, or indirectly through surface primary production, and in several cases this has led to dysoxic or anoxic bottom-water conditions. One example is Drammensfjord (southern Norway, Fig. 4.3), which has a maximum water depth of 124 m, receives about 290 m^3/s water from a river in the innermost part, and is separated from the adjacent Oslofjord system by a sill at 10 m water depth (Alve, 1991, 1995b). From the seventeenth century onwards, the fjord received increasing loads of organic material, first from **saw mills** and later from **paper and pulp mills** until they were shut down in the mid 1970s. Discharges of untreated **domestic sewage** continued until regulations were introduced in the 1980s. Hydrographic investigations showed that the most oxygen-depleted conditions occurred in 1977, 1978 and 1982, when the redox boundary in the middle part of the fjord was positioned at 30–35 m depth in the water column, i.e. the bottom waters had become anoxic.

Benthic foraminiferal analyses in ^{210}Pb-dated cores (two from the northern and two from the southern part) showed that this fjord had not been well ventilated since the time when the Vikings ruled that part of the world some one thousand years ago. The portion of the cores representing sediments deposited at that time contained high-diversity foraminiferal assemblages similar to those living in the adjacent Oslofjord today. Above these sediments, low-diversity and low-abundance assemblages indicated impoverished, stressful conditions lasting for several hundred years, through what we know as the Little Ice Age. In the northern area the transition to black, laminated sediments completely barren of benthic foraminifera shows that the increased load of organic material caused development of predominant anoxia up to 45 m depth in the water column from the middle part of the nineteenth century, and that this had extended upwards to at least 26 m from the 1950s onwards. In the southern area the upward migration of the redox boundary was slower and extended up to a water depth of 70 m by about 1970, because this basin was further away from the effluent sources and closer to the well-oxygenated outer Oslofjord system. A characteristic feature of these assemblages is that, in most cases, they show an upwardly increasing abundance of individuals just before the level at which they were killed off. This increase probably reflects a combination of increased nutrient supply to the foraminifera (directly from the effluents or indirectly from bacteria and, in originally nutrient-limited environments, increased primary production) and decreased predation from other organisms. It required only a small additional change in this highly stressed environment to kill off the foraminifera. A similar trend was seen in sediment cores from the silled, 98-m deep Frierfjord in southern Norway (Fig. 4.3), which, due to increasing discharges of organic material (initially from saw mills and later from paper and pulp industries and sewage), turned anoxic at successively shallower water depths during the 1900s (Alve, 2000). The time resolution in these cores is better than in those from Drammensfjord and Fig. 4.8 illustrates the dramatic faunal changes that occurred in response to the increased organic load (shown by loss on ignition) at 50 m water depth. The weak increase in number of foraminifera/g dry sediment towards the sediment surface suggests slightly improved conditions as a response to legislative limitations on the discharges during the mid 1970s.

The fact that benthic foramininfera can provide information about the positive effects of pollution abatements has been more clearly demonstrated in Drammensfjord. Investigations of the live foraminiferal assemblages from the middle part of the fjord in the late 1980s and early 1990s showed that some of the previously anoxic areas were in the process of recolonisation (Alve, 1995b). The first and most successful coloniser of the previously anoxic sediments was *Stainforthia fusiformis* – exactly the same as suggested for

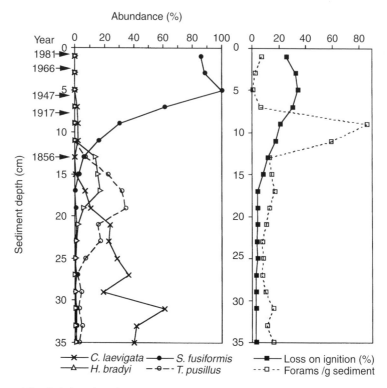

Figure 4.8   Relative abundance of characteristic species, loss on ignition (per cent), and number of benthic foraminifera/g dry sediment in a ^{210}Pb-dated sediment core from Frierfjord, southern Norway. Note the drastic faunal shift which coincides with the peak in foraminiferal density at the transition between the nineteenth and twentieth centuries, the faunal collapse during maximum organic load (as reflected by loss on ignition), and the weak signal of recolonisation following reduced discharges. Generic names: C. = Cassidulina, H. = Haplophragmoides, S. = Stainforthia, T. = Trochamminopsis. (Based on Alve (2000).)

Frierfjord. This is one of the most prominent opportunistic species in NW European waters, flourishing and dominating in normal-marine soft-bottom environments that are commonly exposed to increased nutrient fluxes, to physical disturbance (e.g. trawling) or to occasional anoxia (Alve, 1994).

There are few published records of benthic foraminiferal responses to **aquaculture operations**. However, although sparse, the available data indicate that in originally nutrient-limited environments, there is an inverse relationship between the commercial productivity (i.e. reflecting the supply of organic material to the sediments from fish wastes, excrement and unused fish-feed) and the absolute abundance of foraminifera. This resembles responses seen in environments exposed to other kinds of organic enrichment, with the development of azoic sediments in highly impacted areas where the sediments have turned anoxic, and maximum abundance in the 'hypertrophic zone', where nutrient availability is at a maximum but where the sediments have still not turned permanently anoxic. One example is a study of living (stained) benthic foraminiferal assemblages associated with fin-fish and shellfish operations in coastal Maritime Canada (Schafer *et al.*, 1995). There, most samples which were barren of foraminifera were collected during summer and autumn when anthropogenic and natural organic-matter fluxes were highest. The organic-matter (OM) content ranged between 4 and 26 per cent in the sediment samples studied, and, typically, stations with six or more living species were associated with sediment OM concentrations of less than 14 per cent. As in several other organically impacted environments, *Elphidium excavatum* was one of the most common and persistent species capable of dominating the foraminiferal assemblages directly below fin-fish cages as well as below mussel lines. At a net cage fish farm in the northern Gulf of Eilat/Aqaba in the Red Sea, the sediments underlying the fish farm were substantially different from the nearby areas in terms of organic-matter content, sediment pore-water ammonia and phosphate concentration, and microflora (bacterial mats reflecting the redox status of the sediment). No azoic zone was recorded but the abundances of the seven most common foraminiferal species (which generally made up >83 per cent of the assemblages) were highest in the 'hypertrophic zone', adjacent to the fish cages (Fig. 4.9) (Angel *et al.*, 2000). There was a negative correlation between total abundance (live plus dead) of these species and organic matter in the surface sediment. Negative correlations were also found between the abundances of both total and stained tests of the most abundant species and integrated (0–5 cm sediment depth) ammonia concentrations, whereas no clear indicator species of the organically enriched benthos were identified.

4.3.2   Metal Pollution

Metals are part of nature, and in some areas (e.g. close to exposed mineral ores) the concentrations of certain elements are strongly elevated in the surrounding soil or water as compared with 'normal' values. However, in connection with the development of modern societies, the production of all the facilities we think we need, and the 'buy and throw away mentality', have introduced high concentrations of metals into environments where they do not belong. Metals which typically occur in higher than normal concentrations in anthropogenic effluents include iron (Fe), copper (Cu), zinc (Zn), lead (Pb), tin (Sn), chromium (Cr), cobalt (Co) and mercury (Hg). In low concentrations some of these are essential for living organisms (e.g. Cu and Fe) but in high concentrations they may be toxic, whereas others that are not essential for metabolic activity (e.g. Pb, Sn and Hg) are toxic to cells even at quite low concentrations (e.g. Clark, 1986).

In marine environments, metals are often enriched in bottom sediments. Since many commercially exploited organisms feed on and spend most of

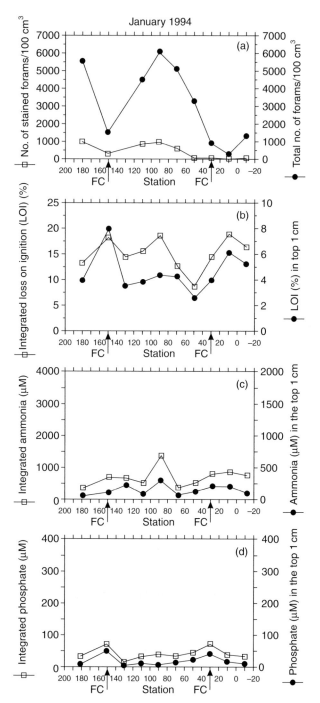

Figure 4.9 Data from samples collected in the northern Gulf of Eilat (Aqaba, Red Sea) in January 1994 along a west to east transect through two cage-fish-farm systems (FC).

their lives in or associated with the sediments, elevated concentrations may potentially affect these organisms and ultimately the humans who eat them. The effects of these contaminants depend on their bioavailability, which in turn depends on a range of physical, chemical and biological factors associated with the sediment. Most polluted areas are exposed to several different kinds of pollutants, and, in most cases, it is difficult or even impossible to isolate the effects of one pollutant from another. In recent years, this has been addressed through multivariate statistical analyses linking a range of faunal and environmental variables.

Mercury, silver and copper are the metals that are most toxic to marine organisms. Because of its widespread use, Cu is one of the most common environmental pollutants. In order to study what effects Cu has on coastal, subtidal assemblages, an *in situ* experiment was conducted at 63 m water depth in the Oslofjord in Norway (Alve and Olsgard, 1999). Sediment Cu-concentrations of >900 ppm caused a change in the living (stained) foraminiferal community structure as compared with control values of 70 ppm. The changes, which were revealed through multivariate statistical analyses of the assemblages from different treatments, were also reflected by increased equitability (a measure of how well specimens are distributed among species) and reduced abundance in treatments with high (967–977 ppm) and very high (1761–2424 ppm) Cu-concentrations (Fig. 4.10). There was no significant decrease in the number of species with increasing sediment Cu-enrichment. At the species level, a significant negative effect could be observed only for *Stainforthia fusiformis* and *Bolivinellina pseudopunctata*.

Analyses of both sediment surface samples and sediment core data from Southampton Water, UK, showed that *Elphidium excavatum* was more common in the live (stained) assemblages than *Ammonia beccarii* and *Haynesina germanica* at the most contaminated stations (max. concentrations in the sediment in ppm dry weight: 1,007 Cu; 160 Cr; 470 Zn; 12 Cd; 281 Pb). The down-core dead assemblages showed that *Ammonia beccarii* had been the most abundant species prior to the introduction of industrial effluents. The degree of tolerance to heavy-metal pollution among the species was ordered as follows: *E. excavatum* > *H. germanica* > *A. beccarii* (Sharifi *et al.*, 1991).

The distribution of benthic foraminifera and trace metals in sediment cores from a Chesapeake Bay estuary, Maryland, USA, document the industrial

Figures along the x-axis represent the sampling stations (in metres from an eastern starting point). (a) Abundance (no. of foraminifera/100 cm³ sediment) of total (stained + unstained) and stained tests of the seven most abundant species; (b–d) geochemistry (loss on ignition, ammonia, and phosphate respectively) of the sediments in cores taken at each sampling station, presented as double y-plots including both integrated (0–5 cm depth) values and concentrations in the top 1 cm of the sediment. (Modified from Angel *et al.* (2000, figs 3b, d, f and 4b). Reproduced with permission from *Journal of Foraminiferal Research*.)

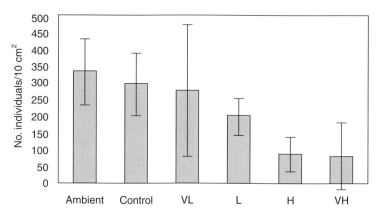

Figure 4.10 Numerical density of live (stained) benthic foraminifera (mean values and SD for replicates) in ambient sea-bed, control, and treatment cores (bulk surface 0–5 cm sediment). VL = very low; L = low; H = high; VH = very high (Cu); for exact values, see text. (From Alve and Olsgard, 1999, fig. 5. Reproduced with permission from *Journal of Foraminiferal Research.*)

and municipal pollution history over more than one hundred years (Ellison *et al.*, 1986). As the concentrations of heavy metals (e.g. Zn, Cr, V) and associated pollutants increased downstream, the distribution of the dominant species, *Ammobaculites crassus*, retreated in the same direction and developed a large population in the lower part of the estuary, probably due to abundant food and lack of competition. With time, as the pollution increased, the populations declined in size and ultimately the whole estuary was nearly barren of foraminifera. Based on the fact that the data point to a recent return of modest-sized populations of this species, it was concluded that this may signal the return to healthier, more natural conditions in the estuary.

4.3.3 Engineering Geological Applications

Foraminiferal biostratigraphy has been shown to be a valuable tool in engineering geology and thereby to address environmental problems. This is due to the fact that the recognition of distinct foraminiferal biofacies can successfully be used to perform correlations between sediment cores or boreholes. This is of particular importance in stratigraphically complex and discontinuous sedimentary deposits. One such example is the Upper Pleistocene estuarine deposits beneath San Francisco Bay, California, USA. There, glacially controlled fluctuations of sea-level over at least the past half million years have resulted in the deposition of alternating alluvial and estuarine sediments which interfinger complexly at the margins of the Bay (Sloan, 1995). Active strike-slip and thrust tectonism along the San Andreas and Hayward faults further complicates the depositional record. The 1989 Loma Prieta earthquake near Santa Cruz caused severe damage about 100 km

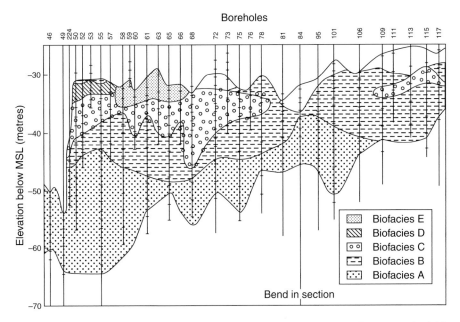

Figure 4.11 Biofacies A–E, all in the same lithological unit, the Yerba Buena mud, which underlies San Francisco Bay at depths of 25–70 m below mean sea-level. (Modified from Sloan (1995), fig. 6. Reproduced with permission from the *Journal of Foraminiferal Research*.)

from the epicentre, accentuated by the fact that substrates such as landfill and Bay mud may amplify seismic waves and locally intensify ground shaking. Major damage to man-made structures built on these substrates, including the collapse of the Cypress Freeway Structure, contributed to loss of human life. Before the freeway could be rebuilt, it was essential to know more about the subsurface sediments. Micropalaeontological studies distinguished five biofacies within the extensive, lithologically uniform deposit of the last-interglacial estuary: the Yerba Buena mud (Fig. 4.11). Of these, biofacies C had a distinct assemblage of microfossils dominated by *Elphidium hannai* and a sand-sized diatom species. Within the occurrence of biofacies C there was a thick layer of oyster shells which gave a good marker in drilling logs. Identification of biofacies C permitted greater resolution of the subsurface stratigraphy than would otherwise have been possible.

Due to the discontinuous nature of the San Francisco Bay margin sediments, it was previously difficult to track contamination plumes in some of the main aquifers. Now this characteristic biofacies of the Yerba Buena mud has also been shown to be particularly useful in delineating the regional hydrogeological conditions in this area, thus aiding the clean-up of polluted groundwater.

4.3.4 Test Deformation: Natural and Unnatural Causes

Test deformations in benthic foraminifera are known to occur in polluted marine habitats as well as in natural, non-impacted environments. An example of the latter is the above-mentioned (section 4.2.1) high abundance of abnormal *Ammonia tepida* tests in a hypersaline inland pool in the Dead Sea area, Israel (Almogi-Labin *et al.*, 1995), and deformed tests are also known from the fossil record. So far, we only have a limited understanding of which environmental parameters cause the development of deformed tests. Suggested causes include: a periodically reduced growth rate as a response to periods with reduced feeding activity; the repair of physically damaged tests; and some chemically stressful conditions. For instance, laboratory experiments have shown that double tests develop spontaneously, but much more commonly under hypersaline than under normal-marine conditions (Stouff *et al.*, 1999). One probable reason is that hypersalinity may inhibit or slow down the pseudopodial activity and thereby slow down dispersal and facilitate fusion of young individuals subsequent to schizogony (asexual reproduction). Other kinds of test deformation include, for instance, double apertures, enlarged apertures, reduced size of one or more chambers, protuberances, twisted or distorted chamber arrangement, and aberrant chamber shape and size. In severe cases, test deformation makes species identification impossible.

In recent years, several investigations have shown that test deformation commonly occurs in polluted environments. For example, in Southampton Water, UK, successively increasing relative abundances of deformed *Ammonia beccarii* tests were found, ranging from none at the 'clean' site to 10–20 per cent at the more polluted sites (Table 4.6) (Sharifi *et al.*, 1991). However, the previously mentioned experiments in the Oslofjord, Norway, showed that not even Cu-concentrations of >2000 ppm induced test deformations beyond the range of abundance normally found in uncontaminated marine sediments (the latter generally have Cu concentrations of <50 ppm).

Table 4.6 Maximum concentrations of some heavy metals in estuarine, surface (top 1 cm) sediments, impact category, and % deformed tests of live *Ammonia beccarii* populations in Southampton Water, UK (based on Sharifi *et al.*, 1991)

Site	Metals in ppm dry weight							Impact category	% deformed *A. beccarii*
	Cu	Cr	Ni	Zn	Co	Cd	Pb		
8	30	32	21	138	22	7.3	55	Clean	0
1	51	42	48	143	32	8	81	Relatively	
3	93	41	38	166	26	7	111	clean	0
7	40	51	36	127	24	10.4	75		
2	122	48	58	396	33	9	128	Relatively	
5	396	88	76	237	52	11	161	polluted	5–10
6	109	80	78	216	55	9	161		
4	1007	160	118	470	71	12	281	More	10–20
9	641	130	140	403	88	14	233	polluted	

This suggests that Cu is not solely responsible for the higher than 'normal' abundances of deformed tests recorded in many polluted areas. On the other hand, it is possible that the fine-grained nature of the sediment used in the experiment, with a mean organic carbon content of 3.2 per cent, caused the Cu to be less bioavailable than would have been the case if it had been performed with more coarse-grained sediments.

4.4 USE IN ARCHAEOLOGY

Between 21 and 10 BC, an all-weather harbour was constructed on the coast of Israel midway between Tel Aviv and Haifa. The harbour had a profound effect on the longshore transport of sand, and the quiet waters in its protection were the site of mud/silt deposition, which is quite unusual on this otherwise high-energy coastline. Archaeological evidence suggested that the harbour was in active use until the mid to late first century AD and that there were problems as early as 66 AD. By 500 AD the harbour was a danger because the breakwaters were submerged and several ships were wrecked there in storms. At present, the original outer breakwaters are submerged 5 m beneath the sea; the subsidence is thought to be due to neotectonism on coast-parallel faults. Now only a small harbour remains, occupying the inner part of the original.

A core taken in the original outer harbour showed a sediment sequence from before the harbour was built, during harbour construction, the active harbour, and post-harbour destruction (Reinhardt *et al.*, 1994). The foraminiferal faunas can be divided into two biofacies representing different environments: first a miliolid and *Ammonia tepida* biofacies like that found in modern low-energy Mediterranean lagoons; and second an *Ammonia parkinsoniana*, *Porosononion granosum* and miliolid biofacies typical of the modern high-energy shelf. At the base is pre-harbour coarse sand with biofacies 2. The rubble from harbour construction was not sampled for foraminifera. The active harbour deposits show alternations of sand and organic-rich muds, the former being attributed to periodic storms and the latter to normal quiet deposition. The muds contain biofacies 1, which is considered to be indigenous. The sands with their associated biofacies 2 are considered to have been transported in during storms. The post-harbour destruction sands have biofacies 2. Combining this palaeoecological evidence with a (somewhat unreliable) radiocarbon date from near the top of the active harbour unit, it is inferred that the harbour area reverted back to normal high-energy shelf conditions probably before 250 AD and no later than 490 AD. Thus, the palaeoecological interpretation based on the sediments and foraminifera gives more detailed interpretation than would be obtained from a standard archaeological excavation.

Estuarine deposits laid down in the fourteenth century at the mouth of the River Fleet in central London, UK, have revealed much of the medieval environment. The succession commences with gravels, overlain by sands, silts and clays. The biological remains present include plants and eggs of parasitic worms from human faeces. This is one of the earliest records of organic pollution (Boyd, 1981). The foraminifera present (especially in the fine-grained sediments), include forms such as *Ammonia beccarii, Elphidium williamsoni, Haynesina germanica, Miliammina fusca* and *Trochammina inflata*, which represent intertidal mudflats and marsh. These may have been indigenous, but there is also the possibility that they were transported in from the macrotidal Thames estuary. In addition, there is an assemblage of marine forms that was clearly transported into the estuary from the adjacent North Sea. Historical records of drought during the summers of 1324, 1325 and 1326 can be matched with an increased inflow of brackish water into the Fleet.

4.5 ROLE IN SEQUENCE STRATIGRAPHY

Sequence stratigraphy is based on the recognition of sequences of genetically related strata which are bounded by surfaces of erosion or non-deposition or their correlative conformities. Sequences comprise systems tracts (ST) such as transgressive (TST), highstand (HST) and regressive (RST). Although sequences are commonly determined from continuous-reflection seismic profiles, they are also determined from boreholes (geophysical records and core samples) and outcrops. Detailed analysis of sediments and microfaunas helps to identify sequence boundaries, parasequences (subseismic scale sequences), and environments of deposition. Unfortunately, the quality of ditch cuttings from commercially drilled boreholes is poor, and limits palaeoecological resolution. Commonly biostratigraphical data are integrated with sequence-stratigraphical data only after the sequences have been mapped (often from seismic data) and this often leads to disagreeing interpretations. To overcome this problem it is better to integrate the biostratigraphical information with the stratigraphical analysis, in order to help to define and map the sequences.

North Island, New Zealand, lies over a subduction zone formed by the subduction of the Pacific plate beneath the Australian plate. Wanganui Basin has an almost complete Plio-Pleistocene stratigraphical record of shelf sediments exposed in outcrop sections which have been studied in detail (Naish and Kamp, 1997). The Rangitikei River valley has a 1-km thick succession comprising a progradational stack of 20 depositional sequences (sixth order, 41 ka duration). Rapid basinal subsidence was matched by similar rates of sediment accumulation. Superimposed on the basin subsidence, the primary control on sedimentation and depositional environments was fluctuations in glacio-eustatic sea-level. Each glacial/interglacial stage couplet is made up of a sequence of transgressive, highstand and regressive systems tracts. Multivariate analysis of the foraminiferal data from outcrop samples defined

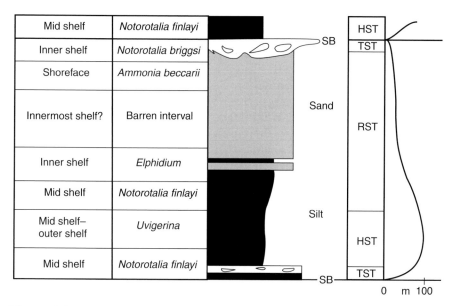

Figure 4.12 Data and interpretation of an asymmetrical-style sequence about 80 m thick from the Plio-Pleistocene of North Island, New Zealand. The right-hand curve is inferred palaeobathymetry. (After Naish and Kamp, 1997.)

13 associations which have been grouped into seven depth-related biofacies ranging from shoreline to outer shelf. An asymmetrical-style sequence shows upward shallowing from mid shelf to shoreface, bounded by unconformities above and below (Fig. 4.12). Palaeobathymetric changes determined from the foraminifera (100–200 m) agree with those obtained from the macrofauna. They are somewhat higher than those calculated from the oceanic oxygen stable-isotope record (50–100 m) which gives a measure of glacio-eustasy, and the difference indicates that, in addition, there was significant subsidence.

4.6 SUMMARY

The case studies show that foraminifera can be used to interpret environmental change, whether natural or anthropogenic, over time-scales ranging from a few years or decades to thousands of years. Natural environmental changes are linked to the climate/ocean system. Because many modern forms were already present in the Neogene, it is possible to make detailed environmental reconstructions that far back in time. Foraminifera are good indicators of changes in salinity, temperature, availability of dissolved oxygen, food supply, eutrophication and changes of sea-level. They can be used to document the effects of various types of pollution (organic, hydrocarbons, heavy metals). In contemporary polluted environments, pre-pollution conditions can be reconstructed from the foraminiferal assemblages preserved in the pre-pollution sedimentary record. Where pollution is diminishing they can be used to monitor recovery, including recolonisation of previously defaunated substrates, or recolonisation by the more sensitive species in areas where the impact did not kill off the entire foraminiferal fauna.

FURTHER READING

The following provide good general background reading:

Alve, E., 1995a. Benthic foraminiferal responses to estuarine pollution: a review. *Journal of Foraminiferal Research*, 25, 190–203.

Lee, J.J. and Hallock, P. (eds), 2000. Advances in the biology of foraminifera. *Micropaleontology*, 46, Suppl., 1–198.

Martin, R. (ed.), 2000. *Environmental micropaleontology*. Kluwer Academic/Plenum Publishers, New York, 481 pp.

Murray, J.W., 1991. *Ecology and palaeoecology of benthic foraminifera*. Longman, Harlow, Essex, 397 pp.

Scott, D.B., Medioli, F.S. and Schafer, C.T., 2001. *Monitoring in Coastal Environments using Foraminifera and Thecamoebian Indicators*. Cambridge University Press, Cambridge, 177 pp.

Sen Gupta, B.K. (ed.), 1999. *Modern foraminifera*. Kluwer Academic Publishers, Dordrecht, 371 pp.

INTERTIDAL FORAMINIFERA AS PALAEOENVIRONMENTAL INDICATORS

W. Roland Gehrels

Intertidal foraminifera are perhaps the toughest foraminifera around! Many species are adapted to harsh environmental conditions, such as low salinity, frequent subaerial exposure and low pH. The sedimentary environments that comprise the intertidal zone (salt marshes, tidal flats and tidal channels and creeks) each harbour their own characteristic foraminiferal assemblages.

Salt-marsh foraminifera are mostly **agglutinated**; i.e. their tests are composed of detrital clastic material kept together by a lining. Agglutinated forms are usually coloured brown or beige when viewed under the light microscope. Unlike calcareous foraminifera, agglutinated species are able to withstand the acidic conditions that commonly occur in salt marshes. In fact, agglutinated tests are usually the only test types that preserve in salt-marsh sediments. Species found on the tidal flats and in the creeks and channels have **calcareous** tests (although some agglutinated forms also occur) and are white or glassy. All intertidal foraminifera are benthic: they live on the sediment surface (epifaunal) or just below it (infaunal). Infaunal foraminifera are particularly common along low-latitude coastlines and may be found alive at depths of up to 1 m below the surface.

There has been some discussion about the **taxonomy** of common intertidal foraminifera. For example, *Trochammina macrescens* (e.g. Scott and Medioli, 1980) constitutes both species *Balticammina pseudomacrescens* and *Jadammina macrescens* (Gehrels and van de Plassche, 1999). Other (likely) synonyms include: *Elphidium williamsoni*, *Elphidium articulatum* and *Cribrononion umbilicatulum*; *Haynesina germanica*, *Haynesina depressula*, *Protelphidium anglicum* and *Protelphidium germanicum* (e.g. Murray, 2000). For reasons of simplicity, and to avoid extensive taxonomic debate beyond the scope of this chapter, species names from case studies correspond to the names presented by the original authors of papers to which reference is made. The reader is referred to these papers if more taxonomic information is required. Figures 5.1 and 5.2 show some of the most common intertidal foraminifera.

The aim of this chapter is to review those techniques that have a clear application in Quaternary **palaeoenvironmental reconstruction**. The large amount

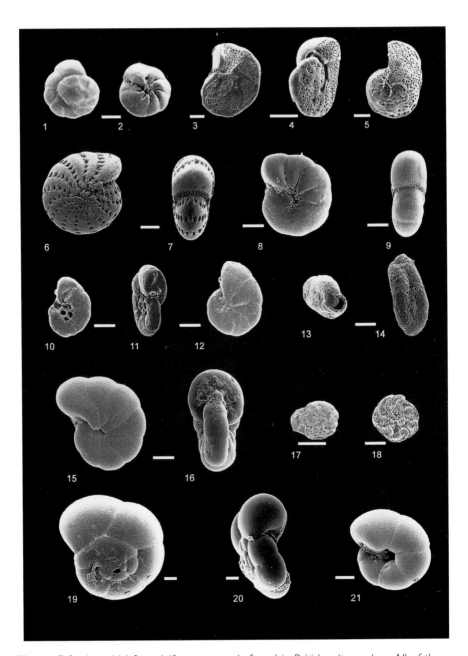

Figure 5.1 Intertidal foraminifera commonly found in British salt marshes. All of these species are also found in North American salt marshes. All scale bars are 100 μm in length. 1–2 *Ammonia beccarii* (Linné), 1 dorsal view, 2 ventral view; 3–5 *Cibicides lobatulus* (Walker and Jacob), 3 ventral view, 4 apertural view, 5 dorsal view; 6–7 *Elphidium williamsoni* Haynes, 6 side view, 7 apertural view; 8–9 *Haynesina germanica* (Ehrenberg), 8 side view, 9 apertural view; 10–12 *Jadammina macrescens* (Brady), 10 dorsal view, 11 view showing multiple apertures, 12 ventral view; 13–14 *Miliammina fusca* (Brady), 13 apertural view,

of literature on the use of intertidal foraminifera as pollution indicators will not be considered, and the reader is referred to Chapter 4 of this volume.

5.1   SAMPLE PREPARATION

5.1.1   Field Sampling

Cores from intertidal environments can be collected using a variety of coring techniques, including a gouge auger, a vibracorer or a piston corer. Hand-held gouges (e.g. Eijkelkamp gouges) are probably the quickest and easiest to use.

In many palaeoecological reconstructions based on foraminiferal evidence, a **uniformitarian approach** is recommended. This means that the relationship between modern foraminifera and the ecological parameter to be reconstructed is investigated first in a field setting that is presumed to be similar to the palaeoenvironment. These data form a **training set** that serves as an analogue of the fossil foraminifera from which the palaeoecological information can then be reconstructed. The case studies in section 5.3 are examples of a uniformitarian approach.

A **Pitman corer** is a handy sampling device for collecting surface samples from saltmarshes and tidal flats. It is an aluminium cylinder, about 8 cm in diameter, with plastic caps on both ends. A tomato-paste tin can be used as an improvised alternative. The sampling device is pushed into the sediment, with the top capped, and the sample is retrieved by pulling back the sampling device while supporting the sample with a knife. The bottom is then capped and both top and bottom caps are labelled. Most commonly a sample is collected that is 10 cm deep. In many studies, in particular those that attempt to reconstruct sea-level changes, it is crucial to survey the heights of the surface samples.

After collection, the samples are placed in a solution of 10 per cent formaldehyde or 98 per cent ethanol, with baking soda (one teaspoon per litre) to buffer the solution, and, if desired, **rose Bengal** (one teaspoon per litre) to stain living foraminifera. Samples may be immersed into solution in the field, but it is often more practical to do this in the laboratory upon return from the field. To avoid sample deterioration, samples should not be left untreated for more than a day, especially when calcareous foraminifera are present in the samples. In the laboratory, a subsample can be obtained if so desired. If there is a substantial infaunal foraminiferal population, the entire

14 side view; 15–16 *Haplophragmoides wilberti*, 15 side view, 16 apertural view; 17–18 *Trochammina ochracea* (Williamson), 17 dorsal view, 18 ventral view; 19–21 *Trochammina inflata* (Montagu), 19 dorsal view, 20 apertural view, 21 ventral view. (Specimens 1–16 from the Erme estuary, Devon, England; 17 from Chezzetcook, Nova Scotia, Canada; 18 from the Taf Estuary, Carmarthenshire, Wales; 19–21 from the Lynher estuary, Cornwall, England.)

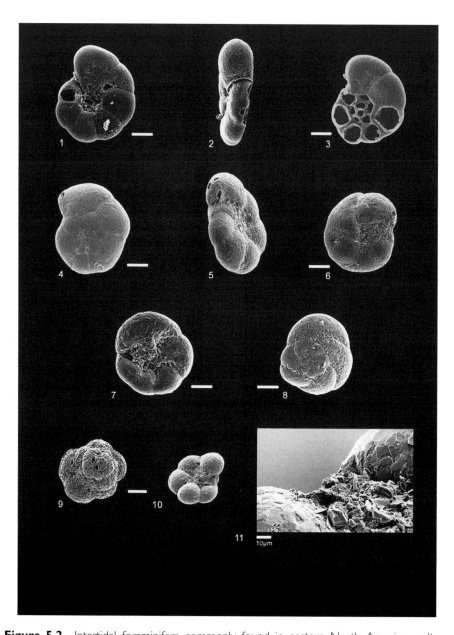

Figure 5.2 Intertidal foraminifera commonly found in eastern North American salt-marshes. These species are rare in British salt marshes. All scale bars are 100 μm in length unless otherwise stated. 1–3 *Balticammina pseudomacrescens* Brönnimann, Lutze and Whittaker, 1 ventral view with supplementary apertures in umbilicus, 2 apertural view, 3 dorsal view; 4–6 *Arenoparrella mexicana* (Kornfeld), 4 dorsal view, 5 apertural view, 6 ventral view; 7–8 *Tiphotrocha comprimata* (Cushman and Brönnimann), 7 ventral view, 8 dorsal view; 9–11 *Siphotrochammina lobata* Saunders, 9 dorsal view, 10 ventral view, 11 detail of aperture. (Specimens 1–3 and 7–8 from Chezzetcook, Nova Scotia, Canada; 4–6 from Watts Island, Connecticut, USA; 9–11 from Wells, Maine, USA.)

sample, 10 cm in depth, should be processed to determine accurately the surface distribution of the living foraminifera (Patterson *et al.*, 1999). If infaunal foraminifera are scarce, it is sufficient to analyse only the top 1 cm. In the laboratory, samples should be stored at a temperature of 4–6 °C.

5.1.2 Sample preparation

The samples should be soaked for at least 24 hours in the solution of rose Bengal and ethanol (or formaldehyde). Shaking the sample will ensure that all living foraminifera take up the rose Bengal stain. The sample is wet-sieved between 63 μm and 500 μm. From the 63 μm sieve, the sample is collected and washed into a beaker. The sediment should settle in the beaker, and the organic material in suspension can be decanted carefully. It is wise to collect the decanted material and check under the microscope that no foraminifera are lost. If the sample is large, it is recommended to use a **wet-splitter** (Fig. 5.3). The wet-splitter is filled with turbulent water and, when the splitter is half full, the sample is added and the splitter is then filled up to the rim with water. The sediment should settle for at least three minutes. This time should be increased if very small or light foraminifera are known to be present, but if the settling time is too long there will be a lot of organic material left in the sample (especially in salt-marsh samples), which will obscure the foraminifera. The water (with suspended fine organic material) is drained via a hose. This residue should be collected in the 63 μm sieve and checked for foraminifera. If foraminifera are present, the settling time should be increased. A plug is pulled out to collect one-eighth of the initial sample and this subsample is collected in a beaker. The sample is again allowed to settle and excess water is carefully decanted before the sample is poured in a counting tray. Alternatively, a pipette is used to collect a portion of the sediment in the counting tray. It is best to spread out the sediment thinly so that the subsample is not too large and foraminifera are not obscured by the large amount of sediment. Usually, intertidal foraminifera samples should not be dried, because the sediment would clump together, making specimen picking very difficult. Also, if the samples are dried, organic linings of foraminifera are lost. These linings are the remains when tests are dissolved and they can provide useful palaeoecological information (Edwards and Horton, 2000). The traditional picking trays are not recommended for wet samples – it is better to use a tray with a spiralled groove, as depicted in Fig. 5.3.

Core samples are prepared in the same manner as surface samples, but staining is only needed to detect infaunal foraminifera living near the top of the cores (if the core is collected from an 'active' environment). A sample volume of 5–10 cm³ usually contains sufficient numbers of foraminifera. When less core material is available, a 2 cm³ sample may suffice.

The counting tray is placed under a binocular microscope. Appropriate magnification is 40–80×. Foraminifera are picked with a fine paintbrush and placed on a pre-glued, labelled microfaunal slide. The entire tray is scanned for foraminifera because some sorting according to size can occur. If

Figure 5.3 Wet splitter and counting tray.

a subsample does not provide sufficient specimens, more sample material is prepared from the raw sample. A total number of 200–300 foraminifera should be counted. A high count is required from a statistical point of view if important indicator species are present in very low numbers (Patterson and Fishbein, 1989). However, if the species diversity is very low and indicator species are present in high abundances, a lower number may suffice. In upper salt-marsh samples the concentration of foraminifera sometimes may be too low to obtain the desired number.

5.2 INTERTIDAL ZONATION

A specific feature common to all intertidal environments around the world is that of intertidal zonation. The concept of intertidal zonation is the key to the suitability of intertidal foraminifera as palaeoecological indicators, as will become clear later in this chapter. Due to the tides advancing and retreating every six hours (approximately), the lower areas of the intertidal zones are more frequently submerged than the higher parts and, conversely, the higher parts of the intertidal zone are more frequently subjected to sub-aerial exposure. The tides also supply most of the nutrients for the animals and plants. The competition for space between species results in distinct bands of biota, with the most adaptable – or the hardiest – species outcompeting other species under the most stressful environmental conditions in the upper parts of the intertidal zone. The most visible examples of intertidal zonation are seaweeds on rocky shores and the vegetation zones in salt marshes. A flooding curve for a salt marsh and its relationship to plant and foraminiferal zonation is shown in Fig. 5.4.

The intertidal zonation of foraminifera was first described by Phleger and Walton (1950) for the Great Marshes in Barnstable Harbor, Massachusetts. They distinguished between four ecological zones ('*Zostera* zone', 'mud flat zone', '*Spartina glabra* zone' and '*Spartina patens* zone'), the latter three zones being intertidal. The *Spartina patens* and *Spartina glabra* zones contain high abundances of *Trochammina macrescens*, *Trochammina inflata* and *Miliammina fusca*, while *Elphidium incertum* dominates the mudflat zone. A vertical zonation of intertidal foraminifera has since been found in many other parts of the world (Fig. 5.5). The species composition of the intertidal foraminiferal fauna varies from place to place, but the agglutinated species *Trochammina inflata*, *Miliammina fusca* and *Jadammina macrescens* appear to have a cosmopolitan distribution (Scott *et al.*, 1990, 1996).

The relationship between benthic foraminifera and water depth is well described in marine environments. In intertidal areas the same relationship exists, but height is often used as the environmental parameter instead of water depth, because the intertidal zone dries out during the low tide and is perceived as 'land'. The relationship between height and foraminifera has been described for many salt marshes around the world, and is best known from the salt marshes in Nova Scotia, Canada.

In Chezzetcook Inlet, Scott and Medioli (1978, 1980) found four distinct zones of foraminifera (Fig 5.5), spanning the vertical range between mean sea-level (MSL) and the highest astronomical tide (HAT). Higher than the HAT, 1 m above MSL, no foraminifera occur. The upper zone IA consists of *Jadammina macrescens* and *Balticammina pseudomacrescens* (grouped by Scott and Medioli as *Trochammina macrescens*) and is found between 0.95 and 1 m above MSL. Zone IB, between 0.75 m and 0.95 m above MSL, is dominated by *Trochammina macrescens* and *Tiphotrocha comprimata*. *Trochammina inflata* and *Miliammina fusca* dominate zone IIA, between 0.55 m and 0.75 m above MSL. The lower zone, zone IIB, is found between MSL and 0.55 m above MSL, and includes the species *Miliammina fusca*, *Ammotium salsum* and *Elphidium williamsoni* (Scott and Medioli's *Cribrononion umbilicatulum*). The four zones in Maine are very similar to those in Nova Scotia (Gehrels, 1994; Fig. 5.4).

In most other coastal areas, two or three foraminiferal zones can be distinguished (Fig. 5.5). It appears that only along the east coast of North America and in England are four zones found. The highest foraminiferal zone usually extends towards the HAT, so that in areas with a large tidal range, the zones are found at high elevations relative to MSL. Foraminifera are precise sea-level indicators where the vertical zones are narrow. Although Fig. 5.5 does not give a complete picture of all studies of foraminiferal intertidal zonation, it is fair to say that many coastlines around the world have not been studied at all and that most studies are from North America.

The vertical zonation patterns of foraminifera as depicted in Fig. 5.5 are qualitative and descriptive in nature. When palaeoenvironmental information is derived from fossil foraminiferal assemblages, it is often useful to

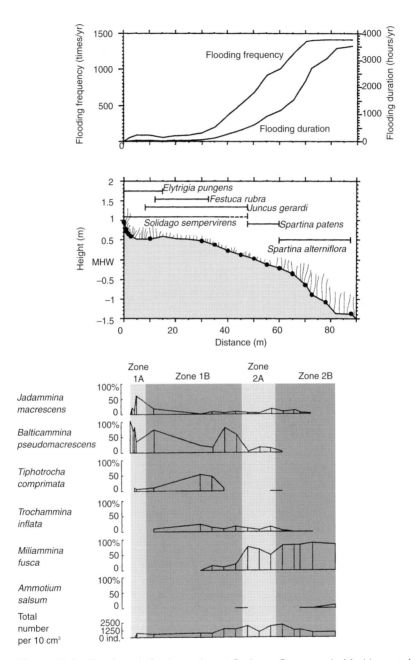

Figure 5.4 The foraminiferal zonation at Sanborn Cove marsh, Machiasport, Maine, in relation to the zonation of plants and the frequency and duration of tidal flooding. The foraminiferal zones (1A, 1B, 2A, 2B) correspond with the zones from Nova Scotia marshes described by Scott and Medioli (1978), except for the fact that *Jadammina macrescens* and *Balticammina pseudomacrescens* are grouped as *Trochammina macrescens* in Nova Scotia. Foraminiferal and plant data are adapted from Gehrels (1994) and Gehrels and Van de Plassche (1999). MHW – mean high water (1.97 m above mean sea-level).

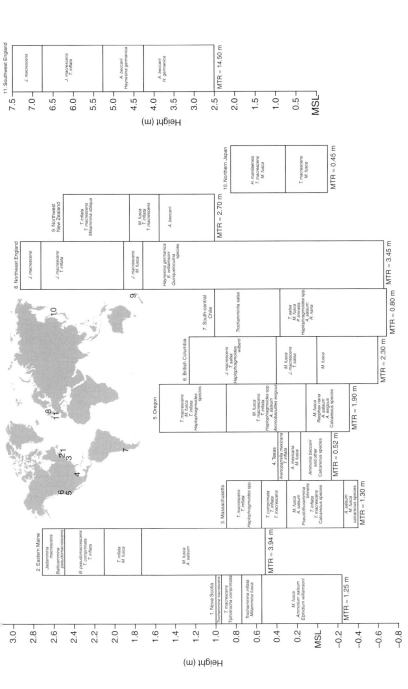

Figure 5.5 Relationship between foraminiferal assemblages and mean sea-level (MSL) for selected salt marshes around the world. (Sources: 1. Scott and Medioli (1980); 2. Gehrels (1994); Gehrels and Van de Plassche (1999), this study (Fig. 5.4); 3. Scott and Leckie (1990); 4. Williams (1994); 5. Jennings and Nelson (1992); 6. Guilbault et al. (1995, 1996); 7. Jennings et al. (1995); 8. Horton (1999); 9. Hayward et al. (1999); 10. Scott et al. (1996); 11. Haslett et al. (1998, 2001). Note the different scale for 11, an area with extreme macrotidal conditions. MTR – mean tidal range.)

quantify relationships between species and variables (Gehrels *et al.*, 2001). A first step to explore the relationship between foraminifera and height in a **quantitative** manner is simply by plotting the species against height. In Fig. 5.6 this is done for three species, (*Jadammina macrescens*, *Miliammina fusca* and *Tiphotrocha comprimata*), for a data-set consisting of 68 surface samples from various salt marshes in Maine. The tidal range varies from marsh to marsh, and the data-set has therefore been normalised to a vertical range of 1 m between MSL and HAT.

It can be seen from Fig. 5.6 that the distribution of salt-marsh foraminifera along the height gradient can take various forms. *Jadammina macrescens* and *Miliammina fusca* show a **linear** distribution. *Jadammina macrescens* has its optimum occurrence near the HAT. Apparently, it can adapt readily to many hours of subaerial exposure – more so than other species. Lower in the marsh, *Jadammina macrescens* coexists with other species. *Miliammina fusca* shows the opposite distribution pattern; its relative abundance increases with decreasing height. *Tiphotrocha comprimata*, on the other hand, does not show a linear distribution pattern. Its optimum occurrence is somewhere near the mean high-water mark. Above and below this level, the species' relative abundance decreases. Therefore, its vertical distribution pattern is best described as **unimodal**.

When a complete data-set of surface foraminifera is analysed, it must first be investigated whether a linear regression model or a unimodal regression model is more appropriate to describe the distribution of all foraminifera

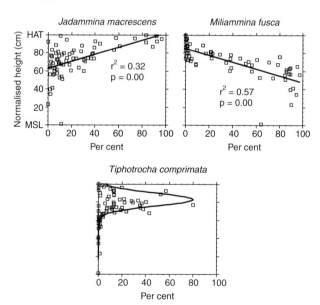

Figure 5.6 Species–height plots for *Jadammina macrescens*, *Miliammina fusca* and *Tiphotrocha comprimata*. The data are from various saltmarshes in Maine, USA, and are expressed as a percentage of the species in each sample. *J. macrescens* and *M. fusca* show a linear distribution along the elevational gradient, whereas *T. comprimata*'s distribution resembles a unimodal pattern. HAT – highest astronomical tide; MSL – mean sea-level. (Data from Gehrels (1994, 2000).)

along the environmental gradient. This can be achieved by **detrended canonical correspondence analysis** (DCCA). DCCA calculates the gradient length for the environmental variables under study, as expressed in standard deviation units (Birks, 1995). If the gradient length is smaller than two standard deviation units, the distribution of the foraminifera is best described as linear. A unimodal regression model is more appropriate for gradient lengths greater than two standard deviation units.

Figure 5.7 shows the results of a **regression analysis** on a set of 68 surface samples from Maine marshes (Gehrels, 2000), using the program CALIBRATE (Juggins and Ter Braak, 1998). The regression was carried out using a unimodal regression model. The regression identifies the optimum height of the foraminifera, as well as their tolerance (which can be interpreted as a standard deviation if the foraminifera are assumed to be normally distributed along the height gradient). The pattern reflects the qualitative zonation pattern of Scott and Medioli (1978, 1980) described earlier in this chapter. *Balticammina pseudomacrescens* and *Jadammina macrescens* (which Scott and Medioli grouped together as *Trochammina macrescens*) are the highest occurring foraminifera. *Tiphotrocha comprimata* and *Trochammina inflata* occur in a middle zone, whereas *Miliammina fusca* has its optimum near the lower end of the sampled gradient.

5.3 RECONSTRUCTIONS

5.3.1 Reconstruction of Sea-Level Changes

5.3.1.1 Salt-marsh Deposits

The concept of foraminiferal intertidal zonation was first applied in palaeo-environmental reconstruction by Scott (1976), who recognised that the narrow vertical zones of the foraminifera, and particularly the specific heights of these zones in relation to sea-level, could be readily used to relocate former sea-levels in cores. In an oft-cited paper, Scott and Medioli (1978) argued that a monospecific assemblage of *Trochammina macrescens*, which occurs in a very narrow zone near the highest astronomical tide mark in salt marshes in Nova Scotia (Canada), could indicate, when found in a core, a sea-level with a precision of ±0.05 m. Although spatial variability may add a degree of imprecision, the value quoted by Scott and Medioli (1978) would make *Trochammina macrescens* arguably the most accurate sea-level indicator in the world.

Because *Trochammina macrescens* is the highest occurring salt-marsh foraminifer living on the salt-marsh surface on the coastlines of the Canadian Maritimes and New England, fossilised *Trochammina macrescens* can be found in abundance near the base of salt-marsh deposits. These deposits are the stratigraphical equivalent of the highest part of the salt marsh, and were formed during the initial stages of the Holocene marine transgression. If the sediments in which *Trochammina macrescens* is found overlie hard substrates, it can be assumed that they are not displaced by the weight of the overlying sediments. These sediments are therefore especially useful for sea-level reconstruction and, when dated, provide **sea-level index points** that can be accurately related to the highest astronomical tide level.

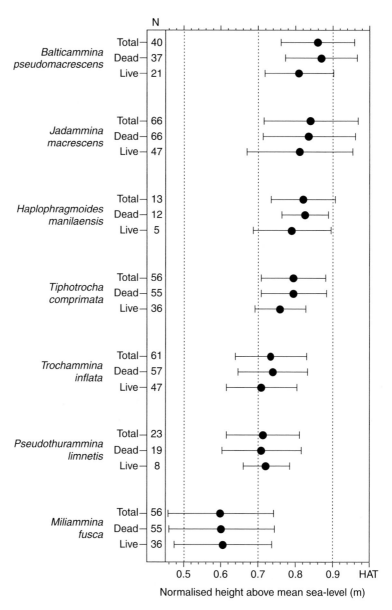

Figure 5.7 Optima and tolerances of foraminifera in Maine salt marshes in relation to their height, calculated from relative species abundances along sampled transects. The heights have been standardised to account for different tidal ranges. Data are shown for live, dead and total (live and dead) species. N – number of samples; HAT – highest astronomical tide. (From Gehrels (2000).)

Sanborn Cove marsh is located in eastern Maine (USA) along the western shore of Machias Bay. The oldest radiocarbon dates on the basal peat overlying the Pleistocene substrate indicate that the salt marsh has been in existence for at least 5000 years (Gehrels and Belknap, 1993). Since its inception, the marsh has gradually built up, keeping pace with the long-term rise in sea-level. A 4 m-thick peat sequence has developed. A core was collected from this peat and analysed for foraminifera (Fig. 5.8).

The foraminiferal assemblages in the core are dominated by *Balticammina pseudomacrescens, Jadammina macrescens, Tiphotrocha comprimata* and *Trochammina inflata*. These are species that thrive in the middle and upper marsh, as can be seen in Fig. 5.4. Near the top of the core, the middle- and low-marsh species *Miliammina fusca* appears. The changes in foraminiferal stratigraphy make it possible to reconstruct the changes in height of the marsh surface relative to mean sea-level. This can be achieved by quantifying the distributions of the foraminifera on the present-day surface of the marsh, as depicted in Fig. 5.7, and applying this information to the core. Each foraminiferal sample in the core is thus assigned a former marsh-surface height. The term **indicative meaning** is commonly used for the height at which the foraminifera once existed. Regression equations are used to calculate the relationships between height and the modern foraminiferal data-set and to calculate the indicative meaning of the fossil foraminifera; these equations form a **transfer function** (see Appendix).

A transfer function can be established for other environmental variables as well, as long as the relationship between the variables and a modern training set of foraminifera is quantified. The indicative meaning of the foraminiferal assemblages in Fig. 5.8 has been reconstructed using flooding duration as the environmental variable. Although in sea-level reconstructions 'height' is the variable of interest, it cannot be considered a true ecological variable, and its relationship with flooding duration (Fig. 5.4) is not linear. In the reconstruction for Sanborn Cove marsh (Fig. 5.8), the transfer function calculates the flooding duration for a given fossil foraminiferal assemblage, and the relationship between height and flooding duration on the modern marsh surface is then used to convert flooding duration into height. More detail can be found in Gehrels (1999).

The indicative meaning is not the only parameter that determines the vertical position of the sea-level index point in the time–depth diagram. Gehrels (1999) calculated the vertical position of a sea-level index point (SLIP) as:

$$SLIP = H - D - I + T - L + A$$

where H is the height of the marsh surface relative to MSL at the site of the core from which the foraminiferal sample is retrieved; D is the depth of the sample in the core; I is the height of the former marsh surface relative to MSL (or indicative meaning); T is the difference between the present mean M_2 high-water level and the former mean M_2 high-water level, relative to present MSL; L is the core compaction at the stratigraphical level of the sample; and A is the **autocompaction** of the peat. Changes in tidal range (T) and autocompaction (A) are the most difficult factors to determine. Over time-scales of less than 1000 years,

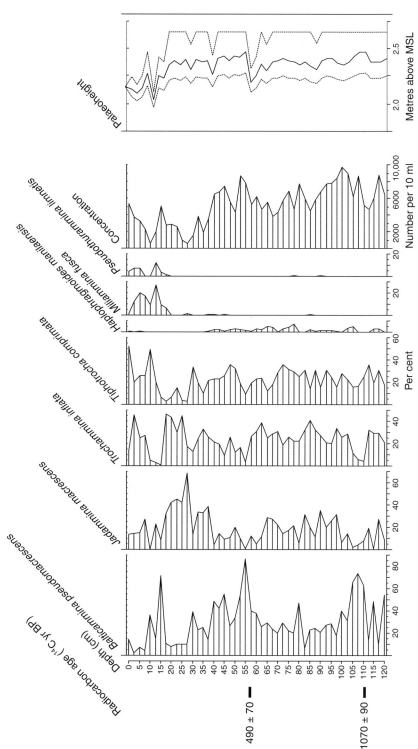

Figure 5.8 Foraminiferal stratigraphy of core SN-VC-1, from Sanborn Cove marsh, Machiasport, Maine. The average error in the reconstruction of the palaeoheight of the marsh surface is +0.23/−0.14 m. MSL − mean sea-level. (From Gehrels (2000).)

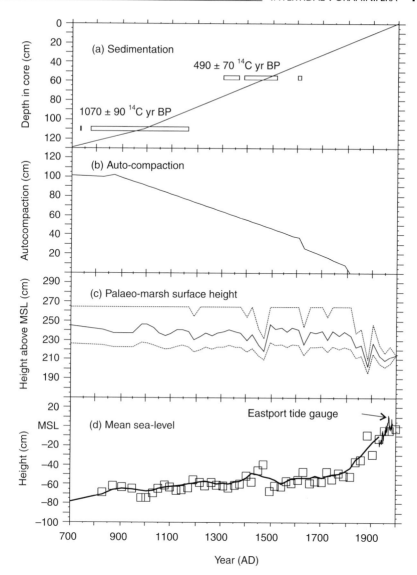

Figure 5.9 The elements that are required to reconstruct a sea-level history (D) for Sanborn Cove marsh from foraminiferal salt-marsh stratigraphy include: rate of sedimentation determined from calibrated radiocarbon dates (a), autocompaction (b) and former marsh-surface height (or 'indicative meaning') (c). Fluctuations in sedimentation rate and changes in tidal range could complicate the reconstruction of sea-level (d). The reconstruction for the past century is validated by a comparison with observations from a nearby tide gauge. MSL – mean sea-level. (Data from Gehrels (1999).)

tidal range may be assumed to be constant, but over Holocene time-scales it is well known that the tidal range in the Gulf of Maine and Bay of Fundy has been amplified substantially (Gehrels *et al.*, 1995). Autocompaction can be quantified by a comparison of ages of basal peat, formed on a hard substrate, with the ages of samples obtained from within the peat section (Gehrels, 1999). Another way to determine autocompaction is by numerical modelling based on geotechnical characteristics (e.g. Pizzuto and Schwendt, 1997; Paul and Barras, 1998).

The core from Sanborn Cove marsh contains two radiocarbon dates, and the ages of the foraminiferal samples are obtained by interpolation between the samples. Ideally, a better-constrained chronology is necessary to resolve any possible change in the accumulation rate of the peat and to ensure that no hiatuses are present in the sequence. The resulting sea-level chronology from the Sanborn Cove marsh core is depicted in Fig. 5.9.

5.3.1.2 Isolation Basins

Salt-marsh deposits are rare in areas of isostatic uplift. Here, sea-level changes can be reconstructed from the marine deposits present in basins that were isolated from the sea at some stage in their uplift history. The basins are commonly lakes or raised bogs, and the marine deposits are buried beneath freshwater lake deposits or terrestrial peat. The stratigraphical contact between the marine and terrestrial deposits marks the isolation event, and this contact is dated to provide a sea-level index point. It is possible that, at some stages during uplift, sea-level rise outstrips uplift and marine waters re-enter the basin. This results in a more complex stratigraphy, with alternating marine and terrestrial deposits. From this type of sequence it is possible to obtain several sea-level index points (e.g. Shennan *et al.*, 1994).

In isolation basin studies, the height at which the index point is plotted in a time–altitude diagram corresponds with the height of the sill that separates the basin from the sea. However, if there is a significant tidal range it is possible to recognise various stages of transgressions and regressions, as demonstrated by the study of Lloyd (2000) at Loch nan Corr in northwest Scotland (Fig. 5.10). The lithology of the core collected from the isolation basin consists of sand overlain by organic-rich sediments. The foraminiferal assemblages show distinct changes. Lloyd (2000) recognised eight **biozones**. The lower zones are rich in calcareous foraminifera from estuarine and shelf environments. The dominant foraminifer in the lower five biozones is *Haynesina germanica*, but in biozone 2 the biodiversity reaches a maximum and many other calcareous foraminifera are also abundant. In biozones 4 and 5 we see an increase in the number of agglutinated foraminifera. *Jadammina macrescens* dominates biozone 6, while *Miliammina fusca* is the last foraminifer to occur before freshwater conditions set in, as indicated by testate amoebae.

The general pattern of water-level change, as reflected by the changes in foraminifera, shows increasing water depth through biozone 1 and the lower half of biozone 2, followed by decreasing water depth through the rest of the sequence. The decreasing water depth is due to the slow isolation process,

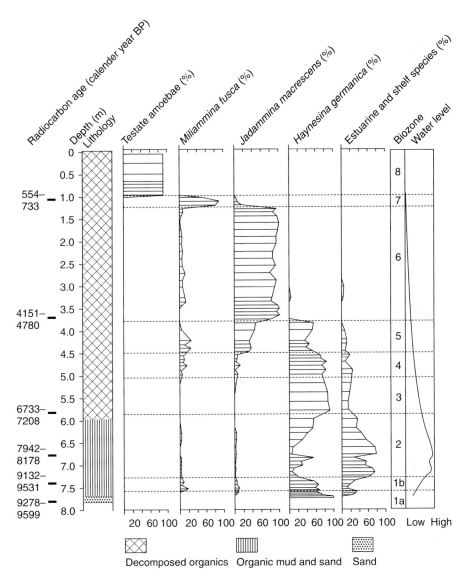

Figure 5.10 Foraminiferal stratigraphy of a core from the isolation basin Loch nan Corr in NW Scotland with calibrated radiocarbon chronology and reconstructed water levels. (Adapted from Lloyd (2000).)

which takes about 7000 years as the rock sill at 2.7 m above mean sea-level passes upwards through the intertidal zone. From the sequence, Lloyd (2000) collected seven sea-level index points. The time of the final isolation, when only the highest tides entered the basin, is dated by the youngest radiocarbon sample, at the boundary between biozones 7 and 8. Because the mean high water of spring tides occurs at 2.59 m above MSL, this sea-level index point is

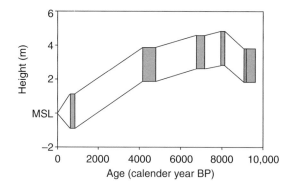

Figure 5.11 Sea-level curve for Loch nan Carr, NW Scotland. MSL – mean sea-level. (Adapted from Lloyd (2000).)

plotted at 0.11 m (the height of the rock sill, 2.7 m, minus 2.59 m) above MSL in the time–altitude diagram. A similar calculation is applied to the other radiocarbon dates to provide five accurate sea-level index points (Fig. 5.11).

5.3.1.3   Co-seismic coastal subsidence

The reconstruction of co-seismic subsidence events using foraminifera applies the same principles as the reconstruction of sea-level changes from salt-marsh deposits. Foraminifera are again used as water-depth (or surface-height) indicators. The obvious difference is that co-seismic subsidence events occur suddenly and the changes in foraminifera are abrupt, and often dramatic. The lithostratigraphy in coastal wetland areas where such events have been recorded commonly shows soils or marsh facies directly overlain by tidal flat facies. Foraminifera and other microfossils, mainly diatoms, can be used to quantify the amount of co-seismic subsidence that has occurred.

An example is shown by the detailed study of Guilbault *et al.* (1996) from Meares Island on the Pacific coast of Canada (Fig. 5.12). The core shows a regressive sequence from middle to upper marsh facies below a sand layer that is interpreted as a tsunami layer deposited during an earthquake event. Above the sand layer, a middle-marsh facies occurs which is, near the top of the core, replaced by an upper-marsh facies. The upper-marsh facies is dominated by the foraminifera *Jadammina macrescens* and *Trochamminita salsa*, whereas the middle-marsh facies also includes *Miliammina fusca*. The tsunami layer includes a significant proportion of subtidal foraminifera, such as *Trochammina ochracea*, *Trochammina nana* and *Eggerella advena*. Guilbault *et al.* (1996) quantified the modern height distributions of the foraminifera in order to establish a transfer function. They were then able to determine that the change in foraminiferal population caused by the seismic event represents a sudden subsidence of 55 cm. Importantly, the quantitative approach used by the authors also made it possible to determine the statistical error on the estimate, in this case 13 cm.

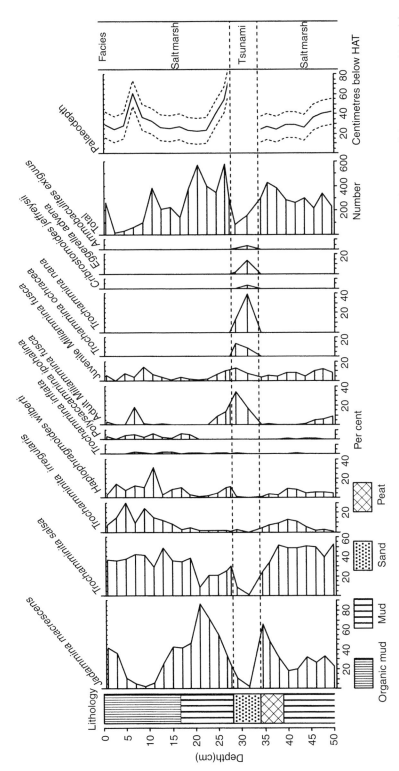

Figure 5.12 Foraminiferal stratigraphy from an exposure on Meares Island, British Columbia, Canada. A sand layer containing transported 'exotic' foraminifera marks a tsunami layer. HAT – highest astronomical tide. (Adapted from Guilbault et al. (1996).)

5.3.2 Reconstruction of Storm Events

The signature of storm events in the foraminiferal stratigraphical record relies on the identification of offshore and marine foraminifera transported into intertidal deposits. Hippensteel and Martin (1999) identified multiple washover events in cores from back-barrier marshes along the South Carolina coast (Fig. 5.13). The authors divided the foraminifera into two groups. The

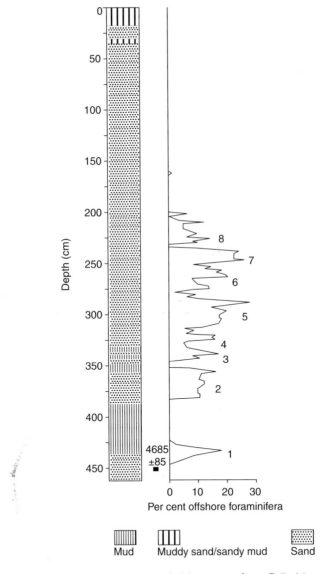

Figure 5.13 Washover events recorded in a core from Folly Island, South Carolina, USA. Eight events are recognised based on peaks in the abundance of offshore foraminifera. (Adapted from Hippensteel and Martin (1999).)

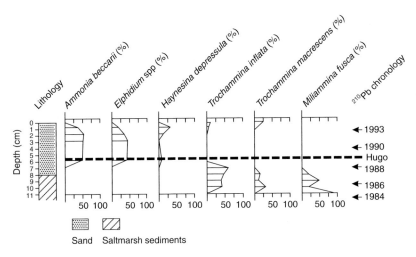

Figure 5.14   Hurricane Hugo recorded in a core from Price's Inlet, South Carolina, USA. (Adapted from Collins *et al.* (1999).)

marsh/estuarine/nearshore foraminifera included *Ammonia* species and *Elphidium* species. The offshore genera, characteristic of the overwash events, included *Bolivina, Buccella, Cibicides, Buliminella, Eponides, Fursenkoina, Nonionella, Quinqueloculina, Rosalina, Virgulina* and planktonic foraminifera. The offshore group also included the reworked Oligocene–Miocene species *Cancris sagra, Hanzawaia concentrica, Saracenaria senni, Siphogenerina transversa, Stilostomella recta* and *Uvigerina calvertensis*. In the core, eight distinct storm events are identified. Importantly, some of the events could not be detected by lithostratigraphical and grain-size analyses, so foraminiferal analyses proved extremely useful.

Collins *et al.* (1999) found a foraminiferal signature of Hurricane Hugo in a short core from a pond located between beach ridges along the southern South Carolina coast. In a sand layer, shallow-marine foraminifera are present in abundance (Fig. 5.14). These species have been washed into the pond from offshore. Before the deposition of the sand layer, a marsh community was present in the pond.

5.3.3   Reconstruction of Tsunami Events

Like storm events, tsunamis can also be recognised in the stratigraphical record from the identification of allochthonous foraminifera. However, on the basis of foraminifera alone, it is difficult to distinguish between storm deposits and tsunami deposits. Other criteria need to be considered, including stratigraphical context, spatial distribution of coarse-grained sediments and sedimentological characteristics.

Clague *et al.* (1999) analysed foraminifera in a core collected from Catala Lake on the west coast of Canada (Fig. 5.15). They identified a gravel layer

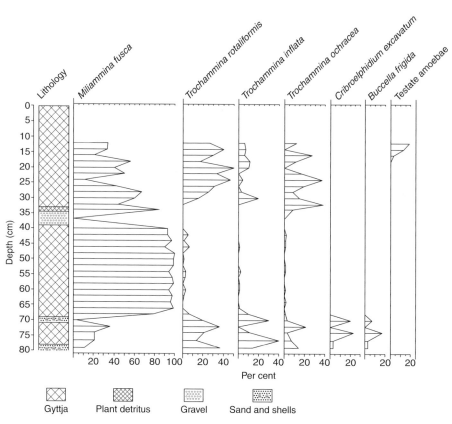

Figure 5.15 A gravel layer, interpreted as a tsunami layer, marks a dramatic change in faunal composition in a core from Catala Lake, British Columbia, Canada. (Adapted from Clague *et al.* (1999).)

in the core, which they suggested was deposited in 1700 AD by a large tsunami caused by a historically documented earthquake. The tsunami layer itself is barren of foraminifera. Below the layer, the fauna consists of an almost monospecific assemblage of *Miliammina fusca*. A much more diverse marsh assemblage appears in the lake after deposition of the tsunami layer, including *Miliammina fusca*, *Trochammina rotaliformis*, *Trochammina inflata* and *Trochammina ochracea*. The authors suggest that this assemblage reflects higher salinity conditions following deepening of the connection with the open sea by the tsunami and, possibly, by co-seismic subsidence.

The tsunami layer from Meares Island (Fig. 5.14) contains foraminifera transported from deeper water. If the sand layer was deposited during a storm, it is difficult to explain the change in height of the former marsh surface that appears to have occurred during formation of the sand layer. A seismic event is likely to have caused the sudden drop of the marsh surface. This makes it plausible that the sand layer is tsunamigenic in origin.

5.4   SUMMARY

For many palaeoenvironmental studies, sampling of modern foraminifera is necessary to create training sets that represent analogues of fossil foraminifera. Cores are easily collected with a hand-held gouge, while modern intertidal foraminifera are best sampled in the field with Pitman corers. In the laboratory, the foraminifera should be counted wet, unlike deeper-water species, because of the high organic content of intertidal samples. Sample preparation techniques are quick and easy.

Palaeoenvironmental reconstructions from intertidal foraminifera rely on the intertidal zonation pattern apparent in their modern distribution. The specific height ranges of the foraminifera are used in reconstructions of sea-level change and co-seismic subsidence. Modern relationships between foraminifera and parameters of interest (e.g. height, flooding duration) should be investigated first, following a uniformitarian approach, before these relationships are reconstructed from fossil foraminifera. Identification of storm layers and tsunami layers is possible because of the incorporation of allochthonous deeper water species. Tsunami layers can be distinguished from storm layers if vertical ground motion can be reconstructed from foraminiferal evidence. It must be kept in mind, however, that reconstructions are more robust if additional microfossil evidence is used, most commonly diatoms. This is especially important when foraminiferal stratigraphical evidence is obscured because of a number of taphonomic problems, including bioturbation, dissolution and mixing of living infaunal and fossil species.

FURTHER READING

De Rijk, S., 1995. *Agglutinated foraminifera as indicators of salt marsh development in relation to late Holocene sea level rise (Great Marshes at Barnstable, Massachusetts)*. Ph.D. thesis, Vrije Universiteit, Amsterdam.
This published thesis is an in-depth study of the ecology, biogeography, taxonomy and sea-level applications of salt-marsh foraminifera. It contains a useful section on various sample preparation and splitting techniques.

Murray, J.W., 1979. *British nearshore foraminiferids*. Academic Press, London.
This is a reasonable identification guide for species common in British intertidal areas. It contains useful background information on the ecology of foraminifera and is ideal for students who have not encountered foraminifera before. Unfortunately, the book is out of print. A taxonomic update is provided by Murray (2000).

Schafer, C.T., 2000. Monitoring nearshore marine environments using benthic foraminifera: Some protocols and pitfalls. *Micropaleontology*, 46, supplement no. 1, 161–169.
This publication discusses the taphonomic problems that need to be assessed when interpreting foraminiferal stratigraphical data, including bioturbation and differential preservation and dissolution of tests.

Scott, D.B. and Medioli, F.S., 1980. Quantitative studies of marsh foraminiferal distributions in Nova Scotia: implications for sea-level studies. *Cushman Foundation for Foraminiferal Research Special Publication*, 17, 1–57.
This provides the first overview of marsh foraminiferal distributions in a regional context. The monograph contains information on sampling methods. Five plates and taxonomic descriptions form the basis for the taxonomy of North American studies of intertidal foraminifera.

Scott, D.B. and Medioli, F.S., 1986. Foraminifera as sea-level indicators. In O. van de Plassche (ed.) *Sea-level research: a manual for the collection and evaluation of data*. Geo Books, Norwich, pp. 435–457.
This provides useful examples of the use of intertidal foraminifera as sea-level indicators in marsh deposits and in isolation basins.

Todd, R. and Low, D., 1981. *Marine Flora and Fauna of the Northeastern United States. Protozoa: Sarcodina: benthic foraminifera*. NOAA Technical Report NMFS Circular 439, US Dept. of Commerce, National Oceanic and Atmospheric Administration, National Marine Fisheries Service, pp. 17–30.
This is a nice identification key to North American benthic foraminifera, recommended for student use.

CHAPTER

6

ENVIRONMENTAL APPLICATIONS OF MARINE AND FRESHWATER OSTRACODA

Ian Boomer

The Ostracoda (or ostracods) are an abundant and diverse group of small crustaceans (most extant species range from 0.3 to 3.0 mm long as adults) with a long fossil record (Ordovician to Recent). They have been successful in colonising almost every aquatic habitat – from deep oceans to mountain springs, and they are even known to live in damp vegetation cover just above the high-water mark along coasts and estuaries.

Ostracods possess a dorsally hinged, bivalved **carapace**, which is laterally symmetrical (Fig. 6.1), and it is the carapace (comprising a left and right valve) that is preserved in the fossil record. Within the carapace are the soft parts (Fig. 6.2), which generally do not fossilise. The valves provide an important source of biogenic carbonate, which can be used for stable-isotope, trace-element geochemistry and ^{14}C studies, while the soft tissue of living species has also been used in genetic investigations.

The ostracods are largely benthonic or nekto-benthonic (the few truly pelagic taxa do not readily preserve as fossils and are not considered further in this work) and therefore do not possess the obvious chronostratigraphical value of planktonic microfossils. However, they are sensitive to changes in a wide range of environmental variables – even slight changes in such parameters can bring about a change in the species composition and structure of living populations. Similar faunal changes in fossil assemblages allow us to reconstruct past environmental change in a wide range of aquatic settings.

The value of ostracods in palaeoenvironmental reconstruction has increased greatly in recent years. This has been encouraged by the production of taxonomic keys, which make Quaternary to Recent species much easier to identify (Athersuch *et al.*, 1989; Henderson, 1990; Griffiths and Holmes, 2000; Meisch, 2000). Advances in analytical techniques and the inception of local and regional databases regarding the modern ecology of non-marine Ostracoda in particular has seen their transition from qualitative to quantitative indicators, although the modern ecology of many species remains

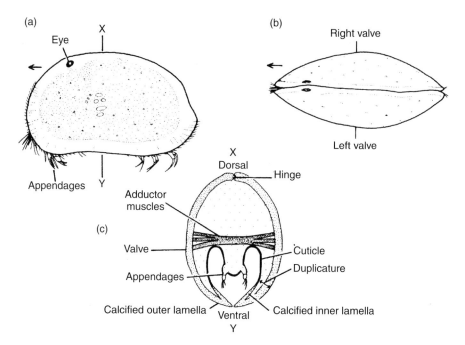

Figure 6.1 General morphology and structure of a podocopid ostracod (arrows indicate anterior direction). (a) left lateral view of carapace (appendages protrude ventrally), (b) dorsal view of carapace, (c) cross-section through carapace (X–Y on Fig. 6.1a). (Reproduced from Athersuch *et al.* (1989), by courtesy of the Linnean Society of London.)

poorly understood. They have proved to be valuable in palaeolimnology, palaeoceanography and sea-level change studies, and many examples of these are given below. A comprehensive range of applications throughout geological time can be found in the proceedings of the International Symposia on Ostracoda (see Further Reading).

6.1 TAXONOMY AND CLASSIFICATION

Modern living species can be taxonomically classified by their soft parts (mainly based on the seven to eight pairs of jointed appendages), whereas fossil taxa are usually classified using characters of the calcareous carapace, particularly the external **ornament**, dorsal **hingement**, **adductor muscle scars** and **inner lamella**. In cases of exceptional preservation it has been possible to describe the soft parts of now extinct species.

The classification of the Ostracoda has been in some flux over recent years, particularly with the ongoing revision of the *Treatise on Invertebrate Paleontology* (Moore, 1961). The most commonly adopted classification

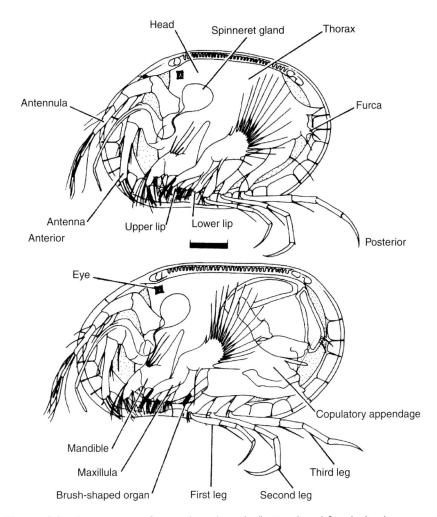

Figure 6.2 Arrangement of appendages in male (bottom) and female (top) carapaces of *Loxoconcha elliptica*. Both viewed from left-lateral view with left valve removed. (Reproduced from Athersuch *et al.* (1989), by courtesy of the Linnean Society of London.)

scheme places them within the Phylum **Arthropoda** (meaning 'jointed-foot') as a Subclass of the Class **Crustacea**, although some authors retain the Ostracoda as a distinct class. Three Orders of Ostracoda are recognised from the Quaternary to Recent: the **Myodocopida**, the **Platycopida** and the **Podocopida**. The three Orders outlined below can be distinguished on the basis of the adductor muscle-scar patterns (see Fig. 6.3 for explanation), and a brief description is given for each. Similarly the four super-families of the Podocopida have distinct muscle patterns, which are briefly described.

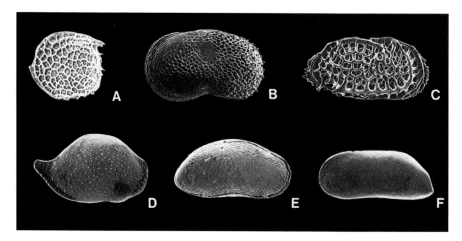

Figure 6.3 SEM illustrations of some of the main ostracod groups. A, *Polycope* sp. (80×). B, *Cytherella* sp. (55×). C, *Bradleya* sp. (45×). D, *Bairdoppilata* sp. (38×). E, *Bythocypris* sp. (47×). F, *Krithe* sp. (57×). All are lateral views of adults from the Cenozoic of the Pacific deep sea, except A, which is from the Arctic Ocean. A, C, D and E right-lateral views, B and F left-lateral.

The Myodocopida (which can reach up to 30 mm in length) are exclusively marine and usually possess weakly calcified carapaces, which aids their planktonic mode of life (suborders Halocypridina and Myodocopina). However, one suborder, the **Cladocopina**, possesses a more heavily calcified carapace and a nekto-benthonic or interstitial mode of life. *Polycope* (Fig. 6.3A) is an important cladocopine genus – round to subround in lateral view – which is usually weakly ornamented or smooth. *Polycope* has been used to reconstruct deep-water-mass changes in the Quaternary of the Arctic Ocean (Jones *et al.*, 1999). Myodocopid muscle-scar patterns can be quite varied, but in the Cladocopina they are generally characterised by a reduced pattern of three small, central, closely spaced scars.

The Platycopida (sometimes included as a suborder of the Podocopida) have moderately to heavily calcified carapaces which are ovate to subovate in lateral view (Fig. 6.3B), and usually display strong valve asymmetry with the right valve overlapping the left. Two of the more common genera are *Cytherella* and *Cytherelloidea*. The Platycopida are marine (or occasionally brackish-water) and benthonic. They are characterised by a biserial, vertically aligned adductor muscle pattern, usually numbering between 10 and 18 scars.

By far the most diverse and abundant order is the Podocopida. Four superfamilies are extant during the Quaternary to Recent interval (**Cypridacea, Bairdiacea, Cytheracea** and **Darwinulacea**). Of these the Cytheracea (Fig. 6.3C and F) is the most abundant and diverse superfamily inhabiting marine and freshwater environments. Shell size, thickness and ornament vary, but these ostracods are distinguished by possessing a vertically aligned row of four or five adductor muscle scars (one or more of which may be subdivided) with up to three frontal scars.

The Cypridacea (Fig. 6.3E), although inhabiting both marine and freshwater habitats, are particularly diverse and abundant in the latter environment. The carapace is usually thin-shelled, and smooth or very weakly ornamented. Muscle-scar arrangement varies considerably within the superfamily, but usually comprises a cluster of variably sized scars.

The Bairdiacea (Fig. 6.3D) are exclusively marine (a few are tolerant of brackish-water conditions) and usually thick-shelled. Ornament varies from smooth to well developed and the typical 'bairdoid' lateral outline is broadly rounded anteriorly, but more acutely so posteriorly. The adductor muscle arrangement comprises a series of elongate scars arranged in a cluster, often as two or more subvertical rows of two to four scars.

Finally, the Darwinulacea are exclusively freshwater (some are tolerant of weakly brackish-water conditions), usually with a thin-shelled, smooth carapace, markedly elongate in lateral view. The adductor muscle pattern consists of a radially arranged rosette of elongate scars.

For a more detailed discussion of taxonomy and classification, see articles by Athersuch *et al.* (1989) and Henderson (1990) which describe the living marine/brackish-water and freshwater ostracods of Britain respectively.

6.2   MORPHOLOGY

The ostracod body consists of a head and thorax with up to eight pairs of appendages (Fig. 6.2), which are variously modified for sensing, locomotion, feeding, respiration or sexual reproduction. In sexually reproducing taxa, the male and female adults also have distinct reproductive organs. The ostracod carapace is a bilaterally symmetrical extension of the living organism, which encompasses the physiological systems (Maddocks, 1992). A chitinous cuticle envelops the carapace, and it is the **outer lamella** (composed of low-magnesium calcite) that is biomineralised on the epidermal cells of the duplicature (folds in the epidermis which surround the ostracod body). This biomineralised shell is what becomes part of the fossil record.

The carapace comprises two laterally symmetrical valves (except for minor differences in hingement and valve closure). The valves are hinged dorsally and thus gape anteriorly, ventrally and posteriorly. The 'soft parts' can be withdrawn inside the carapace for protection. The valves are perforated by **pore canals**, which carry sensilla (external sensory hairs) and may permit chemical exchange between the ostracod organism and its immediate environment. The development of the hinge is of taxonomic importance and varies from weak, little more than a groove, to complex three- to five-part hinges containing teeth, sockets, bars and crenulae (Fig. 6.4).

Another feature of taxonomic importance is the adductor muscle-scar pattern. The scars are situated on the internal surface, roughly in the centre of the valve, although they are sometimes also expressed externally. They mark the point of attachment of bundles of muscle fibres, which facilitate opening and closing of the two valves.

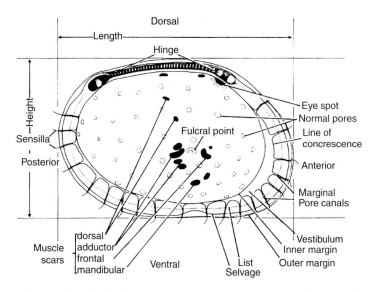

Figure 6.4 Details of the internal structure of the podocopid ostracod carapace. Example is of a female *Loxoconcha elliptica*. (Reproduced from Athersuch *et al.* (1989), courtesy of the Linnean Society of London.)

Although most of the calcification occurs on the outer lamella of the duplicature, the proximal part of the inner lamella is also biomineralised to varying degrees. The **marginal zone** is formed where the inner lamella is in contact, or fused, with the outer lamella. The free or unfused part forms a space or **vestibulum** between the inner and outer lamellae. The development and pattern of contact between these two lamellae as well as the distribution of sensory pore canals through the fused zone are of great taxonomic importance at specific, generic and suprageneric level.

The external carapace morphology of ostracods varies from completely smooth to varying degrees of reticulation, punctation, noding, spinosity and ribbing. These features are usually of taxonomic importance at subspecific, specific and generic level. Many extant and fossil ostracod species exhibit sexual dimorphism, although this is not usually evident until the penultimate or adult stages. In addition to the expected differences in reproductive organs, sexual dimorphism is also evident in the calcified 'hard parts' and may be reflected in carapace size, length/height ratio, tumidity (often to allow brood care in the female) and ornamentation.

6.3 ECOLOGY AND REPRODUCTION

Ostracods reproduce either sexually or asexually (parthenogenesis), and some species have the ability to switch between the two modes under different environmental conditions. Ostracod eggs, once produced, may be

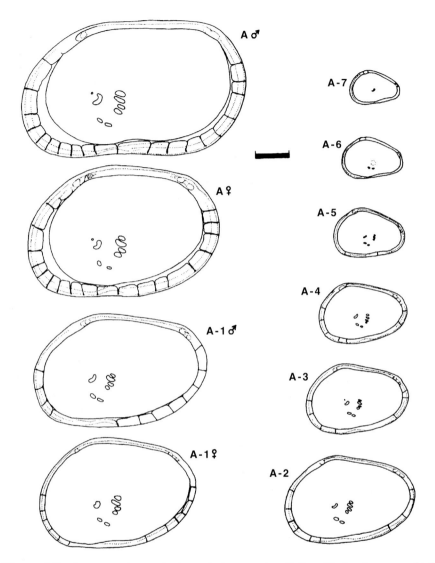

Figure 6.5 Incremental growth stages in *Loxoconcha elliptica*. All drawings are of left valves (scale bar = 100 μm). (Reproduced from Athersuch *et al.* (1989), courtesy of the Linnean Society of London.)

retained within the carapace, deposited in sediments or adhered to a surface where they will develop independently of the parent. A number of taxa in different orders are known to exhibit brood care. As arthropods, the Ostracoda grow incrementally in discrete **moults** or **instars** (Fig. 6.5) (Baltanás *et al.*, 2000; Smith and Martens, 2000). This process, ecdysis, can occur up to eight times during the life-cycle as the individual passes through successive instars. The life-cycle of an individual can be anything from a few weeks through to two years. The number of generations per year is often peculiar to a species (Horne, 1983), but the rate and timing of development are known to be temperature-dependent in many cases. By studying the ostracod population structure (number of valves of each instar) in fossil assemblages (Whatley, 1988) it is possible to determine what taphonomic processes may have occurred *post mortem*. For example, a study of adult to juvenile ratios in subfossil samples across the Alaskan continental shelf (Brouwers, 1988) showed which elements had been transported through the action of bottom currents.

The ostracod organism is particularly sensitive to changes in salinity, temperature, substrate, food supply, pH and dissolved oxygen levels. Undoubtedly salinity is one of the most important environmental variables in determining ostracod species distribution. Consequently, ostracods are commonly used to reconstruct past salinity changes (Neale, 1988). Individual species may be confined to strictly delimited salinity conditions (stenohaline) or be capable of tolerating varying salinity conditions (euryhaline).

As with many other groups of organisms, the global distribution of ostracods is largely temperature controlled, for example, the distribution of shallow-water marine taxa is often latitudinally controlled and the distribution of deep-water species is commonly depth controlled (temperature being a function of depth).

There is often a strong relationship between species distribution and substrate, and this relates to both mode of life and food supply. Detritivore species may show a strong preference for fine-grained sedimentary substrates, while phytal taxa may be more closely associated with algal or plant-rich substrates. Indeed, a wide range of feeding strategies have been recognised, as have a number of plant–ostracod interactions.

6.4 Environmental Applications

The most important characteristic of Ostracoda is their sensitivity to environmental change. Our knowledge of modern species indicates that many species have a well-defined ecological range, and therefore changes in fossil assemblages through time reflect changing environmental conditions, which can be reconstructed if the ecological tolerances of the individual species are known. Similarly, the abundance of each species in an assemblage yields important information on environmental conditions. For example, in modern marginal-marine settings we know that variations in salinity determine the composition and diversity of the Ostracoda and there are well-defined assemblages, each dominated by just two or three species, which represent oligohaline, freshwater to brackish, brackish to marine and fully marine environments (Boomer and Eisenhauer, 2002). Similarly there are some species that prefer muddy substrates, some that like sandy substrates and others which are usually associated with algae.

In some environments, ostracod assemblages are dominated by a single taxon. This is often the case in biologically 'stressed' environments such as hypersaline waterbodies, intertidal settings or reduced-oxygen conditions. In such cases, one species appears to have a physiological (or perhaps reproductive) advantage over its competitors that allows it to dominate the assemblage – sometimes to the exclusion of all other species, but usually to the extent that it constitutes more than 90 per cent of the total assemblage. This can make the interpretation of environmental changes difficult, and it is necessary to examine the relatively small component represented by the 'other taxa'. Slack *et al.* (2000) outlined the potential of these minor elements in distinguishing 'trend' from 'noise'.

6.5 FAUNAL RESPONSE TO ENVIRONMENTAL CHANGE

6.5.1 Biostratigraphy

Little attention has been paid to the biostratigraphical potential of ostracods within the Quaternary. Given the relatively short time-scale of the Quaternary period (in evolutionary terms) there are no well-established global extinctions or originations that are useful in terms of correlation, and Quaternary scientists commonly defer to other chronostratigraphical indicators. Over longer time-scales ostracods have been shown to be important stratigraphical markers, particularly in marine sequences (e.g. Whatley and Coles, 1987; Whatley, 1993). Their stratigraphical potential in Quaternary sequences is more commonly the result of regional changes driven by climatic fluctuations. Quaternary climate-change resulted in broadly latitudinal shifts of species distribution as ostracods sought to remain within their ecological range (usually temperature-related).

A review of Quaternary freshwater ostracod records from Europe was published by Griffiths (1995). Griffiths and Holmes (2000) discussed the biostratigraphical potential of non-marine Quaternary ostracods. They noted, for example, that the genus *Scottia* is represented by a number of similar species, two of which at least are of stratigraphical value: *Scottia browniana* is confined to the Early and Middle Pleistocene, while *Scottia pseudobrowniana* is thought to be restricted to the Ipswichian (Last Interglacial) to Recent interval. One of the most useful stratigraphical applications of ostracods in the Quaternary is the recognition of the transition from the late-glacial through to the Holocene in non-marine environments. This marked climatic change is reflected in the replacement of one ostracod assemblage by another. Late-glacial lake sediments often contain a *candida* fauna (named after *Candona candida*), which is gradually replaced by a more eurythermal assemblage, the *cordata* fauna (named after *Metacypris cordata*), although neither of these two species need necessarily be recorded in either fauna, (see Fig. 6.6). This faunal transition, which reflects climate warming and the development of mature substrates and aquatic macrophytes, was first recognised by Absolon (1973) in Central Europe and has subsequently been recorded throughout NW Europe (Griffiths and Evans, 1995).

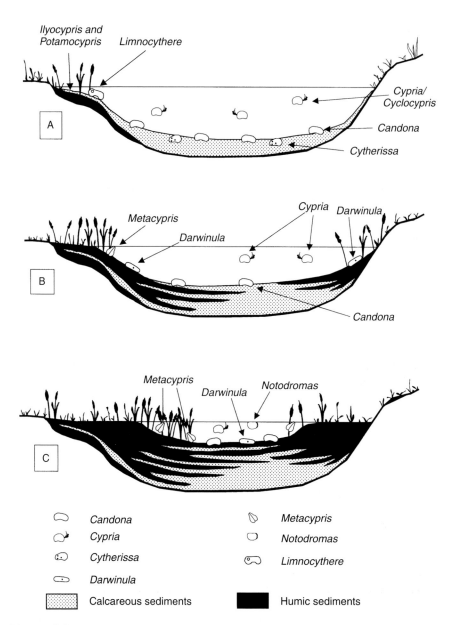

Figure 6.6 Hypothetical model illustrating three stages in the transition of a small European waterbody as the ostracods change from a late-glacial 'candida fauna' to a postglacial 'cordata fauna'. A, late-glacial. B, intermediate. C, post-glacial. (Redrawn from Griffiths and Evans (1995).)

6.5.2 Biogeography

The spatial distribution of any animal or group of animals at a given time (biogeography) is determined by their dependence on particular environmental conditions. Changes in their distribution through time (palaeobiogeography) can tell us about shifting climatic zones or water masses during the Quaternary. Two recent papers discuss the biogeography and palaeobiogeography of marine ostracods from one particular region of the world. The first (Coimbra *et al.*, 1999) describes the modern distribution of ostracods across the Brazilian equatorial shelf and the relationship of the assemblages to faunal provinces elsewhere in the Atlantic. The second paper (Wood *et al.*, 1999) details the changing Oligocene to Recent occurrence of ostracods from the eastern coast of South America, and concludes that modern zoogeographical distributions in the region were strongly influenced by abiotic events such as climate change and plate tectonic movements (expansion of the Drake Passage, closure of Panama), both of which brought about changes in oceanic currents and water-mass temperature.

One of the most intriguing aspects of ostracod biogeography is the establishment of 'isolated marine faunas', either on oceanic islands or on the summits of submerged islands/volcanoes known as guyots. The importance of these faunas lies in the fact that the ostracods, unlike many other aquatic organisms, do not possess planktonic larval stages, therefore their distribution must take place as living, viable eggs, instars or adults. Surface distribution by driftwood or an organic host must be the main route for transport between oceanic islands, but active migration between isolated, submerged guyots is more problematic since the carbonate carapace excludes their presence below the CCD (calcite compensation depth) in the deep oceans. Even the most isolated islands have their own ostracod fauna. Whatley and Jones (1999) described the ostracods from Easter Island – 21 of the 30 species they identified were both new and endemic to the island. They determined, by comparison with assemblages from elsewhere in the Pacific (e.g. Whatley and Roberts, 1995), that the Easter Island ostracods originated from further to the west, possibly during periods of lowered sea-levels (e.g. glacial phases) when a number of now submerged submarine ridges would have been emergent, permitting an 'island-hopping' process.

The distribution of marginal-marine (Grigg and Siddiqui, 1993) and continental ostracods is thought to be partly related to the migration pathways of waterfowl (De Deckker, 1977), particularly since many freshwater taxa have the ability to produce desiccation-resistant eggs which may remain viable for some time before finally reaching a suitable aquatic habitat. Successful dispersal within non-marine ostracods is facilitated by the predominance of a parthenogenetic reproductive strategy (a new population can be established through a single egg or individual). Technological advances in the amplification and study of minute traces of genetic material have added a new dimension to studies of ostracod biogeography, for example, the genetic variability between geographically distinct populations of *Cyprideis torosa* was investigated by Sywula *et al.* (1995). Genetic studies have also opened up questions regarding the evolution of reproductive modes in non-marine ostracods (Martens, 1998; Butlin and Menozzi, 2000), and this has implications for the interpretation of ostracod distribution, migration and colonisation during the late-glacial to Holocene transition.

Undoubtedly the last 500 years of human global exploration and migration have added an anthropogenic element to the distribution of ostracods, and the opportunity for chance distribution of ostracods across hitherto improbable routes has become commonplace. This must be taken into account when investigating the biogeographical distributions of marine species during the last few centuries.

6.5.3   Faunal Response to Pollution

While some ostracod species are sensitive to the most subtle changes in their environment, others are capable of withstanding a wide range of conditions, even to the extent of inhabiting heavily polluted sites. In a number of recent papers the effects of sewage pollution on coastal ostracod faunas has been investigated by Eagar (1999) in New Zealand and a Pacific atoll (Eagar, 2000). These studies have shown that although increasing levels of pollution result in reduced ostracod abundance and diversity, some species are capable of withstanding quite high levels of contamination.

Rosenfeld *et al.* (2000) investigated similar effects in freshwater environments of the Harod River, northern Israel, and concluded that species such as *Heterocypris incongruens* and *H. salina* were capable of withstanding moderate levels of pollution, while *Candona neglecta* and *Cyprideis torosa* are less tolerant. Once again, increasing levels of organic pollutants resulted in low ostracod diversity and abundance. An important conclusion was that although pollution discharge events are, by their nature, often difficult to pick up from occasional water sampling, the ostracod fauna reflects the true state of the aquatic ecosystem. A similar study has also been undertaken on the River Magre in eastern Spain (Mezquita *et al.*, 1999), where the river is subject to both sewage and industrial effluents.

6.6   MORPHOLOGICAL RESPONSE TO ENVIRONMENTAL CHANGE

As aquatic environments change, so the composition of the ostracod assemblage and the abundance of each species reflects this modification. We also know that some species respond to environmental change through the morphology of their carapace. This has been established in living populations where the environmental parameters have been measured, thus allowing a quantitative, or at least semi-quantitative, reconstruction of past environmental changes. Three main forms of morphological variability are recognised (Boomer and Eisenhauer, 2002); the occurrence and development of noding, the shape of sieve-pores, and finally, shell size and ornamentation.

6.6.1 Noding

The occurrence of **nodes** on the external, lateral surfaces of cytherideid species such as *Cyprideis torosa* (Kilenyi, 1972; Vesper, 1972; Vesper, 1975; van Harten, 1996), *Cytherissa lacustris* (Danielopol *et al.*, 1990) and various limno-cytherids (do Carmo *et al.*, 1999) in response to environmental variability has been widely studied; this topic was reviewed by van Harten (2000). The nodes are hollow flexures of the shell and occur in fixed positions, although their number and development may vary (Fig. 6.7). It is generally accepted that this noding is a physiological response to external conditions and not a genetically influenced adaptation as proposed by Kilenyi (1972); however, it is recognised that in many species the position of the nodes is genetically controlled. Although the precise trigger for node development remains unclear, salinity is known to play a part. In *Cyprideis torosa*, for example (a strongly euryhaline taxon), nodes may occur only at relatively low salinity, usually in the range 2–5%. Uncertainty in determining the controlling factor precludes the application of noding as a quantitative palaeosalinometer, although it is a useful indicator of relatively low salinity conditions.

6.6.2 Sieve-pore Shape

All ostracods have pores which perforate the carapace and are thought to act as chemicosensory pathways, allowing the organism to sense its immediate environment. The pores can take a number of forms: they can be simple tubes or more complicated multi-element structures and may be associated with sensilla. One form, the **sieve-type**, is characterised by a perforate plate guarding the external entrance of the pore. A number of studies have described the relationship between salinity and the shape of sieve pores on the lateral surface of *Cyprideis torosa* (Rosenfeld and Vesper, 1977). These pores occur in a number of cytherid families and usually possess a circular outline; however, in *Cyprideis torosa* they vary from round, through oblong, to irregular. Rosenfeld and Vesper determined an inverse relationship between the abundance of round pores and the salinity of the environment. In more

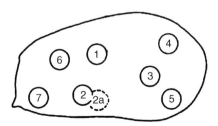

Figure 6.7   Distribution and terminology of nodes on *Cyprideis torosa*. (After Kilenyi (1972).)

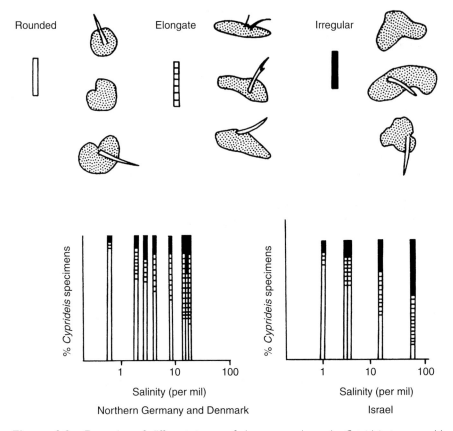

Figure 6.8 Examples of different types of sieve-pore shape in *Cyprideis torosa*, with examples of their occurrence in waters of different salinity. (After Rosenfeld and Vesper (1977).)

saline environments, irregular pore shapes increased in abundance (Fig. 6.8). *Cyprideis torosa* is known from the late Miocene to the present, throughout much of Europe, Asia and Africa, and it may provide a useful tool in reconstructing Quaternary palaeosalinities. The technique has recently been applied to Early Pleistocene assemblages from central Italy (Gliozzi and Mazzini, 1998), where palaeosalinities were estimated to range from freshwater to oligohaline.

6.6.3 Carapace Size and Ornamentation

The size of the carapace and degree of ornamentation within some species is known to vary in response to environmental changes; in some cases there is clearly a seasonal signal. Some species of *Leptocythere* which produce more than one generation per year are known to exhibit seasonal differences in external morphology, with more heavily ornamented species appearing at certain times of the year, which may reflect some degree of temperature control (e.g. *L. porcellanea*, *L. lacertosa*).

Intraspecific variations in shell morphology such as the degree of calcification and the development and nature of ornamentation have been studied, particularly at the transition between fresh and saline water (Carbonel, 1988; Carbonel and Hoibian, 1988; Peypouquet *et al.*, 1988). Carbonel (1988) concluded that although patterns of carapace ornamentation are genetically determined, the degree to which ornament is developed is environmentally controlled. This 'aggradation–degradation' theory, first put forward by Peypouquet *et al.* (1980), proposed that ornament development is determined by the carbonate equilibrium of the water in which the ostracod calcifies its shell. The carbonate equilibrium is strongly influenced by the biological consumption of organic matter (mainly through bacterial decomposition). For example, high levels of bacterial respiration produce carbonate ions, which favour heavier calcification and stronger reticulation. It is possible that studies of fossil assemblages may infer changes in palaeoproductivity and/or detrital organic-matter supply through time (Boomer and Eisenhauer, 2002).

6.7 GEOCHEMISTRY OF OSTRACOD SHELLS

Although precise details of the carapace calcification process are still not fully understood, it has been established that the minerals required to calcify each new shell are sequestered from the ambient water and are not stored within the organism prior to calcification, and neither are they absorbed from the previous instar. Since the whole calcification process takes a matter of hours, the composition of the shell represents a 'snapshot' of environmental conditions at that time (i.e. temperature, chemistry and stable-isotope composition of the host water).

The mineralised ostracod carapace comprises low-magnesium calcite, dominantly $CaCO_3$ but with additional magnesium and strontium atoms (among other elements) substituted within the calcite lattice. It has been established that the uptake of magnesium is largely (but not exclusively) temperature-controlled, while the presence of strontium is determined by its concentration in the host water. By determining the molar concentration of these two elements with respect to calcium, it is possible to quantitatively reconstruct salinity and temperature at the time of calcification. For each genus a **distribution coefficient** must also be determined (through laboratory cultures and/or field collections), which reflects biological control on the uptake of Sr and Mg. De Deckker *et al.* (1999) reviewed the uptake of Mg and Sr in *Cyprideis torosa* through a series of *in vitro* experiments.

The relationship between shell composition and the conditions of the water at the time of calcification can be described by the following equation:

$$K_D[M]_{(T)} = (M/Ca) \text{ ostracod shell} / (M/Ca) \text{ host water}$$

where K_D is the distribution coefficient for an element (M) at a given water temperature (T), and M/Ca are **molar ratios**. Closely related species have very similar distribution coefficients.

Trace-element geochemistry has its greatest potential in environments such as closed basin lakes (e.g. reconstructing salinity variability caused by changes in the precipitation/evaporation ratio within the catchment) and the deep sea (e.g. temperature variability resulting from changes in deep-water circulation/formation (Dwyer *et al.*, 1995; Corrège and De Deckker, 1997; Cronin and Raymo, 1997). By far the majority of shell chemistry research studies are based on non-marine systems and this is reflected in the palaeolimnology section below.

In marginal-marine environments, salinity can change over a wide range within a few hours. This causes problems for shell chemistry studies since the chemistry of the marine water is such that it can overwhelm the fresh-water signal. In tidal settings it is possible that ostracods can select an optimum salinity level at which to begin calcification which may not necessarily be representative of mean salinity conditions at that location (Boomer and Eisenhauer, 2002), and for these reasons shell chemistry studies in marginal marine environments must be treated with some caution.

The calcite ostracod carapace can also be utilised for carbon and **oxygen stable-isotope analysis** using the standard isotope techniques (Lister, 1988; Chivas *et al.*, 1993) to reconstruct past environmental changes. The interpretation of ostracod stable-isotope analyses has been made possible by detailed investigations, such as those by von Grafenstein *et al.* (1999b), into the relationship between shell isotope and water isotope composition in modern lake settings. High-resolution stable-isotope studies of Late Holocene ostracods have permitted a decadal resolution of climate change in central Europe between 15 ka and 5 ka BP (von Grafenstein *et al.*, 1999a). Schwalb *et al.* (1999) used a combined analysis on ostracod and authigenic carbonate stable isotopes to reconstruct Holocene climatic variability in the Chilean Altiplano. Their results indicate changing precipitation levels, with a mid-Holocene dry phase followed by highly variable but increased levels in the Late Holocene. Undoubtedly the application of ostracod shell chemistry in quantitative reconstructions of past salinity and temperature conditions (e.g. Chivas *et al.*, 1985, 1986; Engstrom and Nelson, 1991; Wansard, 1996) has elevated their profile in Quaternary research.

It is possible to undertake both trace-element and stable-isotope analyses on a single ostracod valve. The value of combined analyses on ostracod assemblages was reviewed by De Deckker and Forrester (1988) and Holmes (1996) (Fig. 6.9). Ostracod shell material may also be used to date Quaternary sequences using standard radiocarbon techniques, and the potential of amino-acid racemisation has also been investigated (McCoy, 1988).

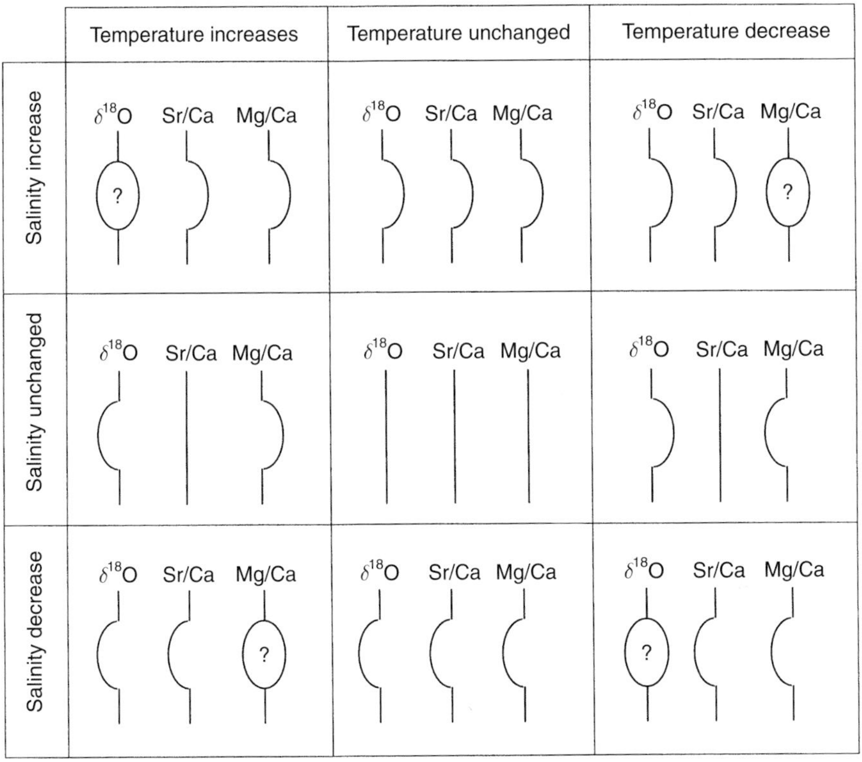

Figure 6.9 The environmental interpretation of relative changes in shell chemistry and stable-isotope composition (after Curtis and Hodell, 1993). Shifts to the right are positive excursions, those to the left are negative. The 'question-marked' curves may respond in either direction. (Redrawn from Griffiths and Holmes (2000).)

6.8 PALAEOCEANOGRAPHY AND QUATERNARY CLIMATE CHANGE

Although ostracods are generally less abundant than foraminifera in marine sequences, their utility has long been recognised. Benson (1988; 1990) and Whatley (1996) reviewed their use in reconstructing deep-sea events and processes. Ostracods have been studied from most of the world's oceans, but the SW Pacific, North Atlantic and the Mediterranean Sea have received most attention, and some recent research has concentrated on the Arctic Ocean. Deep-sea ostracods are distinctly different from shelf and marginal-marine taxa. Within deep-sea assemblages there are pandemic species which occur in all oceans, while other species are restricted to particular oceans or regions (Whatley and Ayress, 1988).

Changes in deep-sea ostracod diversity have been shown to reflect five major palaeoceanographical events from the late Mesozoic to the Quaternary (Benson, 1990). These changes are largely responses to the reorganisation of the global deep-water circulation patterns due to a combination of climatic and tectonic events – the ostracods are responding to changes in the physical and chemical composition of bottom waters. Dingle *et al.* (1989) and Dingle and Lord (1990) first outlined the close association between water masses and ostracod assemblages in their study of Recent and Late Holocene ostracods from the continental shelf off southwest Africa. Jones *et al.* (1999) investigated the ostracods from three Late Quaternary cores in the Arctic Ocean. Comparison with modern core-top samples from the same region permitted a reconstruction of water-mass evolution in the Arctic from isotope stage 7 to the present. They determined that interglacial periods were rich in ostracods, whereas glacial periods were barren. Furthermore, they were able to show that upper Arctic Ocean Deep Water (AODW) could be differentiated from lower AODW by means of the ostracods recovered. They were also able to detect the influence of other water bodies, particularly the relatively warm, saline Arctic Intermediate Water (AIW).

One of the most striking associations between ostracods and water masses is the sometimes dominant occurrence of platycopids within the Oxygen Minimum Zone or OMZ (Whatley, 1991). As **filter feeders** the platycopids continually circulate water across their respiratory surfaces, and this is thought to confer an ecological advantage compared with other taxa in situations where oxygen levels are reduced. Their reproductive mode (internal brood care) also ensures that eggs are also kept oxygenated (a reproductive advantage). This association has been used to interpret palaeo-oxygen levels in the Mesozoic and Palaeozoic, but also has applications in tracing OMZ fluctuations in response to eustatic sea-level change during the Quaternary. The ubiquitous, abundant and diverse deep-sea genus *Krithe* has also been inferred to indicate palaeo-oxygen levels through the variable development of its anterior inner lamella (Peypouquet, 1975); however, the reliability of this technique has been questioned (Whatley and Quanhong, 1993).

Cronin and Raymo (1997) showed that Pliocene deep-sea ostracod diversity in the North Atlantic responded to changes during climatic cycles of 41,000 years which affected bottom-water temperature, surface productivity and nutrient levels. Subsequently, a similar study (Cronin *et al.*, 1999) on the Late Quaternary interval of the same region (isotope stages 7–1), based on ostracod abundance, diversity and trace-element analyses, indicated that high-amplitude fluctuations coincided with the orbital and suborbital scale oscillations. High-diversity samples, characterised by *Argilloecia–Cytheropteron* assemblages, corresponded to interglacial conditions, while glacial intervals were dominated by low-diversity *Krithe* assemblages (Fig. 6.10). It is clear that even deep-sea ecosystems are significantly affected by climatic events over orbital to millennial time-scales.

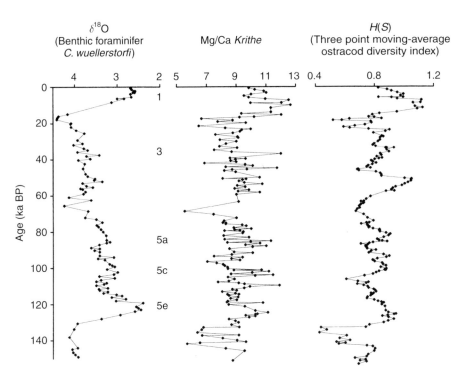

Figure 6.10   A comparison of ostracod diversity and ostracod Mg/Ca ratios with $\delta^{18}O$ from benthic foraminifera. Many of the peaks in ostracod diversity (H(S) = Shannon–Weaver diversity index, three-point moving mean) and trace-element chemistry (based on specimens of *Krithe*) correspond to global climatic warm periods. Oxygen isotope stages 1, 3, 5a, 5c and 5e are marked on the $\delta^{18}O$ graph. (Redrawn from Cronin *et al.* (1999).)

6.9   MARGINAL-MARINE RECORDS OF ENVIRONMENTAL CHANGE

Marginal-marine environments present challenging physiological problems for most organisms, including ostracods, since they may have to survive marked environmental changes over very short periods. Ostracod diversity is often low in these habitats, although abundance may be remarkably high, with one or two species dominating.

Although marginal-marine environments do not always preserve well (Boomer and Eisenhauer, 2002), the global changes in sea-level throughout the Quaternary have resulted in the deposition of sedimentary sequences which, in some areas, yield micropalaeontological records that permit the reconstruction of relative sea-level changes and coastal geomorphology (e.g. Andrews *et al.*, 2000; Mazzini *et al.*, 1999). Neale (1988) reviewed the salinity tolerance of many modern marine and non-marine species and provided a number of examples of the value of ostracods in palaeoenvironmental reconstruction. However, the accuracy of such palaeoenvironmental reconstructions depends on knowledge of modern species autecology.

One application for marginal-marine ostracods in Quaternary research is in reconstructing the deglaciation history of the Scandanavian ice-sheet at the end of the last glacial period. This was associated with high-discharge freshwater pulses into the Baltic and adjacent regions. Hammarlund (1999) recorded the evolution of a marginal-marine site, through ostracod assemblages and stable isotopes, which gradually became an isolated freshwater basin due to isostatic uplift between about 17 ka and 13 ka BP. Majoran and Nordberg (1997) used the stratigraphical record of ostracods in the southern Kattegat to reconstruct environmental changes in this region between about 13 ka and 12 ka BP. The ostracod assemblages are dominated by mixed freshwater and marine elements reflecting the dynamic conditions during this period. Finally, Schoning and Wastegård (1999) investigated the latest glacial ostracods (about 10.5 ka to 10.3 ka BP) from varved glaciomarine sediments in the region. They showed that the ostracod assemblages recorded detailed fluctuations in salinity associated with the latest phase of the deglaciation.

6.10 PALAEOLIMNOLOGY AND OTHER NON-MARINE APPLICATIONS

Much of our evidence for global Quaternary climate change comes from the oceans and ice-cores, but an increasing amount of regional data comes from relatively low-latitude, non-marine settings. There are a number of long-term ostracod records from continental sequences which yield vital information regarding palaeoclimatology and palaeohydrology. Reviews outlining the role of ostracods in reconstructing climate change from palaeolimnological records can be found in De Deckker and Forester (1988) and Forester (1987). The ostracods often constitute just one element of a multidisciplinary approach to reconstructing lake records (including sediment chemistry and isotopes as well as other biotic records such as pollen, diatoms and molluscs, e.g. Barber *et al.*, 1999).

The evolution of lake salinity and solute chemistry (sulphate, bicarbonate and chloride concentration) is strongly affected by changes in hydrology, which is generally controlled by climate. Lake level, salinity and chemistry changes are a function of the balance between water supply (precipitation within the catchment) and water loss (mainly evaporation), the PE (precipitation/evaporation ratio), in closed basin systems (i.e. with no surface outflow). However, the system can also be affected by mineral inputs from sources such as sea spray or dissolution of evaporites in the catchment; similarly, the effects of groundwater must be taken into account (Smith *et al.*, 1997). Modern ostracods from North American lakes (Forester, 1983) indicate that **lake solute type**, rather than salinity, may be more important in determining the faunal composition. Forester concentrated on a number of species, and the modern hydrochemical preferences of two North American *Limnocythere* species (*L. staplini* and *L. ceriotuberosa*) are shown in the top trilinear anion diagrams in Fig. 6.11. Knowledge of the preferred solute chemistry for key species permits the hydrological reconstruction of lacustrine systems whose sediments contain those ostracods and, from such records, regional patterns of past climate change can be reconstructed (Forester, 1986).

The stratigraphical distribution of *L. staplini* and *L. ceriotuberosa* in Quaternary sediments from the Great Salt Lake indicates that the lake chemistry changed from chloride–sulphate dominated to carbonate dominated and back again (see schematic core illustrated in Fig. 6.11). It is also possible to use such data to reconstruct past precipitation and temperature records (Smith and Forester, 1994).

Ostracods have proved to be important palaeoenvironmental proxies in a variety of non-marine environments. This is in no small part due to the ease with which ostracods can be transported between waterbodies (as adults or

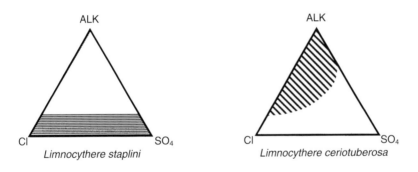

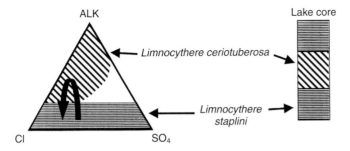

Figure 6.11   The top two trilinear anion diagrams indicate the 'hydrochemical space' that each species occupies, and the bottom two figures show how the changing occurrence of the same species through a core can be used to reconstruct palaeohydrological changes. ALK = total alkaline ions, Cl = total chloride, SO$_4$ = total sulphate. (Redrawn from Griffiths and Holmes (2000).)

eggs via waterfowl or other chance events). Consequently, ostracods are one of the first invertebrate groups to invade newly opened aquatic biotopes. Indeed, it appears that they were some of the first invertebrates to colonise new habitats following the retreat of the ice-sheets and amelioration of periglacial conditions across Northern Europe and North America. The northward retreat of the ice-sheets was followed by the progressive appearance of freshwater ostracod species as they moved out of their more southerly glacial refugia (Griffiths and Horne, 1998). Ostracods in late-glacial to Holocene tufa from the British Isles have been used to reconstruct climatic change and the effects of human forest clearance (Preece and Robinson, 1984; Preece *et al.*, 1984; 1986).

6.11 COLLECTING, OBSERVING AND STUDYING OSTRACODS

Living ostracods are very easy to collect and observe. Collecting equipment need only consist of one or more sieves and some storage containers. Since most adult ostracods fall into the size range 300–1500 μm (0.3–1.5 mm), it is suggested that a large sieve size (>2 mm) be used to exclude large detrital material (sediments and plants), with a smaller sieve size of about 150 μm being used to retain most ostracods. The residue from the finer sieve can then be decanted into the storage containers (preferably wide-necked jars). Since ostracods live in a variety of habitats, it is important to sample from a range of sub-environments. Within a lake, the sediment surface, macrophytes and nekton will yield different assemblages. Similarly, in marine settings, the soft sediments, hard substrate and algae should be investigated separately. To preserve living ostracods in the field the host water should gradually be made up to about 70 per cent ethanol, while the addition of a stain such as rose Bengal will highlight living tissues. Despite their small size, ostracods can be observed in field samples using a hand lens, and some of the larger species, or those with distinctive coloration patterns, can even be identified to species level.

Ostracods are best observed in a petri dish under a binocular microscope at magnifications between 10× and 50×. Both transmitted and reflected light can be used to observe internal and external features respectively; they are easily transferred using a pipette. Freshwater ostracods in particular can be easily cultured in aquaria and require only a small amount of organic material – humus or manure is ideal.

Collecting and preparing fossil ostracod material is similar to that for foraminifera. Since many Quaternary sediments are poorly lithified it is a simple matter to extract the ostracods. Often, a gentle wash in warm water is sufficient to disaggregate the sediment. In the case of more compacted sediments the samples should be air-dried before disaggregation in a 2 per cent H_2O_2 solution. Further details on sample collection and processing methods can be found in Brasier (1980) and Griffiths and Holmes (2000).

6.12   SUMMARY

The success of ostracods in occupying most aquatic habitats, together with their readily fossilised carbonate carapace, make them an invaluable aid to many palaeoenvironmental studies. Faunal and shell chemistry responses to environmental change can be used to qualitatively and quantitatively reconstruct patterns of climate change throughout the Quaternary. Their utility can only be enhanced as we learn more about the ecology of living ostracod species.

FURTHER READING

The proceedings of the International Symposia on Ostracoda (listed below) have been published variously as books and special issues of international journals (the 1997 symposium was published in three separate journals). These are invaluable resources and serve to illustrate the growing importance of ostracods in the earth and environmental sciences.

Boomer, I. and Lord, A.R. (eds), 1999. Marine Ostracoda and Global Change. Theme 2 of the 13th International Symposium on Ostracoda (ISO97). *Marine Micropaleontology*, 37, 227–394.

Hanai, T., Ikeya, T. and Ishizaki, K. (eds), 1988. *Evolutionary biology of ostracoda. Its fundamentals and applications.* Elsevier/Kodansha, Amsterdam/Tokyo, 1361 pp.

Hartmann, G. (ed.), 1976. Evolution of post-Paleozoic Ostracoda. *Abhandlungen und Verhandlungen des Natürwissenschaftlichen Vereins in Hamburg (N/F), Supplement 18/19,* pp. 1–336.

Holmes, J.A. and Horne, D.J. (eds), 1999. Non-marine Ostracoda: Evolution and Environment. Theme 1 of the 13th International Symposium on Ostracoda (ISO97). *Palaeogeography, Palaeoclimatology, Palaeoecology,* 148, 1–186.

Horne, D.J. and Martens, K. (eds), 1999. Evolutionary Biology and Ecology of Ostracoda: Theme 3 of the 13th International Symposium on Ostracoda (ISO97). *Developments in Hydrobiology,* 148, 1–197.

Krstic, N. (ed.), 1979. *Proceedings of the VII International Symposium on Ostracodes – Taxonomy, biostratigraphy and distribution of ostracodes.* Beograd, Serbian Geological Society, 272 pp.

Löffler, H. and Danielopol, D.L. (eds), 1977. *Aspects of ecology and zoogeography of Recent and fossil ostracods.* D.W. Junk, The Hague, 521 pp.

McKenzie, K.G. and Jones, P.J. (eds), 1993. *Ostracoda in the earth and life sciences,* A.A. Balkema, Rotterdam, 724 pp.

Maddocks, R.F. (ed.), 1983. *Applications of Ostracoda.* University of Houston, Houston, 677 pp.

Neale, J.W. (ed.), 1969. *The taxonomy, morphology and ecology of Recent Ostracoda,* Oliver & Boyd, Edinburgh, 553 pp.

Oertli, H.J. (ed.), 1971. *Paléoécologie des ostracodes. Bullétin Centre des Recherches Pau – SNPA Supplement,* 5, 953 pp.

Puri, H.S. (ed.), 1964. *Ostracods as ecological and palaeoecological indicators. Publicazione della Stazione Zoologica di Napoli Supplemento,* 33, 612 pp.

Ríha, J. (ed.), 1995. *Ostracoda and biostratigraphy.* A.A. Balkema, Rotterdam, 454 pp.

Swain, F.M., Kornicker, L.S. and Lundin, R.F. (eds), 1975. Biology and paleobiology of Ostracoda. *Bulletins of American Paleontology*, 65, 687 pp.

Whatley, R.C. and Maybury, C. (eds), 1990. *Ostracoda and global events*. British Micropalaeontological Society/Chapman and Hall, London, 621 pp.

PALAEOCEANOGRAPHIC APPLICATIONS OF PLANKTONIC SARCODINE PROTOZOA: RADIOLARIA AND FORAMINIFERA

Simon K. Haslett

Radiolaria and planktonic foraminifera are holoplanktonic marine protozoa that are mainly found living in the open ocean, rather than in neritic environments. They both belong to the protozoan Phylum Sarcodina, but the Subclass Radiolaria are classified within the Class Actinopoda, while the Order Foraminifera are placed within the Class Rhizopoda. They are often treated separately on taxonomic and skeletal grounds, because radiolaria secrete a siliceous opaline test (shell), while planktonic foraminifera secrete a test composed of calcium carbonate ($CaCO_3$). However, they are very similar microfossil groups in terms of their palaeoecology and palaeoenvironmental applications, and therefore are often treated together in ecological studies (e.g. Michaels *et al.*, 1995; Anderson, 1996). The aim of this chapter is to outline the development of the study of radiolaria and planktonic foraminifera as applied to the reconstruction of past oceanographic environments through the Late Neogene and Quaternary, and to discuss a major controversy currently influencing the use and perceived value of these microfossil groups.

7.1 RADIOLARIA

The Superorder Polycystina is divided into two orders of radiolaria, the Spumellaria (skeleton in the form of a sphere, or derived from a sphere, e.g. ellipsoidal, discoidal, lenticular, spiral), and the Nassellaria (with a skeleton that is bipolar and usually bilaterally symmetrical). Although Cenozoic radiolaria can be studied using both reflected light and scanning electron microscopy, traditionally they have been mounted in Canada balsam and viewed in transmitted light (see below and Fig. 7.1). This allows for examination of internal spicules and skeletal structure, features that are important

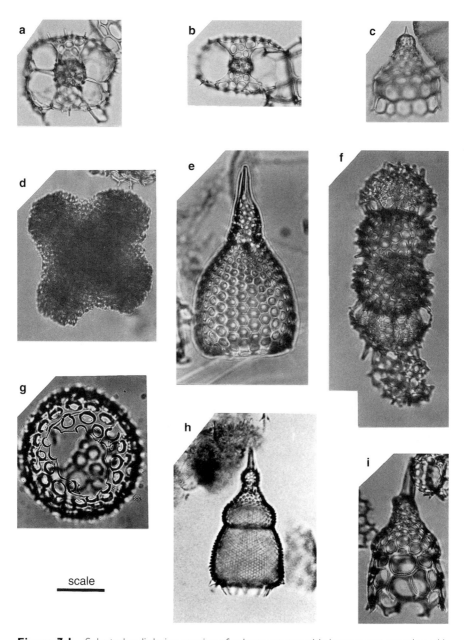

scale

Figure 7.1 Selected radiolarian species of palaeoceanographic importance mentioned in the text. Species employed in the Radiolarian Temperature Index: (a) *Octopyle stenozona*, (b) *Tetrapyle octacantha*, (c) *Cycladophora davisiana*. Other species: (d) *Spongaster tetras*, (e) *Anthocyrtidium zanguebaricum*, (f) *Didymocyrtis tetrathalamus*, (g) *Acrosphaera murrayana*, (h) *Theocorythium vetulum*, and (i) *Lamprocyrtis neoheteroporos*. Scale bar: for (a), (b), (c), (g), (h) = c.100 μm, (d), (e), (f) and (i) = c.75 μm. Specimens (a), (b), (c), (d), (e), (f), (h) and (i) are from the Early Quaternary at ODP Site 677 (eastern equatorial Pacific), while specimen (g) is from the Late Quaternary at ODP Site 658 (eastern tropical Atlantic).

for identification (Kling, 1978; Brasier, 1980; Anderson, 1983; Casey, 1993; Haslett, 1993; De Wever *et al.*, 1994).

Radiolaria inhabit the water column from the surface to depths of many hundreds of metres (Renz, 1976; Kling, 1979; Dworetzky and Morley, 1987; Kling and Boltovskoy, 1995). There are in excess of 500 extant species. Such species diversity is attributable to the niche-specific ecology of many species, and indeed Casey *et al.* (1990) stated that expanding ocean niche-diversity through the Cenozoic was matched by a concomitant increase in radiolarian allopatric speciation and diversity. Therefore, radiolaria are invaluable in the Late Cenozoic as proxy indicators of ocean water-masses, water depth, water temperature, nutrient availability, and physical processes, such as upwelling.

Riedel (1958), comparing differences between low-latitude Pacific and Antarctic radiolarian faunas, was the first to suggest that radiolaria may be useful as palaeoecological indicators. This observation stimulated studies of radiolarian biogeography in the surface sediments of Antarctica (Hays, 1965), the Indian Ocean (Nigrini, 1967), eastern equatorial Pacific (Nigrini, 1968), North Pacific (Nigrini, 1970) and North and South Atlantic (Goll and Björklund, 1971, 1974), in order to establish the distribution of extant species and their relationship to overlying water-masses (Casey, 1971). This aspect of radiolarian research is now focused on gaining an understanding of the nature of radiolaria in the plankton standing stock (Boltovskoy and Jankilevich, 1985; Boltovskoy and Riedel, 1987; Boltovskoy and Alder, 1992), the flux of dead tests through the water column (Gowing and Coale, 1989; Takahashi, 1991; Boltovskoy *et al.*, 1993a; Gowing, 1993; Welling and Pisias, 1998a) and the contribution made by radiolaria to accumulating sediments on the sea-floor (Welling *et al.*, 1992; Boltovskoy *et al.*, 1993b; Welling and Pisias, 1993).

In comparison with radiolaria, planktonic foraminiferal ecology is relatively well known, and their use in reconstructing palaeo sea-surface temperatures (SSTs) is widespread. Their popularity may be because there are relatively few species of planktonic foraminifera compared with radiolaria, and that they are well preserved in Atlantic sediments where radiolaria are often poorly preserved (Sanfilippo, 1987), and where many of the pioneering deep-sea micropalaeontological studies were undertaken. However, in ocean basins where the sea-floor is deeper than the lysocline and calcite compensation depth (CCD) (Libes, 1992), generally *c.*4 km depth, calcareous microfossils such as foraminifera are not preserved and the sediment often comprises 100 per cent siliceous ooze. Also, some oceans, such as the Southern Ocean, are dominated by siliceous plankton. It is in these oceanic and sedimentary environments that the usefulness of radiolaria is apparent (see Abelmann and Gowing, 1997).

Sediment is usually processed for radiolaria through washing with dilute hydrochloric acid (HCl) to remove the $CaCO_3$ component, and then sieved at 63 μm to remove silt and clay-sized particles. The >63 μm fraction is then mounted in Canada balsam on a glass microscope slide and viewed using transmitted light microscopy. Some authors urge that random settling of microfossils on to the slide is required for quantitative studies (see Sanfilippo *et al.*, 1985; Roelofs and Pisias, 1986; Haslett and Robinson, 1991; Welling and Pisias, 1995; Locker, 1996, for various methods). In addition to radiolarian specimens, diatoms (see Chapter 8), sponge spicules (Robinson and Haslett, 1995), and silicoflagellates (Perch-Nielsen, 1985a) are also commonly seen in slides.

7.1.1   Radiolaria as Water-temperature Indicators

Nigrini (1970), in her study of North Pacific surface sediments, statistically identified radiolarian assemblages using recurrent group analysis. This information was applied to a palaeotemperature study of Quaternary sediments of Core V20–130, where Nigrini (1970) constructed a radiolarian temperature number (T_r) for each downcore sample. The T_r was based on an equation developed for diatoms (T_d) by Kanaya and Koizumi (1966):

$$T_r = (X_w / (X_t + X_c) + X_w) \times 100$$

where X_w is the combined abundance of the tropical (warm) assemblage species, X_t is the abundance of the transitional assemblage species, and X_c is the abundance of the subarctic (cold) assemblage species (see also Chapter 8). The resultant T_r record for V20–130 was not entirely successful and did not correspond well with T_d for the same core, although the two did correspond at some important levels (e.g. the last glacial maximum), indicating the potential of the technique. Nigrini (1970) attributed the problem to the position of the core site, in that it occupies a transitional oceanographic setting. However, the use of recurrent group analysis in identifying assemblages may have been misleading, in that the method considers only the presence or absence of a species, which ignores the ecological importance of species abundance. In this way, 17 and 13 species were included in X_w and $(X_t + X_c)$ respectively. Also, the inclusion of transitional assemblage species as a component in X_c of Kanaya and Koizumi's (1966) original equation may have reduced the signal from the cold-water subarctic assemblage. Johnson and Knoll (1974) later applied a T_r, which they called the 'radiolarian climatic index', to Pleistocene cores from the equatorial Pacific, with some success. Nigrini's (1970) study was the first to employ radiolaria as a palaeoceanographic proxy; however, the T_r was not widely used again until quite recently. Recurrent group analysis also has not been employed widely since the early 1970s, although it was used by Johnson and Nigrini (1980, 1982) in a study of radiolarian biogeography of the Indian Ocean.

In 1971, Imbrie and Kipp published their seminal paper on a transfer-function technique, which revolutionised the application of microfossils to palaeoceanographic research, and in particular to the study of sea-surface temperature (SST). It involves analysing surface sediment (Holocene) microfossil faunas (or floras) from the study area and defining the ecological parameters (e.g. SST) for each sample. The number of variables (species) per sample is reduced to a few ecologically meaningful assemblages using factor analysis (Klovan and Imbrie, 1971). A species score matrix is produced indicating the importance of each species to a factor assemblage, and a factor-loading value indicates the contribution made by factor assemblages to each sample, which may be a positive or negative contribution. Factor analysis is then performed on a down-core data-set. The transfer function of Imbrie and Kipp (1971) estimates the SST for each sample, based on their statistical similarity with the Holocene data-set for which SST is known. Factor analysis and the transfer function have become standard micropalaeontological techniques and are widely used by Quaternary radiolarian workers (Moore, 1973, 1978; Sachs, 1973a, b; Hays *et al.*, 1976b; Lozano and Hays, 1976; Dow, 1978; Molina-Cruz, 1977a, b, 1984, 1988; Pisias, 1978, 1979, 1986; Moore *et al.*, 1980; Romine and Moore, 1981; Schramm, 1985; Molina-Cruz and Martinez-López, 1994; Gupta and Fernandes, 1997; Pisias *et al.*, 1997). Many of these studies were contributions to CLIMAP (CLIMAP Project Members, 1976), and used a common taxonomic framework (Nigrini and Moore, 1979).

Factor assemblages are mathematical constructions and as such are objective groupings of species. A wealth of radiolarian ecological information has been derived from the studies cited above, with many of them reaching consistent conclusions concerning particular species. Table 7.1 lists the dominant species that characterise factor assemblages from the main studies of the modern Pacific and its regions. From this summary it is clear that certain species have ecological preferences. For example, the occurrence of *Acrosphaera murrayana* is principally related to coastal upwelling, *Antarctissa denticulata* and *A. strelkovi* are confined to the Southern Ocean around Antarctica, *Botryostrobus aquilionaris* characterises subpolar waters, *Botryostrobus auritus* is dominant along zones of equatorial divergence-driven upwelling, *Cycladophora davisiana* is an important component in cool-water and upwelling assemblages, *Didymocyrtis tetrathalamus* is a characteristic species in tropical/subtropical assemblages, *Pterocorys minythorax* is dominant in both equatorial divergence and coastal upwelling areas, *Spongaster tetras* is dominant only in the western tropical Pacific, and *Octopyle stenozona* and *Tetrapyle octacantha* (often combined) are consistently the dominant species in tropical assemblages. In addition to these important species, the distributions of most radiolarian species counted by CLIMAP Project Members (1976) in the course of their investigations are shown in maps compiled by Lombari and Boden (1985).

The use of factor analysis and transfer functions allows large species datasets to be processed, relating modern to fossil assemblages, yet until the early 1990s there was little independent information available to verify the accuracy of the palaeoenvironmental records being produced using radiolaria. Haslett (1992, 1995a) explored the palaeoecology of a number of these species to establish whether their modern ecology is consistent with

Table 7.1 Radiolaria included in factor assemblages

	Acrosphaera murrayana	Acrosphaera spinosa	Actinomma medianum	Actinomma sp.	Antarctissa denticulata	Antarctissa strelkovi	Botryostrobus aquilonaris	Botryostrobus auritus	Cenosphaera cristata	Cyclodophora davisiana	Dictyocoryne profunda	Didymocyrtis tetrathalamus	Euchitonia elegans/furcata	Eucyrtidium hexagonatum	Giraffospyris angulata	Heliodiscus asteriscus	Larcopyle butschlii	Lirospyris (?) toxarium	Lithelius minor	Octopyle stenozona	Pterocanium aurirum	Pterocorys minythorax	Pterocorys zancleus	Pylospira octopyle	Spongaster tetras	Spongurus sp.	Stylatractus spp.	Stylochlamydium asteriscus	Stylodictya validispina	Tetrapyle octacantha
Pan-Pacific (Moore, 1978)																														
Tropical Factor											✓	✓						✓	✓								✓			✓
Western Pacific Factor						✓	✓	✓	✓	✓	✓																	✓		
Subarctic Factor																		✓												
Transitional Factor		✓										✓						✓												
Antarctic Factor			✓		✓	✓												✓												
East Central Factor																														
Temperate Factor																														
Tropical Pacific (Moore, 1978)																														
Tropical Factor											✓	✓							✓					✓			✓	✓		✓
Western Pacific Factor										✓	✓													✓			✓			
Australian Factor	✓	✓																						✓						
New Guinea Factor																			✓					✓			✓			✓
Eastern Pacific (Romine and Moore, 1981)																														
Tropical Factor											✓							✓			✓						✓			✓
Equatorial Factor (divergence)																			✓		✓									✓
Peru Current Factor (upwelling)																										✓				
Subtropical Factor																✓		✓											✓	✓

Table 7.1 Continued

	Acrosphaera murrayana	Acrosphaera spinosa	Actinomma medianum	Actinomma sp.	Antarctissa denticulata	Antarctissa strelkovi	Botryostrobus aquilonaris	Botryostrobus auritus	Cenosphaera cristata	Cycladophora davisiana	Dictyocoryne profunda	Didymocyrtis tetrathalamus	Euchitonia elegans/furcata	Eucyrtidium hexagonatum	Giraffospyris angulata	Heliodiscus asteriscus	Larcopyle butschlii	Liriospyris (?) toxarium	Lithelius minor	Octopyle stenozona	Pterocanium auritum	Pterocorys minythorax	Pterocorys zancleus	Pylospira octopyle	Spongaster tetras	Spongurus sp.	Stylatractus spp.	Stylochlamydium asteriscus	Stylodictya validispina	Tetrapyle octacantha
Eastern Pacific (Schramm, 1985)																														
Upwelling Factor								✓		✓												✓								✓
East Tropical Factor												✓							✓	✓										
West Tropical Factor																														
Subpolar Factor										✓												✓	✓						✓	✓
Subtropical southeast Pacific (Molina-Cruz, 1977a)																														
Subtropical Factor																										✓				
Equatorial Factor (divergence)														✓		✓		✓	✓	✓	✓	✓	✓	✓				✓		✓
Peru Factor (coastal upwelling)										✓																				
Chile Factor (temperate waters)	✓									✓																				
Backwater Factor										✓																				

their occurrence in the fossil record. Haslett (1992) analysed Early Pleistocene radiolarian faunas from climatic extremes, glacial maxima and minima identified in the $\delta^{18}O$ record of Ocean Drilling Program (ODP) Site 677 in the eastern equatorial Pacific (Shackleton and Hall, 1989). Faunal counts from glacial maxima were combined to produce a composite glacial assemblage, and counts from glacial minima were combined to produce a composite interglacial assemblage. The composite assemblages were ranked and described according to their dominant taxa (dominant taxa comprise ≥ 5 per cent of a composite assemblage). The interglacial assemblage is characterised by *Tetrapyle octacantha*, *Octopyle stenozona* and *Theocorythium vetulum*, and the glacial assemblage by *Cycladophora davisiana*, *Botryostrobus auritus*, *Anthocyrtidium zanguebaricum* and *Hexacontium enthacanthum*. It is inferred that the abundance of these species is influenced by water temperature, and that the assemblages represent warm- and cold-water conditions respectively. This palaeoecological information corresponds closely to the modern distribution of those species listed in Table 7.1, and, of the additional species, *T. vetulum* is now extinct, and *A. zanguebaricum* and *H. enthacanthum* possess modern distributions related to the occurrence of cold SSTs (Lombari and Boden, 1985). This approach is similar to that of Weinheimer and Cayan (1997), who generated warm- and cool-water groups of species in a study of sediments spanning 100 years from the Santa Barbara Basin (off California).

Haslett (1995a) investigated the case of *Spongaster tetras*, which is reported by some authors to be indicative of warm tropical conditions (Moore, 1978; Nigrini and Moore, 1979). Yet doubts about this view are expressed by Anderson *et al.* (1989a, b, c), who conducted laboratory experiments on living cultures of *Spongaster tetras*. Haslett (1995a) compared the Plio-Pleistocene record of this species in the eastern equatorial Pacific with other palaeoenvironmental indicators ($CaCO_3$, $\delta^{18}O$, and $\delta^{13}C$), revealing that *Spongaster tetras* is most commonly present during periods of upwelling, often at isotopically defined glacial maxima, when surface productivity and salinity are high, and SST is expected to be relatively low. Therefore, the palaeoecological significance of *Spongaster tetras* cannot be viewed simply as indicating tropical warm-water conditions, particularly in complex oceanographic settings such as the eastern equatorial Pacific. Furthermore, as other species are being studied in laboratory cultures for their tolerance of environmental parameters, such as *Didymocyrtis tetrathalamus tetrathalamus* (Anderson *et al.*, 1990) and *Dictyocoryne truncatum* (Matsuoka and Anderson, 1992; Sugiyama and Anderson, 1997), then studies should be prepared to evaluate the findings in a palaeoecological context.

In their study of Plio-Pleistocene palaeoceanography of ODP Site 709 in the tropical Indian Ocean, Haslett *et al.* (1994b) constructed 'glacial' and 'interglacial' indices based on the summed percentage abundances of the component species of the glacial and interglacial assemblages of Haslett (1992). These indices were compared with an independently derived radiolarian temperature index (RTI) calculated using the equation of Kanaya and Koizumi (1966). However, the species included in the RTI as warm (X_w) and cold (X_c) water indicators, are selected through factor analysis. Notably, X_w includes *Tetrapyle octacantha* and *Octopyle stenozona*, and X_c *Botryostrobus auritus* and *Anthocyrtidium zanguebaricum*; thus, in composition they are very similar to Haslett's (1992) assemblages from the Pacific. These indices produce a consistent relative SST record that corresponds well with other available palaeoenvironmental proxies ($\delta^{18}O$, $CaCO_3$) for Site 709.

In the Plio-Pleistocene sequences of ODP sites 677, 847 and 851 in the eastern equatorial Pacific, factor analysis again groups *Tetrapyle octacantha* and *Octopyle stenozona* into X_w, while *Cycladophora davisiana* is singled out as the principal X_c species (Haslett and Funnell, 1996). Kennington *et al.* (1999) erected independent SST records for these sites based on a diatom transfer function and, through regression analysis, these data can be used to construct an exploratory calibration set to assign temperature to RTI values (Fig. 7.2, Table 7.2). The resulting temperature records show remarkably consistent trends in sites many kilometres apart (Fig. 7.3).

In summary, two methods of employing radiolaria as SST indicators have had proven success. The radiolarian temperature index based on Kanaya and Koizumi (1966) essentially provides relative SST records, while the transfer-function method of Imbrie and Kipp (1971) is capable of assigning absolute SSTs (°C) to fossil assemblages. The ecology of the relatively small number of individual radiolarian species that have been studied has been derived largely from surveys of their distribution in surface (Holocene) sediments of the modern ocean (Lombari and Boden, 1985), and grouped into

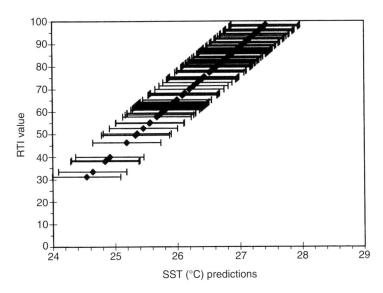

Figure 7.2 Calibration of radiolarian temperature index (RTI) values using diatom-derived sea-surface temperatures of Kennington *et al.* (1999). Error bars are ±0.55 °C

Table 7.2 Sea-surface temperature (SST, °C) corresponding to radioarian temperature index (RTI) values for the eastern equatorial Pacific (calibrated using Fig. 7.2). Accuracy of predicted temperature is ± 0.55 °C

RTI value	SST	RTI value	SST
0.4	24.99	50.9	26.53
1.3	25.01	51.8	26.55
2.2	25.04	52.7	26.57
3.1	25.07	53.5	26.58
4.0	25.09	54.4	26.59
4.9	25.18	55.3	26.60
5.8	25.25	56.2	26.60
6.6	25.26	57.1	26.60
7.5	25.26	58.0	26.62
8.4	25.26	58.8	26.62
9.3	25.26	59.7	26.64
10.2	25.28	60.6	26.65
11.1	25.30	61.5	26.65
11.9	25.31	62.4	26.70
12.8	25.38	63.3	26.75
13.7	25.44	64.2	26.75
14.6	25.44	65.0	26.76
15.5	25.51	65.9	26.77
16.4	25.56	66.8	26.79
17.3	25.61	67.7	26.82
18.1	25.65	68.6	26.82
19.0	25.70	69.5	26.83
19.9	25.70	70.4	26.85
20.8	25.72	71.2	26.87
21.7	25.74	72.1	26.88
22.6	25.76	73.0	26.92
23.5	25.78	73.9	26.92
24.3	25.79	74.8	27.25
25.2	25.80	75.7	27.26
26.1	25.82	76.5	27.27
27.0	25.84	77.4	27.28
27.9	25.85	78.3	27.28
28.8	25.85	79.2	27.33
29.6	25.85	80.1	27.45
30.5	25.87	81.0	27.62
31.4	25.87	81.9	27.72
32.3	25.87	82.7	27.75
33.2	25.89	83.6	27.76
34.1	25.93	84.5	27.77
35.0	25.93	85.4	27.77
35.8	25.97	86.3	27.78
36.7	25.98	87.2	27.79
37.6	26.01	88.1	27.79
38.5	26.05	88.9	27.80
39.4	26.06	89.8	27.80
40.3	26.07	90.7	27.80
41.2	26.09	91.6	27.82
42.0	26.16	92.5	27.82
42.9	26.27	93.4	27.82
43.8	26.36	94.2	27.83
44.7	26.36	95.1	27.84
45.6	26.36	96.0	27.85

Table 7.2 Continued

RTI value	SST	RTI value	SST
46.5	26.39	96.9	27.93
47.3	26.41	97.8	27.95
48.2	26.43	98.7	27.98
49.1	26.44	99.6	28.00
50.0	26.48		

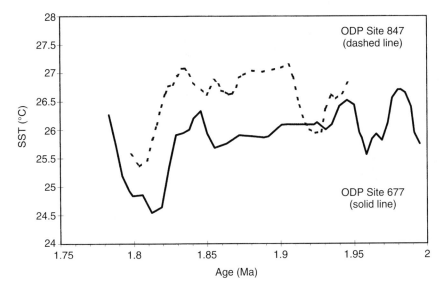

Figure 7.3 Water-temperature reconstructions for ODP Sites 677 and 847 in the eastern equatorial Pacific (raw data are from Haslett, 1994, 1996, and are calibrated using Fig. 7.2 and Table 7.2).

assemblages, aiding down-core comparisons, using a range of statistical techniques such as recurrent group analysis (e.g. Nigrini, 1970) and factor analysis (e.g. Molina-Cruz, 1977a).

7.1.2 Radiolaria as upwelling indicators

Nigrini and Caulet (1992) brought together a number of studies investigating the relationship of Late Neogene–Quaternary radiolaria to upwelling. To accomplish this they examined the stratigraphical distribution of species in cores from the Peru Current, the Oman Margin and the Somalian Gyre, all coastal upwelling centres. They concluded that of the species listed in Table 7.1 as dominant in upwelling factor assemblages, *Cycladophora davisiana*, *Tetrapyle octacantha/Octopyle stenozona*, *Botryostrobus auritus*, *Spongurus* sp., *Stylochlamydium asteriscus* and *Lithelius minor*, their occurrence is not directly related to upwelling. However, they go on to describe an assemblage which characterises Indo-Pacific upwelling (Table 7.3), although not all the 22 component species occur in all areas. They were also able to group species of the upwelling assemblage into three categories, based on the nature of their occurrence in tropical upwelling areas. The first category consists of *endemic upwelling species*, which includes species only found commonly in sediments from upwelling areas; the second category comprises *displaced temperate species*, those species which are common in temperate waters, but not usually found in tropical areas; and the third category consists of *enhanced tropical species*, which are species common in the tropics, but are more abundant and/or more robust in upwelling areas.

Using the upwelling assemblage, Caulet *et al.* (1992) constructed an upwelling radiolarian index (URI) by summing the percentage abundance of component species of the upwelling assemblage. They studied the last 160 ka from core MD85674 from the south Somalian Gyre, and, by comparing the URI with foraminifera, $CaCO_3$, $\delta^{18}O$, and $\delta^{13}C$, were able to recognise some important trends of the glacial/interglacial patterns of upwelling and productivity. Generally, peak URI values were found to be virtually synchronous with $CaCO_3$ maxima and with low $\delta^{13}C$ values in the foraminiferid *Neogloboquadrina dutertrei*. For the eastern equatorial Pacific, Archer (1991a, b) postulated that $CaCO_3$ deposition in pelagic sediments, in the form of calcareous nannofossils and planktonic foraminifera, is controlled by surface-water productivity that – as the URI indicates – is enhanced by upwelling. Also, low $\delta^{13}C$ values from planktonic foraminifera in the northern Indian Ocean are recorded during intense upwelling episodes (Kroon and Ganssen, 1988). Thus, the correlation between peak URI and low $\delta^{13}C$ values, in conjunction with $CaCO_3$ maxima, in core MD85674, suggest that the upwelling radiolarian species described by Nigrini and Caulet (1992) are indeed characteristic of upwelling.

Nearly half the species considered by Nigrini and Caulet (1992) to be characteristic of upwelling were newly described by them, and therefore had not been available for inclusion in previous biogeographical investigations of surface sediments. Thus, for many of these species, there was no modern distributional information with which to verify the upwelling connection. Haslett (1995b) examined 44 Holocene sediment surface samples from the eastern equatorial Pacific upwelling system and constructed a URI for each site. URI spatial variation is mapped and compared with thermocline depth in the region (Fig. 7.4). The depth of the thermocline is an indication of the degree of upwelling that occurs within the water column. In areas of strong upwelling the thermocline lies at a shallow depth. Where upwelling is weak or absent the thermocline extends to greater depths. Haslett (1995b) showed good correlation between the Holocene URI and the present-day thermocline depth in the eastern equatorial Pacific, indicating that the component species of the URI may be regarded as reliable upwelling proxies. Haslett's (1995b) study also demonstrated the potential of the URI for reconstructing upwelling systems over wide geographical areas in the geological record,

Table 7.3 Category, distribution and biostratigraphy of radiolarian species characteristic of areas of upwelling

Category	Species	Biogeography[a]	Stratigraphic range[b]
Endemic upwelling species	Actinomma spp group	Peru and Oman	Miocene–Pleistocene
	Anthocyrtidium rectidentatum	Oman	Pliocene
	Collosphaera sp. aff. C. huxleyi	Peru, Oman, and Somalia	Miocene–Recent
	Cypassis irregularis	Peru, Oman, and Somalia	Pleistocene–Recent
	Dictyophimus infabricatus	NW Africa, Peru, Oman and Somalia	Pliocene–Recent
	Eucyrtidium aderces	Oman	Pliocene
	Eucyrtidium erythromystax	NW Africa, Peru and Oman	Pleistocene–Recent
	Inversumbella macroceras	Peru, Oman and Somalia	
	Lamprocyclas hadros	NW Africa, Peru and Oman	Miocene–Recent
	Lamprocyclas maritalis ventricosa	NW Africa, Peru and Oman	Pleistocene–Recent
	Phormostichoartus schneideri	Peru and Oman	Pliocene–Pleistocene
	Plectacantha cremastoplegma	Peru	Pleistocene–Recent
	Pseudocubus Warreni		
	Pterocanium grandiporus	NW Africa, Peru and Oman	Miocene–Recent
Displaced temperate species	Acrosphaera murrayana group	NW Africa, Peru, Oman and Somalia	Miocene–Recent
	Pentapylonium implicatum	Peru, Oman and Somalia	Miocene–Recent
	Pterocanium aurium	NW Africa, Peru, Oman and Somalia	Miocene–Recent
Enhanced tropical species	Lamprocyrtis nigriniae	Peru, Oman and Somalia	Pleistocene–Recent
	Lithostrobus of L. hexagonalis	NW Africa, Peru, Oman and Somalia	Miocene–Recent
	Phormostichoartus corysforma	NW Africa, Peru, Oman and Somalia	Pliocene–Recent
	Phormostichoartus crustula	NW Africa, Peru, Oman and Somalia	Miocene–Recent
	Pterocorys minythorax	NW Africa, Peru, Oman and Somalia	Miocene–Recent

[a]From Nigrini and Caulet (1992); Haslett (1995c); Zha. et al. (2000).
[b]From Nigrini and Caulet (1992); Haslett (1994, 1996).

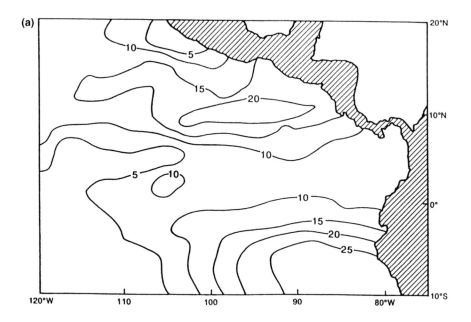

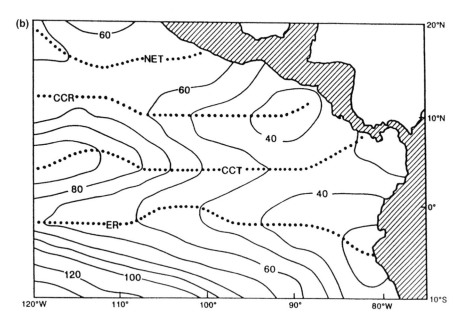

Figure 7.4 Comparison of (a) upwelling radiolarian index (URI%) applied to Holocene surface sediments of the eastern equatorial Pacific; and (b) thermocline depth (m) in the region. ((a) from Haslett (1995c); (b) after Fiedler *et al.* (1991).)

with its application limited only by the availability of well-dated cores, and by the stratigraphical ranges of the URI component species.

The general stratigraphical range of these upwelling species is given in Table 7.3. Most of the species are extant, with two becoming extinct in the Pleistocene. The Pleistocene is the only epoch in which the upwelling species coexist, with fewer species in existence further back in geological time, and with the species that are present displaying increasingly sporadic occurrences and impersistent ranges. Therefore, the Quaternary may be the optimum time period to which the URI can be applied with success (e.g. Caulet *et al.*, 1992; Haslett, 1995b; Vénec-Peyré *et al.*, 1995; Vergnaud-Grazzini *et al.*, 1995; Zhao *et al.*, 2000). For the latest Pliocene and earliest Pleistocene, Haslett *et al.* (1994a, b) constructed upwelling records using the ancestor (*Lamprocyrtis neoheteroporos*) of a common Quaternary upwelling species (*Lamprocyrtis nigriniae*) identified by Nigrini and Caulet (1992). Haslett *et al.* (1994a) found that the percentage abundance of *L. neoheteroporos* correlated reasonably well with $CaCO_3$ preservation (Fig. 7.5), an accepted palaeoproductivity proxy (Archer, 1991a, b), concluding that *L. neoheteroporos*, as the direct ancestor of *L. nigriniae*, could be considered a species indicative of palaeo-upwelling. This conclusion improves the possibility of extending the successful application of the URI to the Pliocene.

Haslett and Funnell (1996) explored fully the application of a URI to the Pliocene–earliest Pleistocene of the eastern equatorial Pacific. Of the 22 species described by Nigrini and Caulet (1992), only *Inversumbella macroceras*,

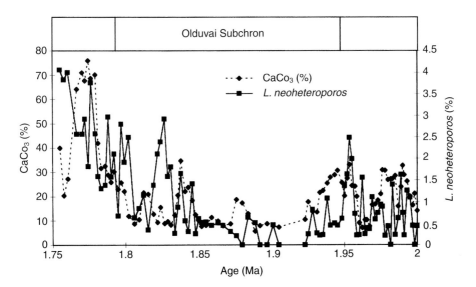

Figure 7.5 Relationship between the Early Quaternary upwelling radiolarian *Lamprocyrtis neoheteroporos* and $CaCO_3$, a palaeoproductivity proxy, through the Olduvai subchron at ODP Site 677, eastern equatorial Pacific. (After Haslett *et al.* (1994a).)

Lamprocyclas hadros, Pseudocubus warreni, Acrosphaera murrayana, Pterocanium auritum, Lithostrobus hexagonalis and *Phormostichoartus crustula* could be expected to be found during the time period under investigation. *Dictyophimus infabricatus* and *Lamprocyrtis neoheteroporos* could also be added to this list, according to the findings of Haslett (1996) and Haslett *et al.* (1994a) respectively. A preliminary survey, however, found *I. macroceras* and *P. warreni* to be absent from the material studied, and *D. infabricatus, L. hadros* and *P. crustula* to be extremely rare. Only *A. murrayana, L. hexagonalis, L. neoheteroporos* and *P. auritum* were found to be present in significant numbers. These four species were included in the URI, and produced upwelling records consistent with other proxy indicators (Funnell *et al.*, 1996). Based on these URI records Haslett and Funnell (1996) attempted to construct a palaeothermocline depth record for ODP sites 847 and 850 in the eastern equatorial Pacific by converting URI values to palaeothermocline depths according to the relationship between the spatial variation of Holocene URI values and present-day thermocline depths in the region (Haslett, 1995b). The results are again consistent with other proxies and do not deviate excessively from the present-day thermocline depth at each site (Fiedler *et al.*, 1991).

Nigrini and Caulet (1992), in describing the upwelling radiolarian assemblage, only investigated sites in the Indo-Pacific region. Haslett (1995c) made a preliminary survey of radiolaria in Late Pleistocene material from ODP Site 658 located in a tropical North Atlantic Ocean coastal upwelling cell, offshore of Cap Blanc, northwest Africa. A number of elements of Nigrini and Caulet's upwelling assemblage were encountered, representing the first record of the assemblage in the Atlantic Ocean. The species encountered by Haslett (1995c) included *Acrosphaera murrayana, Lamprocyrtis nigriniae, Lithostrobus hexagonalis, Pterocanium auritum* and *Pterocorys minythorax*. The recognition of this assemblage in the Atlantic has allowed a URI to be constructed for the Late Quaternary of ODP site 658 (35–0 ka) (Zhao *et al.*, 2000), which has been compared successfully with a number of other palaeoceanographic proxies. In addition to those species encountered by Haslett (1995c), Zhao *et al.* (2000), in their more extensive study, also encountered *Dictyophimus infabricatus, Eucyrtidium erythromystax, Lamprocyclas hadros, Lamprocyclas maritalis ventricosa, Phormostichoartus caryoforma, Phormostichoartus crustula* and *Pterocanium grandiporus*.

The recognition of an upwelling radiolarian assemblage and the construction of the URI provides the palaeoceanographic community with a potent tool for reconstructing temporal and spatial aspects of both coastal and oceanic upwelling systems in the Late Neogene and Quaternary. The technique compares well with other micropalaeontological upwelling proxies, but is particularly well suited to tropical upwelling systems where opal content is usually high and well preserved (Lisitzin, 1971; Davies and Gorsline, 1976), and also in areas where deposition occurs below the calcite compensation depth. Radiolaria also have the advantage of being relatively large, compared to diatoms, coccoliths and dinoflagellates, allowing more rapid faunal analysis.

7.2 PLANKTONIC FORAMINIFERA

Planktonic foraminifera (Fig. 7.6) belong to the Superfamily Globigerinacea of the Suborder Rotaliina. They construct a $CaCO_3$ test arranged as a series of coiled chambers, often globular. The history of research into planktonic foraminifera is similar to that of radiolaria and other microfossil groups, in that the nineteenth century witnessed the erection of a taxonomic framework, followed by taxonomic refinement during the first half of the twentieth century. Application to the petroleum exploration industry, as a means of dating the stratigraphy of marine sediments, was well established by the 1950s. The advent of deep-sea drilling in the 1950s prompted their use as palaeoceanographic indicators, and associated modern ecological and biogeographical studies ensued (e.g. Bé and Tolderlund, 1971; Bé, 1977). The application of planktonic foraminifera to palaeoceanographic studies spans three principal techniques: (1) variations in test morphology, (2) analysis of faunas and assemblages, and (3) because the $CaCO_3$ test is readily dissolved in HCl, the geochemistry of the test can be analysed.

7.2.1 Test Variation in Planktonic Foraminifera

It is recognised that test size varies through time and space in certain species of planktonic foraminifera, either due to evolutionary trends (Huber *et al.*, 2000) or to ecological adaptations. For example, water temperature appears to influence test size in *Orbulina universa* (Bé *et al.*, 1973) and *Globigerina bulloides* (Malmgren and Kennett, 1978), and temperature and salinity changes appear to affect *Globigerinoides ruber* (Hecht, 1976). An interesting study, from an upwelling area in the Arabian Sea, examined the test-size of four species (*Globigerinoides ruber*, *Globigerinita glutinata*, *Globigerina bulloides* and *Neogloboquadrina dutertrei*) through a sediment sequence spanning the last 19,000 years (Naidu and Malmgren, 1995b). The results show that these species were larger during a period spanning 11,000 to 5000 years ago. Independent evidence suggests that this period was characterised by strong upwelling, therefore, the test size of these species may be considered a useful indicator of upwelling intensity.

Another test variable that has palaeoenvironmental value is that of coiling direction. For example, *Globorotalia truncatulinoides* predominantly coils to the left (sinistral) in colder water at high latitudes (Bé and Tolderlund, 1971; Bé, 1977), but is right-coiling (dextral) in warmer waters (Jian *et al.*, 2000). The same is true of the better-known species *Neogloboquadrina pachyderma*, which again displays sinistral coiling in polar and subpolar waters, but dextral in warmer waters. This species has become an important palaeoceanographic indicator in polar to temperate regions (e.g. Bond *et al.*, 1993); however, it is also becoming important in tropical and subtropical waters that are anomalously cold or sourced from high-latitude waters, such as upwelling areas

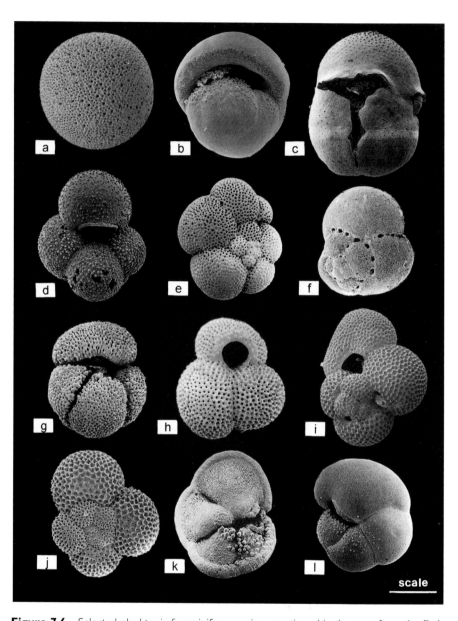

scale

Figure 7.6 Selected planktonic foraminifera species, mentioned in the text, from the Early Quaternary at ODP Site 709B (tropical Indian Ocean). (a) *Orbulina universa*, scale bar = 200 μm, (b) *Pulleniatina obliquiloculata*, view of aperture, scale bar = 200 μm, (c) *Sphaeroidinella dehiscens*, scale bar = 231 μm, (d) *Globigerina spp.*, umbilical view, scale bar = 75 μm, (e) *Neogloboquadrina dutertrei*, view of spiral side, scale bar = 200 μm, (f) *Candeina nitida*, view of spiral side, scale bar = 150 μm, (g) *Globigerinoides conglobatus*, umbilical view, scale bar = 231 μm, (h) *Globigerinoides ruber*, umbilical view, scale bar = 136 μm, (i) *Globigerinoides sacculifer*, view of spiral side, scale bar = 200 μm, (j) *Globoquadrina hexagona*, view of spiral side, scale bar = 150 μm, (k) *Globorotalia tumida*, umbilical view, scale bar = 270 μm, (l) *Globorotalia scitula*, umbilical view, scale bar = 176 μm (see Haslett and Kersley, 1995, 1997).

and the cold currents that move along the western coasts of Africa and South America respectively. Ufkes *et al.* (2000) report a number of major abundance peaks of sinistral *Neogloboquadrina pachyderma* in a sediment core spanning the last 420,000 years from the Walvis Ridge offshore of southwest Africa (Fig. 7.7). The authors variously attribute these peaks to an intensification of the nearby Benguela coastal upwelling system and the propagation of unmixed parcels of cold water from the African coast out into the South Atlantic.

Other species with significant test variations include the surface-dwelling *Globigerinoides sacculifer*, which possesses a large additional chamber (called a sac) in waters greater than 20 m deep (Bé *et al.*, 1985),

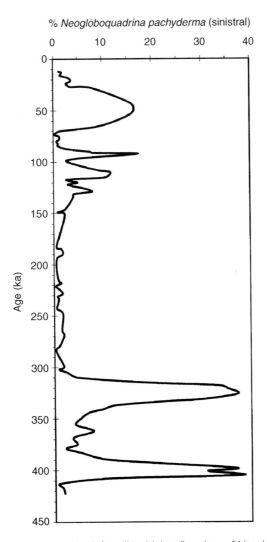

Figure 7.7 The occurrence of the left-coiling (sinistral) variety of *Neogloboquadrina pachyderma* in Quaternary sediments from Walvis Ridge (offshore southwest Africa). (Data from Ufkes *et al.* (2000).)

and the test colour of *Globigerinoides ruber* may be white or pink, depending on environmental conditions.

7.2.2   Planktonic Foraminiferal Faunas/Assemblages

Many studies of planktonic foraminiferal ecology have, as with radiolaria, highlighted the preference many species display for certain environmental conditions. For example, Table 7.4 indicates the water-temperature preferences of a selected number of species estimated by comparing the distribution of microfossils in oceanic surface sediment with overlying water temperatures

Table 7.4   The distribution of selected planktonic foraminifera species in relation to oceanic provinces and sea-surface temperature (°C). (Information from Bé and Tolderlund, 1971.)

Planktonic foraminifera species	Oceanic provinces (and sea-surface temperature)					
	Polar (0.5 °C)	Subpolar (5–10 °C)	Temperate (10–18 °C)	Subtropical (18–24 °C)	Tropical (24–30°C)	
Neogloboquadrina pachyderma	Abundant	Common	Few			Polar assemblage
Globigerina bulloides	Few	Abundant	Common			} Subpolar assemblage
Globorotalia scitula		Rare	Rare			
Globorotalia inflata		Few	Common	Few		} Temperate assemblage
Globorotalia truncatulinoides		Few	Common	Common	Rare	
Globigerinita glutinata		Rare	Few	Common	Few	
Neogloboquadrina dutertrei			Few	Common	Few	
Orbulina universa			Few	Common	Few	Subtropical assemblage
Globigerinella siphonifera			Few	Common	Few	
Globigerinoides conglobatus			Rare	Common	Few	
Globigerinoides ruber			Common	Abundant	Common	
Globorotalia menardii/tumida			Rare	Few	Few	(Sub)tropical assemblage
Pulleniatina obliquiloculata			Rare	Few	Few	
Globigerinoides sacculifer			Few	Common	Abundant	
Sphaeroidinella dehiscens					Rare	
Candeina nitida					Rare	Indigenous tropical species
Globoquadrina hexagona					Rare	

(Bé and Tolderlund, 1971). Large-scale studies, such as this, prove extremely useful in eliciting general relationships between microfossil occurrences and environmental parameters. However, surface sediments comprise average assemblages that may have accumulated over many years, and thus short-lived environmental signals will be masked. More recently, numerous studies have been undertaken on living communities and the **flux** of tests to the sea-floor (e.g. Conan and Brummer, 2000; Kincaid *et al.*, 2000), and these enable an analysis of faunal variation in response to annual and seasonal oceanographic changes. For example, Sautter and Thunnell (1991a) use sediment traps to investigate seasonal flux variations in the San Pedro Basin, off-shore California. They are able to map the maximum test production for a given species against hydrographic conditions (Table 7.5), suggesting that *Globigerina bulloides* is characteristic of spring upwelling conditions, *Neogloboquadrina dutertrei* dominates in the post-upwelling period, and that *Globigerinoides ruber* characterises the warm, stratified waters of the early summer.

Information on modern species occurrences allows fossil assemblages to be interpreted in terms of oceanographic variables. This may be achieved statistically using factor analysis (see section 7.1.1), a modern analogue technique (MAT) (e.g. Martinez *et al.*, 1999), and/or through the application of a transfer function (Imbrie and Kipp, 1971). Hendy and Kennett (2000) use abundance data, a MAT and a transfer function in their study of Late Quaternary palaeoceanography of the Santa Barbara Basin (California). They find that during cold stages (stadials), assemblages are dominated by *Neogloboquadrina pachyderma* (sinistral) and *Globigerinoides glutinata*, which suggest a strong subpolar influence, while warmer stages (interstadials) are characterised by *Neogloboquadrina pachyderma* (dextral) and *Globigerina bulloides*, which they interpret as representing a more subtropical influence. Similar studies have investigated upwelling (Marchant *et al.*, 1999), palaeoproductivity (Cayre *et al.*, 1999) and sea-surface temperatures (Pflaumann and Jian, 1999).

Table 7.5 Times and/or conditions of maximum abundance of planktonic foraminifera species in the San Pedro Basin, offshore California. (Information from Sautter and Thunell, 1991a.)

Species	Dominant season/hydrographic condition
Globorotalia theyeri	Winter (cold, stratified)
Globigerina digitata	Winter (cold, stratified)
Globigerinita glutinata	Early spring (warming, stratified)
Globigerinella aequilateralis	Early spring (warming, stratified)
Neogloboquadrina pachyderma	Early spring (warming, stratified)
Orbulina universa	Early spring (warming, stratified)
Globigerina quinqueloba	Early spring (warming, stratified)
Globigerina bulloides	Upwelling (cold)
Globoquadrina hexagona	Upwelling (cold)
Neogloboquadrina dutertrei	Post-upwelling (warming, stratifying)
Globigerinoides ruber	Early summer (warm, stratified)

Furthermore, the ratio between planktonic and benthic foraminifera occurring in a sample is a crude tool for estimating depositional setting. As discussed in section 3.1.6, offshore settings are characterised by abundant planktonic species, while benthic species increase in abundance shorewards. This has been demonstrated recently by Schmuker (2000) in a study off the coast of Puerto Rico using samples spanning a depth range of 432–4700 m and distance of up to 95 km from the shelf break.

7.2.3 Geochemistry of Planktonic Foraminiferal Tests

The geochemistry of the $CaCO_3$ (calcite) planktonic foraminiferal test is interesting from an environmental view, for a number of reasons. First, the amount of calcite used in test construction may vary according to environmental conditions, so that specimens of the same species may be weakly or strongly calcified. For example, Sautter and Thunnell (1991b) suggested that less-calcified forms of *Globigerina bulloides* are indicative of upwelling conditions. The degree of calcification also influences the preservation potential of tests, as weakly calcified forms are more likely to suffer dissolution as they sink through the water column to the sediment surface. However, acidic bottom waters or sediment substrate may also dissolve tests following deposition. Dissolution may be influenced by changing environmental parameters, and some authors measure dissolution using a ratio between intact and broken specimens. For example, Funnell *et al.* (1996) measured intact versus broken tests of *Globorotalia tumida*, and found that during severe dissolution events there is an associated increase in benthic foraminifera and other more dissolution-resistant planktonic species. Therefore, care must be taken when interpreting assemblages where dissolution appears to have influenced preservation.

Foraminiferal tests also contain traces of other elements, which are incorporated during calcification, many of which are incorporated in relation to environmental variables. For example, experiments on laboratory cultures of *Globigerina bulloides* and *Orbulina universa* have demonstrated that the ratio between magnesium and calcium (Mg/Ca) in the test is temperature dependent, so that the Mg/Ca ratio increases by 8–10 per cent for every 1 °C rise in water temperature (Lea *et al.*, 1999). This has been explored in deep-sea cores, where a comparison between the Mg/Ca ratio in *Globigerinoides sacculifer* and other SST proxies shows that the Mg/Ca ratio does indeed reflect changes in SST (Nurnberg *et al.*, 2000). The application of other geochemical ratios is being investigated, such as cadmium (Cd/Ca ratio) which appears less straightforward (Mashiotta *et al.*, 1997; Rickaby and Elderfield, 1999). Radiolaria also take up trace elements in test silicification; however, SiO_2 is problematic to dissolve in the laboratory, whereas $CaCO_3$ dissolves easily in weak hydrochloric acid (HCl) to release the trace elements for analysis.

In addition to trace elements, the isotopic composition of $CaCO_3$ tests can be investigated. Stable isotopes of oxygen (^{16}O and ^{18}O) and carbon (^{13}C) are among the most widely used in palaeoceanographic studies, but others may also be important, such as neodymium isotopes that appear to be influenced by continental weathering and rates of oceanic circulation (Vance and Burton, 1999). It is known that a number of planktonic foraminifera species, but not all, incorporate isotopes in equilibrium with the surrounding seawater, so that they faithfully record the isotopic composition of the water during calcification of their tests.

Oxygen isotopes in foraminifera have been studied for some time, with Emiliani (1955) doing much to stimulate research in the application of isotopes to palaeoceanography (Sirocko, 1996). Broad variation in the ratio of $^{16}O/^{18}O$ reflects changes in global ice volume, as oceans become depleted in

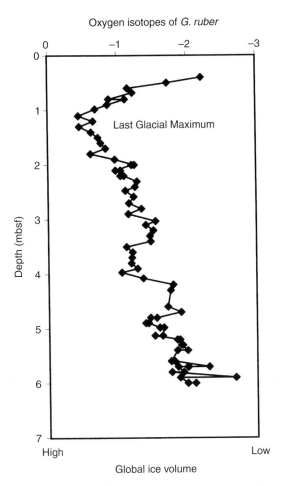

Figure 7.8 The Late Quaternary $\delta^{18}O$ record of *Globigerinoides ruber* at ODP Site 677 in the eastern equatorial Pacific, clearly showing the Last Glacial Maximum; mbsf = metres below sea-floor. (Data from Shackleton and Hall (1989).)

^{16}O during glacial stages, because, being a lighter isotope, it is preferentially evaporated from the oceans and then precipitated and, during glacials, stored in continental ice sheets. During interglacials, ^{16}O returns to the sea via rivers to maintain a higher ^{16}O/^{18}O ratio (Williams *et al.*, 1998).

Shackleton and Hall (1989) provide a high-resolution δ^{18}O record for the Quaternary, based on analysis of *Globigerinoides ruber* at ODP Site 677 in the eastern equatorial Pacific; the Late Quaternary section is shown in Fig. 7.8. Because global ice-volume changes are considered to be related to predictable changes in the Earth's orbit (**Milankovitch Cycles**, see Hays *et al.*, 1976a), it is possible to use the periodicities of these orbital changes (*c*.23,000, 41,000 and 100,000 years) to date the δ^{18}O curve, through a technique known as **astronomical tuning**. This has been performed on the ODP Site 677 Quaternary record (Shackleton *et al.*, 1991), and subsequently a Pliocene record has been calibrated (Shackleton *et al.*, 1995a, b). These time-scales derived from foraminiferal isotope studies have been adopted widely, providing a 'standard' Quaternary chronology (e.g. Funnell, 1995).

In addition to using stable isotopes to establish global palaeoenvironmental conditions, it has been shown that more rapid and less extreme variations in isotopic composition reflect changing oceanographic parameters. For example, Sautter and Thunell (1991b) found in a flux study (San Pedro Basin, California) that seasonal variations in δ^{18}O of *Orbulina universa* reflected fluctuations in water temperature above the thermocline throughout the year,

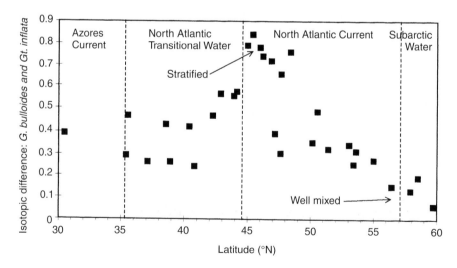

Figure 7.9 The δ^{18}O difference between *Globigerina bulloides* (shallow-water species) and *Globorotalia inflata* (deep-water species) in relation to latitude and ocean currents in the northeast Atlantic. The water column is interpreted as being well mixed where the difference is minimal (e.g. Subarctic Water), but stratified where there is a more marked difference (e.g. near the southern boundary of the North Atlantic Current). (Data from Ganssen and Kroon (2000).)

and that $\delta^{18}O$ in *Globigerina bulloides* documents near-surface temperatures during upwelling (see Table 7.5). Furthermore, Ganssen and Kroon (2000) in a study of surface sediments from the North Atlantic, suggest that $\delta^{18}O$ in *Globigerinoides ruber* and *Globigerinoides trilobus* reflect summer surface-water conditions, *Globigerina bulloides* is associated with the spring bloom (and therefore, enhanced productivity), *Globorotalia inflata* and *Globorotalia truncatulinoides* reflect deep-water (100–400 m) temperatures, and that $\delta^{13}C$ in *Globigerina bulloides* correlates with surface-water phosphate values and, therefore, may represent a useful palaeonutrient indicator. Interestingly, they also state that the difference in $\delta^{18}O$ between *Globigerina bulloides* (a shallow-water species) and *Globorotalia inflata* (a deep-water species) indicates the degree of water-column mixing, so that the difference is minimal where the water column is strongly mixed, and marked where the column is well stratified (Fig. 7.9). A similar approach has been developed in the Pacific using a near-surface (<20 m deep) variant of *Globigerinoides sacculifer*, and the thermocline-dweller *Neogloboquadrina dutertrei* (Farrell *et al.*, 1995).

7.3   THE CLIMAP CONTROVERSY

The CLIMAP (Climate/Long Range Investigation Mapping and Prediction) Project was an early attempt at reconstructing Quaternary SSTs using proxies such as radiolaria and planktonic foraminifera. The aim of CLIMAP was to reconstruct SSTs at the time of the Last Glacial Maximum (LGM) 18,000 years ago. Microfossils were studied in surface sediments, and their occurrences were related to the SST of the overlying water-mass. A transfer function (Imbrie and Kipp, 1971, see section 7.1.1) was then employed to apply SSTs to LGM sediments (CLIMAP, 1976). In general, the results indicated that SSTs were colder at higher latitudes, as would be expected, but that low-latitude SSTs were not appreciably different to modern temperatures. This was certainly the case for reconstructions of the tropical Pacific that relied heavily upon radiolarian data (Moore, 1978; Moore *et al.*, 1980). Therefore, it may be inferred from these data that the SST gradient from low to high latitudes was much steeper during the LGM.

The CLIMAP study was significant in that it provided data that could be employed in pioneering palaeoclimate simulation models (e.g. Kutzbach and Wright, 1985), and that it also provided a model that could be tested through empirical investigations, so stimulating further research. Subsequent research in the tropical Pacific region included investigations of altitudinal changes in mountain environments (e.g. Hawaii) and water-temperature records derived from fossil coral reefs, among others (Rind and Peteet, 1985; Guilderson *et al.*, 1994; Stute *et al.*, 1995). These studies suggest that tropical temperatures were between 3 °C and 6.5 °C cooler than present. Evidence for cooler temperatures from mountain sites is at first sight not entirely incompatible with CLIMAP's reconstructions, as an increase in altitudinal temperature gradients at the LGM could explain the discrepancies. However, cooler reef-derived temperatures do suggest that CLIMAP's reconstructions are inaccurate. Unfortunately, this has led to a view that radiolaria (and to a lesser degree planktonic foraminifera) are not useful water-temperature indicators and that their palaeoceanographic applications are limited. A re-evaluation of the radiolarian contribution to CLIMAP suggests a number of reasons why inaccurate SSTs were reconstructed:

- In the 1970s radiolarian taxonomy was less developed than at present, so that some ecologically important species and subspecies that have since been described (e.g. Nigrini and Caulet, 1992) were not then available. Also, subsequent laboratory culture experiments and detailed comparative palaeoceanographic studies have contributed to refining species–environment relationships (see sections 7.1.1 and 7.1.2).
- The method employed (i.e. factor analysis, see section 7.1.1) could only include generally abundant taxa (for example, comprising no less than 2 per cent of an assemblage). Therefore, uncommon species, which may be ecologically significant (biogeographically restricted) on their own, were grouped into larger artificial taxonomic counting groups, resulting in less-detailed environmental information being generated.
- The processes involved in the deposition of microfossils from the living fauna (**biocoenosis**) to fossil assemblage (**thanatocoenosis**) was poorly understood. It is now appreciated that assemblages occurring in bottom sediments are sometimes different from faunas living in the overlying water column: being affected by current transport, flux variations, time-averaging in the sediment, and dissolution. Welling and Pisias (1998b) have suggested that radiolarian assemblages being deposited in eastern Pacific upwelling regions are very similar to their living counterparts and therefore accurately reflect water temperatures and other oceanographic parameters, but elsewhere in the oceans the differences between fossil assemblages and living faunas are more marked, so that palaeoceanographic inferences are less secure.

With improved understanding of taxonomy, ecology and oceanographic processes, coupled with methodological developments, it is very likely that a re-investigation of LGM palaeoceanography in some areas, using planktonic sarcodine Protozoa, would yield results compatible with other proxies, indicating that deficiencies of past studies should not foreshadow future applications.

7.4 SUMMARY

The range and precision of applications of planktonic sarcodine Protozoa in palaeoenvironmental studies has advanced since the 1950s. As water-temperature indicators, radiolaria and planktonic foraminifera have proved useful proxies, through semi-quantitative techniques, such as the radiolarian temperature index (RTI), or through more fully quantitative and numerical methods, such as through the use of factor analysis and transfer functions. The same is true for their use as proxies of upwelling conditions – with the construction of an upwelling radiolarian index (URI). Study of the geochemistry of planktonic foraminiferal tests is possible due to their easily dissolved $CaCO_3$ composition. Magnesium/calcium and isotopic ratios are proving very useful in answering a range of palaeoceanographic questions. The validity of palaeoceanographic results has been questioned in the past (see section 7.3), but they now appear to be as reliable as those obtained by other proxies. However, further work is required on modern and living faunas, both in the natural environment and in laboratory cultures, in order to define more accurately the associated environmental parameters.

FURTHER READING

Anderson, O.R., 1983. *Radiolaria*. Springer-Verlag, New York, 355 pp.
A useful introduction to living and fossil radiolaria.

Brummer, G.J. and Kroon, D. (eds), *Planktonic foraminifers as tracers of ocean-climate history*. Free University Press, Amsterdam.
A useful collection of relevant case studies.

Hemleben, C., Spindler, M. and Anderson, O.R., 1989. *Modern planktonic foraminifera*. Springer, New York, 363 pp.
A comprehensive introduction to Quaternary planktonic foraminifera.

Kennett, J.P. and Srinivasan, M.S., 1983. *Neogene planktonic foraminifera: a phylogenetic atlas*. Hutchinson Ross, Stroudsburg, PA, 265 pp.
A very useful identification guide to planktonic foraminifera.

THE ENVIRONMENTAL APPLICATIONS OF DIATOMS

Kevin Kennington

Diatoms (Class Bacillariophyceae) are microscopic (1–1000 μm in size) unicellular algae that inhabit almost every aquatic environment on Earth. They are characterised mostly via their cell walls, which are composed of opaline silicate – being made up of two overlapping valves that together constitute the **frustule**. As with other siliceous microfossil groups (e.g. the **silicoflagellates** and **radiolarians**), diatoms are often well preserved in sedimentary sequences and have become invaluable for unravelling past environmental change in both freshwater and marine systems.

Diatoms can be free-floating (**planktonic**), attached to macrophytes (**epiphytic**) or other organisms (**epibiontic**), attached to sand or silt particles (**epipsammic**), or to rock surfaces (**epilithic**). They occupy a diverse range of niche spaces (Fig. 8.1) and often have very limited environmental ranges within which they can survive. Consequently, distinct diatom communities have been used to define different habitat types, and have provided a useful tool for scientists interested in pollution studies.

The diatoms are first found as fossils in Middle Cretaceous rocks. However, since these Cretaceous diatoms occur as assemblages containing numerous species, it is possible that they developed earlier, perhaps in the Late Palaeozoic, and existed as naked protoplasm or were not fossilised because of taphonomic processes or perhaps because early forms were too lightly silicified. The centric diatoms were the first to appear in the fossil record, with the pennate forms not appearing until the late Palaeocene. The earliest freshwater species date from the Eocene and were pennate in form. An understanding of the requirements of extant forms is invaluable in assessing environmental change over time. The age of the sedimentary sequence being examined is important and will dictate how much information can be determined regarding such changes, since fewer extant forms may be found. Any assessment of the environmental requirements of extinct forms must be made with a great deal of caution. The use of diatom microfossils in interpreting past environmental perturbations becomes more problematic the further back in time

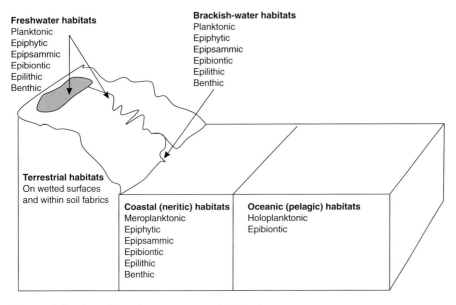

Freshwater habitats
Planktonic
Epiphytic
Epipsammic
Epibiontic
Epilithic
Benthic

Brackish-water habitats
Planktonic
Epiphytic
Epipsammic
Epibiontic
Epilithic
Benthic

Terrestrial habitats
On wetted surfaces
and within soil fabrics

Coastal (neritic) habitats
Meroplanktonic
Epiphytic
Epipsammic
Epibiontic
Epilithic
Benthic

Oceanic (pelagic) habitats
Holoplanktonic
Epibiontic

Figure 8.1   Aquatic environments occupied by diatoms.

the researcher tries to study. It is not surprising, therefore, that most publications reporting on diatom palaeoecology deal with the Cenozoic.

The purpose of this chapter is to introduce the student to the principles of diatom palaeoecology and how these have been applied in the assessment of environmental change in the fossil record. The first section covers a basic introduction to diatom taxonomy, while the last two sections look into the use of diatoms for interpreting changes in freshwater and marine habitats respectively.

8.1   DIATOM TAXONOMY

The purpose of this section is to give a brief introduction to the principles of diatom taxonomy. It is only through the proper identification of diatom species that correct interpretations of the palaeoecological significance of fossil assemblages can be made. References to individual species are only made in order to illustrate the major morphological features of the diatom frustule, since it would be well beyond the scope of this introduction to describe the major taxa. For more information on the components of the diatom frustule, an excellent review of the major genera can be found in Round *et al.* (1990). Other texts describing specific diatom taxonomy are Hendey (1937, 1964), Patrick and Reimer (1966–75), Barber and Haworth (1981), Medlin and

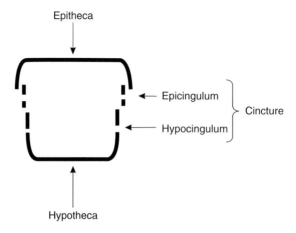

Figure 8.2 The components of a diatom frustule.

Priddle (1990), Snoeijs (1993), Snoeijs and Vilbaste (1994), Snoeijs and Potapova (1995), Snoeijs and Kasperovičienė (1996) and Snoeijs and Balashova (1998). Information on diatom taxonomy is also spread throughout the scientific literature, and may be found in an array of journals (e.g. *Diatom Research, Beihefte zur Nova Hedwigia, Ocean Drilling Program Reports*, etc.).

The taxonomy of diatoms is based primarily on the characteristics of their siliceous frustule. Broadly speaking, the diatoms can be split into two morphological groups, the round diatoms (**centrales**) and the bipolar diatoms (**pennales**), but this is not a strict division, since both round pennate diatoms and bipolar centric diatoms exist. Information regarding the terminology of the diatom frustule can be found in Anon (1975) and Ross *et al.* (1979). The pennate diatoms can be divided further into the raphiated and araphiated forms (based upon the presence or absence of a longitudinal 'slit' or **raphe** on the frustule). The major characteristics of both centric and pennate forms are discussed separately below.

The diatom frustule resembles a petri dish and is composed of four basic parts (Fig. 8.2). The upper part of the valve (**epitheca**) is the oldest part of the frustule and sits over the lower part of the valve (**hypotheca**). These parts of the frustule are separated by a series of girdle bands known collectively as the **cincture**.

8.1.1 Components of the Centric Diatom Frustule

The centric diatoms are for the most part planktonic, that is, they spend the majority of their life floating freely in the surface waters of the sea, lakes or rivers. Indeed there may be a selective advantage to having a circular shape

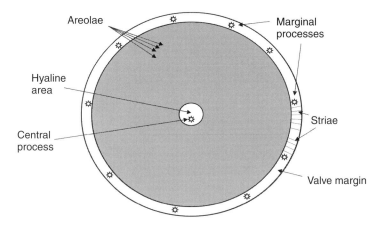

Figure 8.3 Stylised drawing of a centric diatom, showing some key diagnostic features of the valve face.

in such an environment, since this offers a relatively high surface area to weight ratio compared with pennate forms and may help to reduce sinking rates. Not all centric diatoms are round, however, and many other geometrical shapes occur.

The taxonomy of the centric diatoms relies heavily upon the characteristics of the valve face (Fig. 8.3). Pore-like openings (**areolae**) are found running across the valve face, and these are one of the diagnostic features used by diatomists to separate genera. These patterns can be separated into several broad groups. Radial areolation patterns always show areolae running from the centre of the valve face towards the valve margin. Tangential areolation, on the other hand, runs from one side of the valve face to the other. Other characteristics of the centric diatom frustule include the presence of small tube- or sac-like openings on the valve surface. These openings are known as **fultoportulae** and **rimoportulae** (Fig. 8.4). There are several different forms of both these opening types, and some can be species-specific. Other morphological components that aid identification include spines (**setae**), **striae** and hyaline areas.

8.1.2 Components of the Pennate Diatom Frustule

As mentioned earlier, the pennate diatoms can be split into the raphiated and araphiated groups. The araphid group is probably the most difficult group to identify, since the valve structure is rather simple and the valves small and linear. Correct identification of the araphid group is difficult under the light microscope, and many features can only be discerned using the scanning electron microscope. Both pennate forms share morphological features such as striae and areolae. A major component of pennate taxonomy is an assessment of the number and distribution of these structures along the

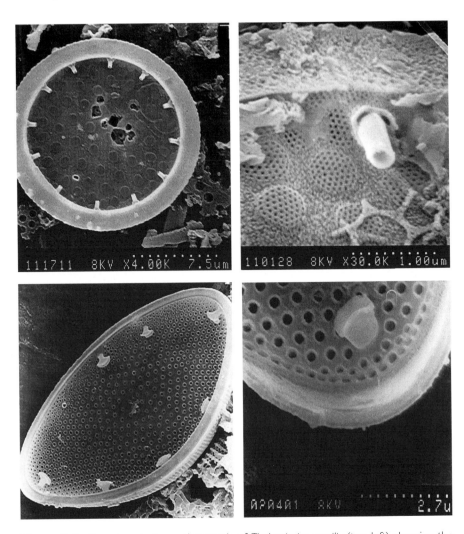

Figure 8.4 Scanning electron micrograph of *Thalassiosira gracilis* (top left) showing the internal arrangement of fultoportulae or strutted processes. Close-up (top right) of fultoportula in *T. gracilis*. SEM of *Hemidiscus cuneiformis* (bottom left) showing arrangement of rimoportulae or labiate processes. Close-up (bottom right) of individual rimoportula in *H. cuneiformis*.

length of the valve, as well as the presence and location of processes and raphe structures (Fig. 8.5).

8.2 Diatoms in Palaeolimnology

Palaeolimnological studies utilising diatoms to describe past environmental changes have developed rapidly over the past few decades. Such studies have

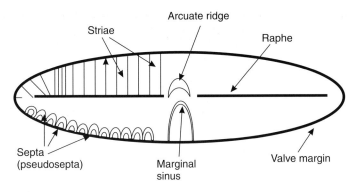

Figure 8.5 Stylised drawing of a pennate diatom, showing some key diagnostic features.

proved to be important tools in the assessment and management of lacustrine systems (Battarbee, 1999). The biological components of a lake are responsive to changes operating within the lake catchment; consequently, any alteration in the processes operating in the catchment over time may be recorded in the sedimentary record. Such factors may include alterations to the nutrient status of the lake, or alterations to the lake pH via pollution or transformations of the lake biota in response to climate change. A major contribution to palaeolimnological interpretation of diatom assemblages has been the application of **calibration sets** (also known as training sets). The development of a calibration set involves the assessment of diatom assemblages across a gradient of differing environmental parameters (both physical and chemical). These assemblages can then be tied numerically to different physicochemical parameters and used to interpret past environmental conditions operating within the lake catchment. Such a technique is referred to as a 'transfer function' (see Appendix).

8.2.1 Diatoms and Lake Eutrophication

Eutrophication, the process of enrichment of a waterbody with plant nutrients, is one of the greatest threats to lacustrine environments. This process can be a result of the gradual change of a lake via natural conditions, or through anthropogenic discharges to watercourses and to the atmosphere. Figure 8.6 shows some of the ways in which nutrients enter lakes. Since lakes by their very nature are enclosed waterbodies, these nutrients can – if not assimilated by the lake flora – increase to levels that are detrimental to the ecosystem as a whole. Responses of a lake to eutrophication include increased productivity of both phytoplankton and macrophytes, which can result in decreased water clarity and water quality. Increased productivity can also lead to the deoxygenation of bottom waters and to the development of anoxic sediment, especially in stratified lakes. This then has a knock-on

Figure 8.6 Pathways by which nutrients and other pollutants enter lacustrine waters: (A) via natural processes, e.g. geological weathering, deglaciation; (B) via atmospheric rain-out, e.g. nitrogen in rainwater from the burning of fossil fuels, etc.; (C) via forestry-management practices: for example, deforestation may lead to increased runoff of nutrients into streams entering the lake; (D) via rivers and streams: a major transport pathway for carrying pollutants into the lake from its catchment area; (E) from municipal wastes, e.g. sewage waste-water, storm-drains, etc.; (F) from agricultural and aquacultural practices, e.g. nitrate and phosphate fertilisers; and (G) via industrial sources such as mining operations, quarrying, etc.

effect throughout the food web of the lake, as anoxic conditions become a barrier to benthic flora and fauna.

Diatoms have been widely used as indicators of lake trophic status and productivity (Stoermer *et al.*, 1996). Diatoms are highly sensitive to changes in nutrient concentrations, because individual taxa have different tolerances for nutrients. The effect of differing nutrient-salt concentrations (mainly NO_3, NO_2, NH_3, PO_4 and SiO_2) upon individual species can be assessed either by laboratory experiments or from direct assessment of diatom assemblages across an array of lakes with differing environmental conditions. Problems exist with both approaches, for it can be argued that experiments using cultured diatoms may not be truly representative of ambient lake conditions, while the direct assessment of a suite of lakes requires a great deal of time and resources. Limnologists have adopted a sort of 'half-way house' technique in recent years. This technique relies on the use of **mesocosms** – large containers filled with lake water of differing nutrient (and other) regimes (Fig. 8.7). Mesocosm experiments allow the limnologist to assess the effect of nutrient enrichment upon the diatom flora in conditions close to those found across a suite of 'natural' lakes.

Figure 8.7 An experimental mesocosm set-up at the University of Liverpool's Ness gardens. The use of such experimental set-ups allows for the assessment of different environmental parameters upon diatom populations. (Photograph courtesy of D. Wilson and B. Moss, University of Liverpool.)

A number of methods have been employed to interpret the trophic history of lakes. One of the simplest techniques adopted was a simple ratio of the centrales to pennales (Nygaard, 1956). This technique was based on the fact that nutrient-poor (**oligotrophic**) lakes contain mainly raphiated pennate diatoms associated with epiphytic, epipelic and epipsammic communities. The ratio becomes very high in large eutrophic lakes, owing to increased suspended matter and increased numbers of centric diatoms (Stockner and Benson, 1967). Stockner (1971) continued the theme of using such ratios and developed the araphid/centric ratio to document levels of lake **eutrophication**. This technique also had its limitations in that it was not applicable to shallow lakes and bogs, owing to the predominance of benthic and heavily silicified diatoms which inhabited such environments and were independent of the trophic status of the lake (Stockner, 1971). The use of such ratios allowed palaeolimnologists to infer the trophic history of such waterbodies, but this approach was still only qualitative, and a lot of information was lost owing to the need to clump diatoms into a few broad categories.

Agbeti and Dickman (1989) developed a diatom inferred trophic index (DITI) by examining the diatom flora from sediment-surface samples across 30 lakes representing a continuum from oligotrophic to eutrophic conditions. The DITI was used in a multiple regression analysis of contemporary phosphorus and chlorophyll-*a* concentrations across all the lakes, and showed significant correlations between the index and these variables. Such linear regression models also involved the clumping of various diatom taxa according to information gleaned from the literature, and as such were over-simplistic in their approach. Modern quantitative techniques have been developed which, instead of relying upon ratios or indicator species, utilise the ecological requirements of common diatom species to quantify changes in lake trophic status (Stevenson *et al.*, 1991; Birks, 1994). These methodologies rely upon the development of calibration sets, which in turn depends upon the assessment of extant diatoms normally collected from sediment surface samples. This has the advantage over sets derived from living diatoms collected from benthic or planktonic habitats, in that sediment surface samples contain diatoms from all habitats in the lake. (Nevertheless, one must be aware of other problems, such as bioturbation and reworking of older sediments.) The diatom assemblages collected from sediment-surface samples are then statistically related to the overlying water chemistry, enabling the discrimination of distinct community groups that have a direct relationship with specific chemical parameters. These community–chemical relationships can then be used in multiple-regression analyses to interpret the fossil record from core material (Anderson *et al.*, 1990).

8.2.2 Diatoms and Lake Acidification

During the early 1970s, Scandinavian scientists noted changes in lake pH across Norway and Sweden. These changes were attributed to '**acid rain**', which it was claimed was having a major impact upon the ecology of such lakes (e.g. by altering the balance between acid-sensitive and acid-tolerant species at different trophic levels). One of the most noticeable impacts of this acidification process was the decline in certain fish stocks within these lakes (e.g. the salmonids). The cause of lake acidification became a hotly debated topic during the following decades, and several factors were highlighted as causative mechanisms for lake acidification. Some scientists claimed that acidification was a natural process (e.g. via glacial retreat) or was a response to land-use changes (e.g. deforestation) operating in the catchment of the lake (Pennington, 1984; Engstrom *et al.*, 2000). However, it soon became apparent that although in certain instances lake acidification could be attributed to natural and catchment-management processes, the major contributor to this environmental problem was acid deposition. What was not known, however, was how long such processes had been affecting these systems. Reliable data on the long-term record of lake pH from monitoring programmes were sparse, and scientists turned to the sedimentary record to try to reconstruct the pHs of lakes across the world.

Diatoms have long been known to be good indicators of pH, and the scientific literature relating diatom distribution to pH is extensive (Hustedt, 1937–39; Flower, 1986) and several good reviews have been published subsequently (Battarbee, 1984; Charles and Smol, 1988; Round, 1990, 1994; Battarbee *et al.*, 1999). The ecological assessment of a lake's pH status and of its pH history has been assessed primarily through the analysis of its diatom flora. One of the earliest such systems was that produced by Hustedt (1937–39), who divided diatom communities into five categories according to the pH tolerance of individual species.

Hustedt's pH classification:

1. alkalibiontic: species occurring at pH values of >7;
2. alkaliphilous: species occurring at pH of about 7, with their widest distributions at pH >7;
3. indifferent: species with equal occurrences on both sides of pH 7;
4. acidophilous: species occurring at pH about 7, with their widest distribution at pH <7;
5. acidobiontic: species occurring at pH values <7, but with an optimum distribution at pH <5.5.

Hustedt's system was readily adopted by palaeolimnologists, and other classification schemes evolved from Hustedt's work (Nygaard, 1956; Meriläinen, 1967; Foged, 1969; Renberg and Hellberg, 1982). Although the aforementioned systems aided the limnologist in interpreting lacustrine alkalinity and pH, there were limitations to these approaches. Any interpretation of the diatom assemblage required the accurate assessment of any individual taxon's pH range and optimum, and these tolerances were often misinterpreted leading to the misclassification of a species' pH class. Meriläinen (1967) also showed how seasonality can have large effects on a lake's pH, especially in shallow productive lakes with natuarally high alkalinity, and recommended that pH measurements of lakes used in 'calibration sets' should be standardised by using values from the autumn turnover period when stratification breaks down and waters in the lake become mixed once more.

The various pH classification schemes developed since Hustedt's (1937–1939) original work went some way to enabling the classification of a lake's pH status through time. More recent methodologies have, however, enabled more accurate reconstructions to be made. These methodologies involved the use of calibration sets obtained from lakes across a gradient of differing pHs. The earliest calibration sets used in limnology were relatively small and their application was sometimes geographically limiting. A more modern approach has been the development of integrated calibration sets where a number of smaller calibration sets are combined, which enables more reliable estimates of pH to be made and can be used over a wider geographical area (see Stevenson *et al.*, 1991; Cameron *et al.*, 1999).

8.3 Diatoms in Marine Environments

8.3.1 Diatoms and Sea-Level Change

The coastal zone provides a high diversity of depositional environments, each with their own unique character. The diatom communities recorded from such environments may also be unique, and can therefore provide a reliable record of changes in coastal environments. As with any interpretation of the fossil record, attention must be paid to the dynamic character of the coastal zone under study, since misinterpretations can easily be made. One major problem of such studies is the discrimination of diatoms that lived at the place of deposition (**autochthonous**) from those transported to the depositional environment by wind or water (**allochthonous**). Information on the taphonomy of diatom communities can allow for the interpretation not only of the depositional environment in question, but also of the environment of the surrounding area (Vos and de Wolf, 1988). Notwithstanding such limitations, the high abundances of diatoms found in coastal environments, together with factors such as niche specificity and the sensitivity of such organisms to varying environmental factors, have meant that diatoms have played a key role in interpreting past sea-level changes (Palmer and Abbot, 1986; Shennan *et al.*, 1994, 1995a, b; Denys and Baeteman, 1995; Shennan *et al.*, 1999, 2000).

As mentioned previously, it is important to be able to separate autochthonous from allochthonous diatom communities. Vos and de Wolf (1988) demonstrated several methods of separating such communities for coastal areas of The Netherlands. Vos and de Wolf's methods employed criteria related to the diatoms (e.g. the composition of different ecological assemblages, the succession of different ecological groups within the sedimentary sequence, the occurrence of rare taxa and the percentage of broken diatom valves, etc.) and other non-diatom related criteria (e.g. lithology, sedimentary structures, palaeogeographical location). This allowed for the classification of ecological groups, and their distinct sedimentary environments to be established (e.g. subtidal, intertidal, supratidal areas, etc.). Certain species were included in more than one group in Vos and de Wolf's halobian classification if they inhabited more than one area (e.g. *Auliscus sculptus* can be either planktonic or epipsammic) or if they had a wide salinity range (e.g. *Cocconeis placentula*).

Analysis of the diatom flora and comparison with altering lithologies within coastal sequences has shown that diatoms can give a good description of marine **regressions** and **transgressions**. Further dating of the lithological boundaries enables the identification of the extent and nature of environmental change. Such changes can be local to that site (e.g. from storm-surge events and tsunamis) or a result of larger global environmental changes (e.g. sea-level rise). Hemphill-Haley (1995) showed how fossil diatoms from stratigraphical sections along the Niawiakum River in Washington indicated the occurrence of an earthquake-induced tsunami some 300 years ago, an interpretation which would have been difficult to make by studying the lithology alone.

Contemporary studies on sea-level changes using the fossil record rely upon the use of **sea-level curves**. Sea-level curves are constructed by charting the relative difference in marine transgressions/regressions over time, after taking into account any subsequent alterations such as isostatic recovery (Tooley, 1982; Lambeck *et al.*, 1998) or sediment compaction. Sediments accumulating in coastal basins isolated from the sea during marine regressions have been shown to provide excellent deposits to study relative sea-level change (Shennan *et al.*, 1993, 1994, 1995a, b, 1999, 2000). The occurrences of altering marine, estuarine, saltmarsh and freshwater diatom communities in sediments recovered from such isolation basins in western Scotland (Fig. 8.8) have aided the development of glacio-hydro-isostatic models for the region (see Shennan *et al.*, 2000). Isolation basins are not, however, the only envi-

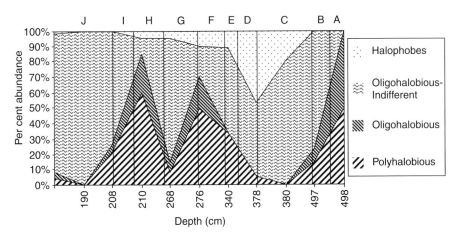

Figure 8.8 Down-core per cent abundance of diatoms with differing salinity tolerances from an isolation basin, Loch nan Eala, Scotland. Note the incursions of saltwater-tolerant species (polyhalobious) representing times of elevated sea-level during the Holocene. (After Shennan *et al.* (1994).)

ronment in which past sea-level changes can be detected; saltmarshes and offshore deposits can also yield valuable information from the preserved diatom flora (see also section 5.3.1).

8.3.2 Diatoms and Sea-Ice Cover

Diatoms are abundant, diverse and well preserved in high-latitude sediments compared with those beneath the mid-ocean gyres. As such, these sediments from the high latitudes can provide a useful proxy for unravelling records of past oceanographic and climatological change. As with the use of diatoms in palaeolimnology, the development of diatom calibration sets from the Arctic and Antarctic oceans has aided palaeoceanographic interpretation of these waters in recent years (Fenner *et al.*, 1976; Pichon *et al.*, 1987; Zielinski and Gersonde, 1997; Cunningham and Leventer, 1998; Zielinski *et al.*, 1998). The provision of these detailed surface-sediment diatom-assemblage distributions has enabled the palaeoceanographic community to document the changing environmental conditions of the Arctic and Antarctic throughout the Quaternary.

Cunningham *et al.* (1999) studied several cores from the Ross Sea, Antarctica, and analysed the diatom assemblages therein. The authors were able to determine that, during the Late Holocene, conditions in the Ross Sea became less stable and more wind mixed. They further noted a gradual decrease in the percentage abundance of diatoms normally associated with sea-ice (e.g. *Fragilariopsis curta*) and an increase in species normally associated with more open-water conditions (e.g. *Thalassiosira gracilis, Fragilariopsis obliquecostata*). These transitions noted in the diatom sedimentary record reflected a change in the interaction between the Ross Sea **polynya** (extensive areas of open water maintained throughout the winter, and which tend to occur more or less in the same place each year) and increasingly cooler seasonal spring temperatures. Such an alteration to the spring ice-melt would mean the breaking-up and carrying away of sea-ice before melting, subsequently releasing sufficient numbers of ice-associated diatoms to seed the planktonic population. In a similar study, Leventer *et al.* (1993) showed how both the **Little Ice Age** and the **Medieval Warm Period** appeared to be times of greater open-water conditions in the Granite Harbour area of Antarctica. Leventer *et al.* suggested that increased wind strength and warmer temperatures were responsible for a reduction in annual sea-ice during these periods. A key species in their interpretation was *Fragilariopsis cylindricus*, a diatom most often associated with subsea ice and ice-edge conditions (Garrison *et al.*, 1987; Kang and Fryxell, 1992).

Kaczmarska *et al.* (1993) provided a very useful diatom index that was applied to detect latitudinal oscillations of the winter sea-ice cover of Prydz Bay, Antarctica, during the Brunhes Palaeomagnetic Chroneity (0.783 Ma). Kaczmarska *et al.* used a ratio of two variant forms of the chain-forming diatom *Eucampia antarctica* (var. *antarctica* and var. *recta*). These two variant forms are morphologically very similar, but they are known to form colonies

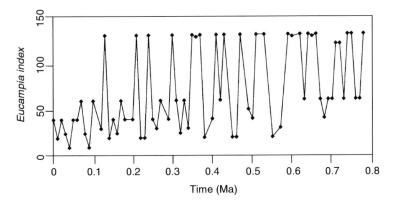

Figure 8.9 The ratio of two variant forms of *Eucampia antarctica* as used to form the *Eucampia* Index for the Kerguelen Plateau. Note the lower values found after 0.4 Ma, which were interpreted as an indication of increased winter sea-ice cover. (After Kaczmarska *et al.* (1993).)

of different lengths – *E. antarctica* var. *antarctica*, which is common in subpolar waters, forms long chains, while *E. antartica* var. *recta* forms shorter colonies and is regarded as a truly polar form. A straight ratio of the total number of terminal to intercalary valves of the two variants was made by Fryxell (1991), who reported a ratio of 1:95 for the subpolar variant *antarctica* and 1:2 for the truly polar variant *recta*. The stratigraphical assessment of this index from Prydz Bay by Kaczmarska *et al.* (1993) showed oscillations corresponding to frequencies of the Earth's obliquity cycles (Milankovitch, 1941; Crowley and North, 1991). The results (Fig. 8.9) showed a decrease in values of the *Eucampia* Index from 1:120, a ratio characteristic of *Eucampia antarctica* var. *antarctica*, to a *Eucampia* Index of 1:10, which is more characteristic of *E. antarctica* var. *recta*. This decrease, which started around 0.4 Ma, was interpreted as being indicative of increasing winter sea-ice cover in Prydz Bay during this time (Kaczmarska *et al.*, 1993).

8.3.3 Diatoms and Palaeotemperature Reconstruction of the Oceans

The development of sediment-surface-derived diatom calibration sets has enabled micropalaeontologists to link distinct floral and faunal communities with contemporary oceanographic parameters. Planktonic **foraminifera** have traditionally been used to map sea-surface temperature (SST), either by direct comparison between their biogeographical setting and modern-day SSTs (CLIMAP 1976, 1981, 1984), or via the use of **oxygen isotopes** (Shackleton and Opdyke, 1973, 1976, 1977). The use of diatoms in reconstructing the SST records of the oceans has received less attention than studies utilising foraminifera. However, diatoms have become invaluable tracers of the SST changes of the oceans through time, especially in regions where calcareous fossils are poorly preserved (e.g. Antarctica). The previous section outlined how diatoms can be used to trace variations in sea-ice cover, which of course is a function of temperature, but they have also been used to reconstruct the sea-surface temperature record of other latitudes (Kanaya and Koizumi, 1966; Barron and Baldauf, 1989; Koizumi, 1989; Koc Karpuz and Schrader, 1990; Schrader and Koc Karpuz, 1990; Barron, 1992).

One of the earliest studies to use diatoms to reconstruct SST records was that of Kanaya and Koizumi (1966). This study examined the diatom data from surface sediments in the North Pacific, and grouped taxa that occurred together on a regular basis. In so doing, these authors developed an 'index of affinity', which allowed them to identify species groups, the areal extent of these groups then being used to show meaningful patterns in terms of the present oceanographic setting. Kanaya and Koizumi applied these species groups in a simple algorithm (see below) to develop a diatom temperature curve for the North Pacific:

$$T_d = (X_w/X_c + X_w) \times 100$$

where X_c and X_w are frequencies of cold- and warm-water forms in a random count of 200 diatom specimens made for a sediment sample (from Kanaya and Koizumi, 1966).

More recent studies utilising diatoms to reconstruct SSTs have relied heavily upon the use of sediment-surface calibration sets. In one such study, Koc Karpuz and Schrader (1990) analysed 104 surface-sediment samples and two piston cores from the Greenland, Iceland and Norwegian Seas. The authors incorporated the data in a Q-mode factor analysis and defined six significantly different floral assemblages. Transfer functions were then applied to these species groupings relating to both winter (February) and summer (August) surface-water temperatures. A down-core assessment of palaeotemperature changes in waters overlying the Iceland Plateau during the Holocene was then calculated (see Fig. 8.10). The palaeotemperature trends revealed that, in the Iceland Sea, higher than present winter and summer SSTs existed from the base of the core until approximately 4000 years ago. After that time the diatom palaeotemperature record shows a steady decrease in SST before a slight increase in temperature is recorded in the uppermost (0–2.5 cm) sediments, a time corresponding to approximately 0–500 years ago. Koc Karpuz and Schrader (1990) concluded that these data indicate a strong influx of warmer Atlantic waters on to the Icelandic Plateau until around 4000 years ago, then, as the influx of warm waters diminished, they were replaced by cooler Arctic waters during the Late Holocene.

8.3.4 Diatoms and Palaeoceanography of Equatorial Upwelling Systems

Upwelling systems are by their very nature some of the most productive regions found in the marine realm. Equatorial winds blowing from east to

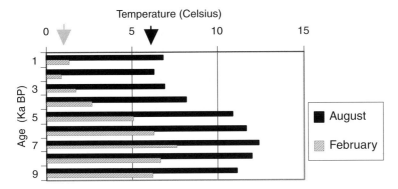

Figure 8.10 Diatom palaeotemperature record of the Iceland Sea during the Quaternary. Note the relative cooling that began around 4000 years BP, which was attributed to an influx of cooler arctic waters at this time. Present-day winter and summer temperatures are indicated by the grey and black arrows respectively. (Drawn using data from Koc Karpuz and Schrader (1990).)

west along the Earth's continental landmasses cause deep water to be upwelled along the eastern margins of the continents. These deep water-masses are preferentially enriched with nutrient salts, which they bring to the surface, stimulating primary production in these regions. The high levels of primary production within upwelling areas have been shown to be dominated by phytoplankton such as diatoms and **coccolithophorids** (De Mendiola, 1981). Such high levels of bioproduction lead to high rates of pelagic sedimentation, creating the potential for high-resolution records of temporal and spatial environmental changes of the equatorial ocean. Diatoms have been successfully employed to trace changes in productivity (Abrantes, 1988; Schrader and Sorkness, 1990, 1991; Schrader 1992a, b, c; Schrader *et al.*, 1993) and sea-surface temperatures of upwelling regions (Pokras, 1987; Kennington *et al.*, 1999).

In a study of the Peruvian upwelling system, Schrader and Sorkness (1991) employed a transfer-function technique to reconstruct productivity changes for the past 400,000 years. Schrader and Sorkness (1991) suggested that the decreases and increases in palaeoproductivity during this time interval were not in phase with the general glacial–interglacial cycles. They also noted that the largest swings in the magnitude of palaeoproductivity seemed to occur across oxygen-isotope stage boundaries and that the highest productivity was characterised by good diatom preservation in laminated sediments with high abundances of *Skeletonema costatum* and *Delphineis karstenii*. Lowest palaeoproductivity was characterised by bioturbated sediments with poor diatom preservation and high abundances of *Azpeitia nodulifer* and *Fragilariopsis doliolus*.

Kennington *et al.* (1999) used the diatom calibration set of Schrader *et al.* (1993) to interpret the palaeotemperature record of the eastern Equatorial

Pacific during the Olduvai magnetic subchroneity (1.75–2.0 Ma). These authors were able to show that the early and late Olduvai was characterised by decreased sea-surface temperatures and increased primary productivity, while the mid-Olduvai (1.80–1.90 Ma) appears to have been a period of weaker upwelling, higher SSTs and lower bioproductivity. Kennington *et al.* (1999) also demonstrated the use of diatoms from different oceanic realms in interpreting palaeoceanographic processes. The authors were able to identify and separate neritic and pelagic species of diatoms from a suite of cores across the eastern equatorial Pacific. These species groupings were substituted into a neritic/pelagic diatom ratio (NPDR):

$$NPDR = (N/N + P) \times 100$$

where N = per cent abundance of neritic species, and P = per cent abundance of pelagic species.

The results showed that high NPDR ratios were well correlated with increased palaeoproductivity and decreased SST estimates for the overlying waters (Fig. 8.11). These results suggested that episodes of intensified upwelling during the Olduvai subchron entrained more neritic diatoms into the westward-flowing equatorial currents and deposited them further offshore than in less-intensified upwelling periods. The authors went further, and suggested that the offshore transport of neritic diatoms was a function

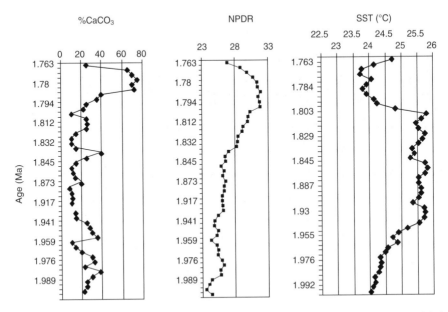

Figure 8.11   Estimated sea-surface temperature (degrees Celsius), productivity (%CaCO$_3$) and trade-wind strengths in the equatorial Pacific (c.1.75–2.0 Ma). For an explanation of the NPDR, see the text. (From Kennington *et al.* (1999).)

not only of surface current regimes but also of variations in the strengths of the trade winds responsible for driving the Peruvian upwelling system.

Laminated sediments underlying the eastern equatorial Pacific Ocean have been linked with increased palaeoproductivity in the region. Kemp and Baldauf (1993) reported on vast diatom mats deposited between 15 and 4.4 Ma. These mats were almost entirely composed of *Thalassiothrix longissima*, a diatom that has been widely used as an indicator of upwelling and increased bioproductivity (Fig. 8.12). The exact processes governing the development of these large mats, which may extend hundreds of miles, is not at present known and they have been observed only a few times (Villereal, 1988; Yoder *et al.*, 1994). Pike and Kemp (1999) suggest that the occurrence of these monospecific assemblages within the sedimentary fabric are a result of blooms settling out of the water column during the autumnal breakdown of water-column stratification.

Analysis of such laminated or **varved** sediments can provide important information on interannual and seasonal variations in production and/or climate of the overlying waters (Sancetta, 1995). Truly varved sediments are characterised by alternating light and dark laminae, together representing one year's deposition. The light lamina is usually dominated by diatomaceous microfossils deposited during the season of maximum production, and can therefore be an indication of increased nutrient availability. The dark lamina, in contrast, results from sediments deposited during periods of reduced bioproduction, or as a result of high rainfall over land or an increase in the deposition of lithogenic material.

Figure 8.12 Laminated or varved sediments (A) retrieved from equatorial waters often contain monospecific mats of diatoms such as *Thalassiothrix* species (B and C). Such assemblages can provide important information on interannual and seasonal variations in productivity and climate.

8.4 Summary

The study of diatoms and more particularly of diatom taxonomy came of age during the Victorian era. Initially these studies were restricted to freshwater and coastal habitats and were mainly descriptive in nature. The inception of the Deep Sea Drilling Project and its successor the Ocean Drilling Program have extended these studies into the deeper ocean realm and have enabled the scientific community to make a meaningful assessment of the Earth's hydrological and climatological perturbations. Such studies are important not only on the local scale, as in the management practices of a particular lake, but also on the global scale. The oceans, which after all cover 70 per cent of the Earth's surface, are extremely important for the assimilation of carbon dioxide, a gas that has the potential to alter radically the climate of the Earth.

Modern micropalaeontological methods incorporating the use of diatoms have advanced rapidly over the last 30 years or so. The development of the electron microscope has enabled the finer details of the diatom frustule to be revealed, and, in so doing, has increased our understanding of the diversity of this group of organisms. More powerful multivariate statistical methods have also been developed during the last 20 years; this – combined with advanced technologies in the field of computing and the development of larger and more detailed diatom calibration sets – has enabled diatomists to assess the fossil record quantitatively and make more accurate assessments of past environmental change.

Further Reading

Battarbee, R.W., 1986. Diatom analysis. In B.E. Berglund (ed.) *Handbook of Holocene palaeoecology and palaeohydrology*. Wiley, New York and London, pp. 527–570.

Round, F.E., Crawford, R.M. and Mann, D.G., 1990. *The diatoms: biology and morphology of the genera*. Cambridge University Press, Cambridge, UK, 747pp.

Stoermer, E.F. and Smol, J.P. (eds.) 1999. *The diatoms: applications for the environmental and earth sciences*. Cambridge University Press, Cambridge, UK, 469 pp.

Werner, D. (ed.), 1977. *The biology of diatoms*. Oxford: Blackwell Scientific Publications, 498 pp.

ENVIRONMENTAL APPLICATIONS OF CALCAREOUS NANNOFOSSILS

Richard W. Jordan

Most marine sediments are composed of siliceous or calcareous biogenic particles in varying concentrations. Calcareous oozes or carbonate-rich sediments cover roughly 50 per cent of the total sea-floor worldwide (Berger, 1976), with calcareous nannofossils being the main component (Bramlette, 1958). Their small size (c.2–25 μm in diameter), abundance, provincialism, morphological diversity and rapid evolution make them invaluable tools for biostratigraphy and palaeoenvironmental reconstruction. Bramlette and Riedel (1954) were the first to recognise this potential, and the results from subsequent Deep Sea Drilling Project (DSDP) cruises allowed the pioneering coccolith workers to construct biostratigraphical schemes for much of the Late Mesozoic and Cenozoic. The main focus of these early investigations was to utilise coccoliths for coarse age determinations, with low-resolution sampling over long time-scales. In this way, the approach was geared mostly towards the oil industry. However, with the start of the Ocean Drilling Program (ODP), more effort was put into high-resolution sampling with the aims of enhancing the biostratigraphical schemes and looking for smaller time-scale changes in the coccolith record to complement those of other microfossil groups, biomarkers and isotopes. Partly in response to the public's concern about the global environment, more funds became available for projects related to Quaternary palaeoceanography (i.e. interglacial/glacial cycles and human impact over the last million years). In the first part of this chapter, background information is provided on the taxonomy and morphology of modern coccolithophorids and the factors affecting phytoplankton communities. In the latter section, two applications of calcareous nannofossils to palaeoenvironmental analyses are provided as case studies.

9.1 TAXONOMY AND CLASSIFICATION

Calcareous nannofossils include two components, coccoliths and nannoliths, which may be unrelated but are formed of the same mineral, calcite. Fossil coccoliths show shared ultrastructural characteristics which allow us to infer that they are derived from bona fide coccolithophorids, while many nanno-liths are perhaps more akin to the internal skeletons or calcified tests of dinoflagellates. Here, the two groups will be treated together, because they are often found in the same sample, and the preparation methods used by most fossil dinoflagellate workers involve hydrochloric acid, which destroys all calcium carbonate microfossils. In particular, one group of calcified dinoflagellates (the thoracosphaerids) will be treated in detail.

For convenience and in order to keep within the confines of this book, the following classification will be limited to Quaternary and modern genera. Table 9.1 represents a synthesis of current taxonomy (for more detailed schemes see Jordan and Green, 1994; Young and Bown, 1997). For obvious reasons, the

Table 9.1 Classification of Quaternary and modern coccolithophorid genera

Division Haptophyta Hibberd ex Edvardsen and Eikrem

Class Prymnesiophyceae Hibberd emend. Cavalier-Smith

Order Coccolithales E. Schwarz emend, Edvardsen and Eikrem

Family Braarudosphaeraceae Deflandre
Braarudosphaera Deflandre

Family Calciosoleniaceae Kamptner
Anoplosolenia Deflandre
Calciosolenia Gran
Alveosphaera Jordan and Young

Family Calyptrosphaeraceae (holococcoliths)
Anthosphaera Kamptner emend. Kleijne
Calicasphaera Kleijne
Calyptrolithina Heimdal
Calyptrolithophora Heimdal
Calyptrosphaera Lohmann
Corisphaera Kamptner
Daktylethra Gartner
Flosculosphaera Jordan and Kleijne
Gliscolithus R.E. Norris
Helladosphaera Kamptner
Homozygosphaera Deflandre
Periphyllophora Kamptner
Poricalyptra Kleijne
Poritectolithus Kleijne
Sphaerocalyptra Deflandre
Syracolithus (Kamptner) Deflandre
Zygosphaera Kamptner emend. Heimdal

Family Ceratolithaceae R.E. Norris
Ceratolithus Kamptner

Family Coccolithaceae Poche
Calcidiscus Kamptner
Coccolithus E.H.L. Schwarz
Cruciplacolithus Hay and Mohler
Cyclolithus Kamptner ex Deflandre
Hayaster Bukry
Oolithotus Reinhardt
Umbilicosphaera Lohmann

Family Helicosphaeraceae Black emend. Jafar and Martini
Helicosphaera Kamptner

Table 9.1 Continued

Family Hymenomondaceae Senn
  Hymenomonas Stein emend. Gayral and Fresnel
  Ochrosphaera Schussnig

Family Papposphaeraceae Jordan and Young
  Pappomonas Manton and Oates
  Papposphaera Tangen
  Trigonaspis Thomsen

Family Pleurochrysidaceae Fresnel and Billard
  Pleurochrysis Pringsheim

Family Pontosphaeracae Lemmermann
  Pontosphaera Lohmann
  Scyphosphaera Lohmann

Family Rhabdosphaeraccae Haeckel
  Acanthoica Lohmann ex Lohmann
  Algirosphaera Schlauder emend. R.E. Norris
  Anacanthoica Deflandre
  Cyrtosphaera Kleijne
  Discosphaera Haeckel
  Palusphaera Lecal emend. R.E. Norris
  Rhabdosphaera Haeckel

Family Syracosphaeraceae
  Calciopappus Gaarder and Ramsfjell emend. Manton and Oates
  Coronosphaera Gaarder
  Michaelsarsia Gran emend. Manton *et al.*
  Ophiaster Gran emend. Manton and Oates
  Syracosphaera Lohmann

Genera incertae sedis
  Alisphaera Heimdal emend. Jordan and Chamberlain
  Balaniger Thomsen and Oates
  Calciarcus Manton *et al.*
  Canistrolithus Jordan and Chamberlain
  Ericiolus Thomsen
  Florisphaera Okada and Honjo
  Gladiolithus Jordan and Chamberlain
  Jomonlithus Inouye and Chihara
  Polycrater Manton and Oates
  Quaternariella Thomsen
  Turrilithus Jordan *et al.*
  Tetralithoides Theodoridis emend. Jordan *et al.*
  Umbellosphaera Paasche
  Vexillarius Jordan and Chamberlain
  Wigwamma Manton *et al.*

Order Isochrysidales Pascher emend. Edvardsen and Eikrem
  Family Noelaerhabdaceae Jerkovic
  Emiliania Hay and Mohler
  Gephyrocapsa Kamptner
  Pseudoemiliania Gartner
  Reticulofenestra Hay *et al.* emend. Gallagher

Division Dinoflagellata (Bütschli) Fensome *et al.* (= Pyrrhophyta Pascher)
  Class Dinophyceae Pascher
  Order Thoracosphaerales Tangen

Family Thoracosphaeraceae Schiller emend. Tangen
  Thoracosphaera Kamptner

scheme is restricted to those taxa that produce coccoliths and thus have a potential fossil record. The Order Thoracosphaerales contains only one family and one genus, and is distinguished from other calcareous dinoflagellates in that there are no thecal plates, the calcareous stage is vegetative (not a cyst) and the cells are coccoid (Fensome *et al.*, 1993).

9.2 Morphology

9.2.1 Coccolith Morphology

The morphology of extant coccolithophorids and of calcareous nannofossils has been reviewed extensively by Jordan *et al.* (1995) and Young *et al.* (1997), respectively. To simplify matters, only terms used frequently (i.e. those associated with common taxa) will be mentioned here.

Coccoliths can be separated into two main types, with very different structures: holococcoliths and heterococcoliths (Fig. 9.1). Holococcoliths are made up of tiny rhombohedral or hexagonal microcrystals, identical in size and shape, and are thought to be produced extracellularly between the cell membrane and an organic layer called the 'skin'. Despite their apparent simplicity, holococcoliths exhibit a diverse range in shapes (see Kleijne, 1991). However, the holococcoliths produced by modern species are small, delicate structures and are easily dissolved in the water column and at the sediment/water interface, and therefore are normally absent from Neogene and Quaternary fossil assemblages. Holococcoliths have been found in sediments dating back to the early Jurassic, and in the late Cretaceous and Palaeogene some larger, more robust holococcoliths are common (see e.g. Perch-Nielsen, 1985a, b; Burnett, 1998).

Heterococcoliths are produced inside the cell within vesicles of the Golgi body. Within each vesicle, a single heterococcolith begins as a circular or oval ring of crystals upon an organic base-plate. The crystals of this proto-coccolith ring then 'grow' and extend in as many as three directions to form a variety of shapes and sizes. The crystals may grow outwards horizontally to form structures such as shields and flanges, or into the central area to form grills, bridges and central structures, or vertically to form walls, central tubes and collars. However, in some members of the Coccolithaceae it appears that the proto-coccolith ring also grows downwards and that the ring becomes embedded in the developing coccolith (Young, 1992, 1993; Young *et al.*, 1999). To add to the complexity, it is now known that the original crystals in the ring are of two alternating types, called V- and R-crystal units, essentially oriented vertically and radially, respectively (Young *et al.*, 1992, 1999; Didymus *et al.*, 1994). In each species these units are responsible for producing different parts of the coccolith, although their contributions are not always equal. In the case of *Emiliania huxleyi*, the V-crystal unit is so small that it is almost invisible (Young *et al.*, 1992). Heterococcolith structures seem to be more complex than those of holococcoliths, and this is

Figure 9.1 Holococcoliths and heterococcoliths. Electron micrographs of a typical holococcolithophorid (*Periphyllophora mirabilis*, from the western Mediterranean, left two specimens) and a typical heterococcolithophorid (*Calcidiscus leptoporus*, culture specimen, right two specimens). Scale bars – one micron. Note that the upper images are enlargements of the specimens in the lower images.

perhaps largely due to the degree of cellular control on the heterococcolith calcification process. Although heterococcoliths are generally more robust than holococcoliths, not all heterococcoliths are regularly recorded in fossil assemblages. Delicate structures like spines, tubular processes and thin walls are often broken during the sedimentation process, and so the most commonly preserved heterococcoliths are placoliths. Placoliths are double-shielded coccoliths, with each shield composed of wedge-shaped elements. The two shields allow the placoliths to form tightly interlocking coccospheres, and it is for this reason that most intact coccospheres in the Tertiary fossil record belong to species in the Coccolithaceae and Noëlaerhabdaceae.

Parke and Adams (1960) showed from culture studies that one holococcolithophorid, *Crystallolithus hyalinus*, was in fact an alternate life-cycle stage of a heterococcolithophorid, *Coccolithus pelagicus*. Subsequently many further examples of holo–heterococcolith pairings have been documented through observations of combination coccospheres (e.g. Kleijne, 1991; Thomsen *et al.*, 1991; Cros *et al.*, 2000). These combination coccospheres contain both holococcoliths and heterococcoliths and are thought to represent

Figure 9.2 *Thoracosphaera heimii* – the most common extant calcisphere. *Thoracosphaera*, and most other calcispheres, are dinoflagellates, not coccolithophores, but they are often studied together.

cells undergoing life-cycle transitions. The available evidence suggests that most and perhaps all holococcolithophorids are life-cycle stages of hetero-coccolithophorids. As a result the traditional taxonomy which places such forms in discrete genera and the Family Calyptrosphaeraceae is intrinsically artificial. However, for palaeontological purposes this is still probably the only practical approach.

9.2.2 Morphology of Calcispheres

Most calcispheres encountered in living or Quaternary fossil assemblages belong to the genus *Thoracosphaera* (Karwath *et al.*, 2000), and so may be termed thoracospheres (Fig. 9.2). Thoracospheres represent a calcified vegetative stage (i.e. they are not cysts) in the life-cycle of calcareous dino-flagellates (Tangen *et al.*, 1982; Inouye and Pienaar, 1983). In living speci-mens, the wall is composed of polygonal calcite crystal elements arranged between two organic membranes. The shell may possess an aperture (open-ing), which if present is normally covered by an operculum (lid), although it has been shown that this structure normally forms in mature cells prior to the release of swarmers or non-motile daughter cells. Prior to cell division, the cell contents emerge from the shell (which retains the two membranes) through the aperture, quickly divide into two non-motile daughter cells (called aplanospores) and presumably recalcify. Alternatively, the cell con-tents may emerge from the shell as a biflagellate, which divides into two new cells that later develop into *Gymnodinium*-like naked motile stages (called planospores).

9.3 METHODOLOGY: SAMPLE COLLECTION AND PROCESSING

Suitable equipment, quantitative methodologies and advanced preparation techniques are discussed in detail in Winter *et al.* (1994) and Bown and Young (1998), while Young and Ziveri (1999) describe procedures for esti-mating mass fluxes from coccolith counts. However, for routine study, nanno-fossils are usually examined using light microscopy on smear slides; these can be prepared rapidly and simply.

For soft sediments, use a toothpick (flat-sided toothpicks are best) to take a small amount of sample. Place the sample on a cover-slip and make a sus-pension with a drop of buffered (pH 10–11) distilled water. Thoroughly mix the suspension, and then, with the toothpick, make a thin, even smear across the cover-slip. Make sure that unwanted large particles are at the side of the cover-slip, and then place the cover-slip on a hot-plate until dry. Scrape off the large particles, and affix the cover-slip (smear-slide down) on to a labelled clean glass slide, using an optical mounting medium (UV curing media are widely used).

For hard sediments, a small amount of sample needs to be crushed into a powder using a pestle and mortar, or the sample can be scraped from a cleaned rock surface. The above procedure can then be followed.

9.4 ENVIRONMENTAL APPLICATIONS

9.4.1 Factors Affecting Plankton Communities

Coccolithophorids and thoracospheres are phototrophic organisms, i.e. they both rely on photosynthesis as their main nutritional mode, and so they often live together within the photic zone, and thus may be affected by the same environmental conditions. There are a number of environmental parameters that are known to affect plankton communities in general, on both spatial and temporal scales. Some of these factors may be localised effects and may only occur on a microscale, while others may be much more widespread, affecting phytoplankton on the mesoscale. The order in which they are listed is not significant; however, it should be remembered that many of these factors are interrelated and thus it is often difficult to separate out which factors are causing the changes in the community structure. To keep this section on environmental factors to a minimum, each factor has been discussed in terms of the relevant literature on coccolithophorids and thoracospheres where possible. On a further note, some marine phytoplankton groups (including diatoms and chrysophytes), as well as many dinoflagellates, produce resting stages, spores or cysts when conditions become unfavourable for continued growth. However, at present, there is no evidence to suggest that either coccolithophorids or thoracospheres produce such structures (although some authors have suggested or illustrated such things in the past), so it seems likely that these two groups of organisms have evolved different strategies when escaping or resisting unfavourable environmental conditions.

9.4.2 Factors Affecting Living Coccolithophorids and Thoracospheres

9.4.2.1 Water Temperature

Water temperature has often been thought of as one of the most important factors affecting phytoplankton populations. This may be due in part to the fact that during many of the earlier investigations only depth, temperature and salinity measurements were taken. Detailed on-board chemical analyses and physical oceanography were not routinely carried out until much later into the twentieth century. Using their extensive data-set of coccolithophorid biogeography, McIntyre and Bé (1967) showed that most common species had relatively narrow temperature ranges, although these ranges were extended by subsequent workers (Braarud, 1979; Okada and McIntyre, 1979). As these authors showed, temperature plays a major role in controlling the largest scale distribution of species, largely defining broad latitudinally related biogeographical zones. However, on smaller scales the species temperature ranges are less informative, as no oceanic waterbody has annual temperature fluctuations covering the entire range (1–30 °C). What they are actually showing is the total range of water temperatures that each species (the sum of all its genotypes and ecotypes) can grow in, rather than the temperature fluctuations experienced by a discrete population. For instance, subtropical water temperatures do not vary greatly (perhaps by less than 5 °C) through the year, and even in glacial times the estimated temperatures were only about 2 °C lower. Thus, for subtropical populations, away from the coast or major frontal boundaries, temperature change is perhaps unimportant, even for lower photic zone (LPZ) communities, although the depth of the LPZ may shift up and down marginally with the seasons. Of course, for temperate and subpolar communities, temperature plays a more prominent role, especially in the formation and breakdown of the seasonal thermocline. Coastal species probably have to withstand the greatest fluctuations as the shallow waters are prone to greater temperature fluctuations than offshore waters.

Temperature experiments on cultured algae have shown that the growth temperature can make a difference. Mjaaland (1956) and Fisher and Honjo (1991) showed that temperature affects the growth rates of *Emiliania huxleyi* clones, while Karwath *et al.* (2000) showed a similar effect in *Thoracosphaera heimii* clones. Wilbur and Watabe (1967) reported that the length and width of *E. huxleyi* coccoliths decreased with increasing temperature above 18 °C, and that in general the length of the upper elements decreased, while the width increased above 18 °C.

9.4.2.2 *Macronutrients*

All phytoplankton require certain nutrients for their growth and biochemical reactions, and these are usually separated into those that are needed in large quantities (termed macronutrients) and those needed in smaller quantities (termed micronutrients; see below). Macronutrients are usually considered to be nitrate, phosphate and silicate, although other nitrogenous compounds like nitrite and ammonia may be important. Nitrate and phosphate are required by all phytoplankton for synthesis of proteins, while silica is required by diatoms and chrysophytes for production of their siliceous skeletons. These three macronutrients are only found at micromolar concentrations in sea-water, and in surface waters they are frequently depleted to vanishingly low concentrations. The low levels in the surface waters are due to the uptake by phytoplankton and bacteria, followed by removal of phytoplankton biomass to deeper water through their digestion by zooplankton and settling out in faecal pellets. In these situations, nutrient recycling, whereby the nutrients from phytoplankton cells are returned to the water by bacterial degradation before they can be removed from the photic zone, and nitrogen-fixation, in which phytoplankton possess nitrogen-fixing cyanobacterial symbionts to provide nitrogen for the host, become important. Nonetheless, macronutrients may become limiting, producing oligotrophic conditions. In these conditions, biomass levels are usually low, and communities are dominated by specialised (K-selected) species adapted to such conditions. While phytoplankton abundances are low in these conditions, diversity may be very high. Other phytoplankton species not adapted to such conditions may be stressed, die out or be forced to sink to deeper levels to maintain a slower growth rate, or to 'hibernate' as spores or cysts. Many coccolithophorids are K-selected, and coccolithophorids as a group achieve their highest relative abundances within phytoplankton communities in oligotrophic conditions. Typical oligotrophic coccolithophorids include: *Umbellosphaera*, *Discosphaera*, holo-coccolithophorids and many *Syracosphaera* species; in addition, the remarkably adaptable species *Emiliania huxleyi* is often abundant in oligotrophic conditions.

Nutrient availability can be increased in a number of ways, by:

1. winter mixing, whereby more nutrient-rich waters below the nutricline are mixed with the impoverished surface waters when the seasonal thermocline is destroyed;
2. storm mixing;
3. upwelling, whereby deeper waters flow upwards to replace surface waters;
4. by input of nutrient-rich river waters.

Although these are different oceanographic conditions, they have the common effect of introducing nutrient-rich waters to the surface and thus creating eutrophic conditions in which phytoplankton can flourish. Such eutrophic waters typically show high phytoplankton abundances but relatively low diversities, with communities being dominated by a limited suite of opportunist, R-selected species. In eutrophic environments coccolithophorids are typically outcompeted by diatoms and form a relatively minor component of the total community. Nonetheless, certain coccolithophorid species (notably many placolith-bearing species) are adapted to such conditions, and, since productivity is much greater in eutrophic conditions, a very large proportion of total coccolith production occurs under such conditions.

Many scientists consider that silicate availability controls diatom growth. When the supply runs out at the end of the spring bloom, the diatoms sink out of the surface waters, often aggregating together, bound by mucilage excreted by the cells due to nutrient stress (Smetacek, 1985). However, other phytoplankton, especially flagellates that do not require silicate for their cell walls, may be controlled by either phosphorus or nitrate limitation (Tyrrell, 1999). The differential effect of such controls means that as nutrient levels decline from eutrophic through mesotrophic toward oligotrophic, conditions become unpredictable and very different phytoplankton assemblages may develop under superficially similar conditions.

The effect of trophic levels on coccolithophorid ecology is discussed by Brand (1994) and Young (1994), and detailed case studies are given by, for example, Hagino *et al.* (2000), Takahashi and Okada (2000) and Kinkel *et al.* (2000).

Nutrient stress caused by low or high concentrations of certain nutrients may affect phytoplankton in different ways. For instance, in cocco-lithophorids, nitrate is essential for growth and calcification; however, at high nitrate concentrations calcification is inhibited, but only at unnatural concentrations like 1000 μM NO_3^- (Wilbur and Watabe, 1963; Nimer and Merrett, 1993), and nitrate-limited algae are more likely to be photo-inhibited (Prezelin *et al.*, 1986; Rhiel *et al.*, 1986). The growth rate of *E. huxleyi* appears to be independent of the N/P ratio in the medium, with little or no response from the cellular N/P ratio, N/C ratio and P/C ratio (Sakshaug *et al.*, 1982), suggesting that *E. huxleyi* is not affected by nutrient stress (Merrett *et al.*, 1993). However, Egge and Heimdal (1994) and van Bleijswijk *et al.* (1994) showed that *E. huxleyi* cell numbers increased in enclosures with a high N/P ratio as compared with those with a low N/P ratio. Likewise, in a chemostat experiment, Paasche and Brubak (1994) showed that phosphorus limitation enhanced calcification in *E. huxleyi*, and caused cells to produce >100 coccoliths/cell. In contrast, Båtvik *et al.* (1997) found that cell numbers increased in the enclosure with low N/P ratio, and that the average distal length of the coccoliths became smaller, while in the high N/P ratio enclo-sure the coccoliths became larger. Nutrient depletion has been invoked for the malformation of coccoliths in natural populations (Okada and Honjo, 1975; Kleijne, 1990, 1991), and although malformation occurred in 10 per cent of the coccoliths in the enclosure experiment, there was no direct evidence to support the above hypothesis. The observations of Okada and Honjo (1975) and Kleijne (1990, 1991) were from marginal seas where other factors could have been involved, such as lower salinity during the downwelling season (NW monsoon) as speculated in both papers.

Some haptophytes, including coccolithophorids, combine phagotrophy with photosynthesis (called mixotrophy), in which they use their haptonema to capture bacteria and other small particles (Green, 1991; Kawachi *et al.*, 1991; Jones *et al.*, 1994; Kawachi and Inouye, 1995). It has been shown that increased bacterivory is linked to phosphate limitation (Nygaard and Tobiesen, 1993), and thus could be important for those organisms living in oligotrophic environments.

9.4.2.3 *Micronutrients*

Brand (1991) demonstrated that open-ocean phytoplankton were less dependent on iron than coastal species and cyanobacteria, and suggested that the former were more likely to be controlled by phosphate limitation. More recently, it has been shown that heterotrophic bacteria play an import-ant role in iron uptake, constituting as much as 20–45 per cent of the total biological uptake (Tortell *et al.*, 1996). An iron fertilisation experiment in the equatorial Pacific showed that pennate diatoms benefited mostly from the introduction of dissolved iron, despite their low abundance in the flagellate-dominated pre-fertilisation community (Coale *et al.*, 1996).

9.4.2.4   Salinity

Most coccolithophorids and thoracospheres live in the open ocean, in temperate, subtropical and tropical surface waters, and so are adapted to living in salinity concentrations of 32–37 ‰. Furthermore, it is true to say that coccolithophorid species diversity reaches a maximum in stratified oligotrophic environments, where salinity values are generally high due to evaporation. In general, low salinity values are found in the polar regions – due to melting ice, in the equatorial zone – due to rainfall, and in coastal areas – due to the discharge of river water. In all three of these areas, coccolithophorid diversity is relatively low, and in the case of coastal and polar regions the communities are specialised (i.e. benthic or partially calcified, respectively). At present, only one coccolithophorid species, *Hymenomonas roseola*, is known to inhabit freshwater environments. It is also important to note that in high-salinity areas, due to extreme evaporation, coccolithophorid diversity is also restricted. In the laboratory, Fisher and Honjo (1988/1989) found that clonal cultures of *E. huxleyi* could not grow at 45 ‰, in contrast to the earlier findings of Mjaaland (1956), while Braarud (1951), working with a culture of *Pleurochrysis carterae*, also managed to maintain growth at this salinity value. In recent years, a number of experiments with cultured coccolithophorids have shown that salinity changes affect the cells in different ways. In the case of *E. huxleyi*, Paasche *et al.* (1996) showed that coastal clones could grow at 12 ‰, while offshore clones required higher salinities. Green *et al.* (1998) showed that decreasing the salinity of the medium from 34 ‰ to 24 ‰ caused an increase in the calcification of certain parts of the coccolith, while at low salinities (14–16 ‰), distortion or even gross malformation of the coccoliths took place. In general, the coccolith dimensions and element number decreased with decreasing salinity. However, it was noted that some clones had different tolerances, suggesting the existence of genetic variation, which are perhaps related to the environments from which they were originally collected.

9.4.2.5   Light Penetration and Quality

As coccolithophorids and thoracospheres are photosynthetic organisms, they require light to fix carbon. However, sunlight is affected by various factors in the atmosphere before it reaches the sea surface, such as absorption, reflection and scattering by clouds and dust particles. Once it reaches the ocean, it is absorbed again, this time by the sea-water, algal pigments and particulate and dissolved organic matter. Light comprises various wavelengths, which can penetrate the sea-water to different depths. In general, red light is absorbed first, and blue light penetrates furthest. The amount of light reaching the sea surface is termed the surface irradiance and given a value of 100 per cent. The depth at which about 1 per cent of this surface irradiance remains, the 1 per cent light level, may be important when identifying the so-called 'shade flora' (Sournia, 1982). Most phytoplankton live in the upper photic waters, but may become photo-inhibited close to the surface. *Emiliania huxleyi* is unusual in that it appears to be uninhibited by high light levels, which may account for its success at outcompeting other species when it forms blooms (Nanninga and Tyrrell, 1996). Experiments have shown that some, if not all, phytoplankton cannot grow in continuous light, and thus appear to need some time in the dark (Brand and Guillard, 1981). Others require light of a certain intensity and wavelength for optimum growth, and this is reflected in the diversity of photosynthetic pigments used by marine phytoplankton. With respect to the intensity and duration of daylight, both seasonal and latitudinal differences affect the composition of the phytoplankton assemblage.

It is well established that detached coccoliths, if in large enough numbers, cause significant back-scattering of light due to the high refractive index of calcite, and so are visible by orbiting satellites. Tyrrell *et al.* (1999) have even suggested that these coccoliths may raise the temperature of the surface waters and cause greater stratification.

9.4.2.6 Turbulence

The ocean surface is constantly moving, due to wind-driven circulation and frictional stresses caused by the shearing of different water masses. The coastal areas are also affected by wave and tidal action and by upwelling. It has been shown that phytoplankton tolerate turbulence to varying degrees (Margalef, 1978; Estrada and Berdalet, 1997). In general, diatoms grow in turbulent waters, while flagellates require calmer waters. This simplistic approach holds true for most situations, with diatoms dominating coastal waters, upwelling areas, frontal boundaries and periods of changeable weather (like in spring and autumn), and flagellates dominating calmer summer stratified waters and permanently stratified gyre waters. Microscale turbulence may be important in some geographical areas, but in the subtropical gyres the existence of discrete vertical communities is perhaps indicative of a lack of physical disturbance or that the phytoplankton have evolved mechanisms to combat sinking, or rising into an undesirable part of the photic layer, or being removed from the sunlit zone.

9.4.2.7 Water Depth

Water depth alone is probably not a factor affecting phytoplankton populations; however, many parameters change with water depth, including light availability, temperature, salinity and nutrient supply. As a result, in most environments, a strong depth stratification of phytoplankton communities is developed. In addition, water depth may be important for those species that require a benthic stage or resting stage (e.g. spores, cysts and 'resting' cells). Most resting stages and benthic stages still require light or suitable conditions, i.e. they are still respiring and in some cases dividing. Furthermore, these stages are only part of the life-cycle and thus they must return to the planktonic stage at some later date. Many coastal species produce resting stages following the onset of unfavourable conditions, and thus sink to the surface sediments to await the return of more favourable conditions. Therefore, it is essential that they do not settle in waters too deep to be resuspended into the surface waters. Experiments have shown that the spores of some *Chaetoceros* species can remain viable for up to 1–2 years. Those species living in open ocean waters, which need to sink out of the surface waters, probably go no further than the pycnocline or 1 per cent light level where there are abundant nutrients, and at a depth from which they can easily return. These survival strategies may be the reason why some species can consistently recur in the same area each year. Such species are usually the bloom-forming plankton that dominate spring and autumn communities. Often called 'opportunists', they are really perfect planners, ensuring that they are numerically superior even while they are waiting, and/or relying on their faster growth rates to outcompete their rivals when the nutrient supply in the surface waters is replenished.

In summer stratified waters and permanently stratified gyre waters, the presence of discrete photic-layer communities suggests that water depth may play some role in the vertical distribution of phytoplankton. In general, the upper photic layer may occupy the top 100 m, the lower photic layer is usually found below 100 m, and sandwiched between these two layers is a thinner middle-photic layer. However, the boundaries of these photic layers are defined by various physicochemical factors, which vary with depth but are not constrained by it. So the depth at which each photic layer boundary exists may change daily, seasonally or in response to environmental conditions.

9.4.2.8 Bloom Termination

It is known that viruses affect marine phytoplankton and bacterial populations (Proctor and Fuhrman, 1990; Suttle *et al.*, 1990; Suttle and Chan, 1995) and may cause changes in their composition and diversity (Bratbak *et al.*, 1993). More specifically, it has been noted that blooms of *E. huxleyi* were sometimes succeeded by increases in the abundance of a morphologically homogeneous viral population (Bratbak *et al.*, 1993, 1996). Emiliani (1993) speculated that these types of virus may lead to background extinctions, by targeting microplankton species which are both numerous and widespread.

9.4.2.9 Toxins and Chemical Warfare

Certain non-coccolithophorid haptophytes produce toxins, which can have drastic effects on the marine community (Underdal *et al.*, 1989; Maestrini and Granéli, 1991; Moestrup, 1994). However, less is known about the effects of other chemicals released by haptophyte (and other phytoplankton) cells within the microenvironment (Kellam and Walker, 1989). One such extracellular product is acrylic acid, which is produced during the synthesis of dimethyl sulphide (DMS) following the cleaving of its precursor, dimethyl sulphoniopropionate (DMSP). Acrylic acid first came to the attention of marine biologists when Sieburth (1960) demonstrated that the compound, produced by *Phaeocystis* blooms, acted as a bactericide, effectively sterilising the guts of penguins that were feeding in the same waters. Huntley *et al.* (1983) reported that *Proteroceratium reticulatum*, a dinoflagellate that produces acrylic acid, could suppress copepod feeding by as much as 60 per cent. Although the presence of the dinoflagellate did not discourage the zooplankton from eating other species, it suggests that acrylic acid may be acting as a feeding deterrent in the microenvironment, thereby allowing one species to remain ungrazed. *E. huxleyi* is known to produce DMSP (Keller, 1988/1989; Levasseur *et al.*, 1996) and coccolithophorid-dominated blooms have been reported as major sources of DMS and DMSP (Malin *et al.*, 1993). Wolfe *et al.* (1997) have shown that *E. huxleyi* can deter protozoan herbivores (like *Oxyrrhis marina*) when their cells are lysed following initial grazing, and acrylic acid is formed when DMSP is cleaved by the released enzyme, DMSP lyase.

9.4.2.10 Grazing Pressure

Microzooplankton (e.g. tintinnids) and macrozooplankton (e.g. copepods) readily ingest coccolithophorids, a fact corroborated by the coccolith-laden faecal pellets and coccolith-covered tintinnid loricae that accumulate in sediment traps underlying productive zones. However, much debate exists as to whether these animals use particle selectivity or merely eat what is available and abundant. Both pellet contents and loricae usually consist of coccoliths from more than one species, and in some cases are mixed with the skeletons, tests and frustules of a variety of microplankton types. As inferred from above, zooplankton grazing must be an important control on phytoplankton numbers and composition, although some species may be utilising preventative measures through the release of extracellular products.

9.4.3 Calcareous Nannofossils and Past Environments

The case studies below briefly describe two current key topics in coccolith research. It has been difficult to interpret past environmental conditions using fossil assemblages, since the most abundant coccoliths are produced by species that are cosmopolitan and tolerant to a wide range of conditions.

Recent advances have come from intensive study of fine-scale variation in particular species and better appreciation of the ecology of particular species (e.g. Bollmann *et al.*, 1998; Takahashi and Okada, 2000). *Florisphaera profunda*, due to its restricted ecological preferences and dominance in underlying subtropical/tropical sediments, has since become a particularly useful tool. In parallel with such work based on floristic analysis, geochemical palaeoproxies based on coccolithophorid-derived material have become increasingly important, notably alkenone palaeothermometry.

9.4.3.1 *Florisphaera profunda*

Florisphaera profunda (Fig. 9.3) was first described by Okada and Honjo (1973) from the central Pacific, as a new species inhabiting and characterising the 'lower euphotic layer'. This layer was later referred to as the lower photic zone (LPZ). The LPZ is characterised by low light levels (<1 per cent to 4 per cent of the surface irradiance according to Okada and Honjo, 1973), by temperatures 5 °C (or more) lower than at the surface, higher nutrient concentrations (i.e. below the nutricline), and appears to exist just below the deep chlorophyll maximum (DCM) (Winter *et al.*, 1994; Jordan and Chamberlain, 1997). The LPZ is usually a permanent feature of subtropical gyres, but may develop in well-stratified waters in equatorial and temperate regions during the summer months. It has also been demonstrated that the vertical distribution of *F. profunda* is controlled by the depth of the nutricline (Molfino and

Figure 9.3 *Florisphaera profunda*. Left – a complete coccosphere of this unusual, deep-dwelling species. Right – a disintegrating coccosphere. Scale bars – one micron. The individual plates of *F. profunda* are enormously abundant in sediments, but with their simple morphology for a long time they were overlooked. The abundance of *F. profunda* relative to other coccoliths essentially provides an indication of lower vs. upper photic zone productivity and has provided high-quality records of environmental fluctuations on Milankovitch frequencies.

Table 9.2 Inferred correlations of *Florisphaera profunda* with environmental conditions

Proxy	F. profunda(%)	References
Water depth	Increases with increasing water depth	Okada (1983) Ujiié et al. (1991)
Distance from the shore	Increases with increasing distance	Okada (1983)
Periods of intensified winds (shallow nutricline)	Low	Molfino and McIntyre (1990a, b), McIntyre and Molfino (1996)
Sea-water turbidity	Decreases with increasing turbidity	Ahagon et al. (1993)
Upwelling	Low	Jordan et al. (1996), Okada and Matsuoka (1996), Zhao et al. (2000)
Weak upwelling	High	Broerse et al. (2000), Andruleit et al. (2000)
Increased primary productivity (during sapropel formation):		
(a) In the DCM	High	Castradori (1993)
(b) In the shallow pycnocline	Low	Castradori (1993)
Palaeoproductivity	Low	Beaufort (1996) Henriksson (2000)
Heinrich events	Low	Jordan et al. (1996)
	High	McIntyre and Molfino (1996)
Glacial periods	Low	Okada and Wells (1997)
	High	Okada and Matsuoka (1996)

A wide range of different conditions may favour, or inhibit, deep photic zone species, and, consequently, different specific effects have been invoked in different cases; the common thread, however, is that it is an indicator of low productivity in surface waters. In the case of Heinrich events and glacial periods, these correlate variously with high or low productivity in different oceanographic settings.

McIntyre, 1990a, b). Given its apparent preference for stratified waters, it is not surprising that it is rare or absent in turbulent waters, for example, in coastal or upwelling areas. Strong turbulence causes mixing and thus destroys the barriers protecting the specialised LPZ flora (often referred to as the 'shade flora'; Sournia, 1982). In addition to those parameters, the top of the LPZ may also be associated with a pycnocline, a region of maximum density change. Turbulence in shallow waters also causes an increase in turbidity, and thus *F. profunda* abundance negatively correlates with water depth, distance from the shore (Okada, 1983) and water transparency (Ahagon *et al.*, 1993). However, Okada (1983) noted that, unlike the nominate variety, *F. profunda* var. *elongata* failed to show a clear correlation with water depth, although the reasons for this are still unclear, as the two varieties usually coexist in the LPZ. It should be noted that Okada (1983) used the same data-set to correlate the *F. profunda* abundance with water depth and distance from the shore, and so the correlations resembled each other.

High and low abundances of *F. profunda* in the sediments have been proposed as proxies for a number of parameters (Table 9.2). As *F. profunda* is known to extend back to at least the middle Miocene (Young, 1998), these proxies could provide valuable information on low-latitude palaeoceanography, in particular, for much of the last 5 million years. Although *F. profunda* dominates the LPZ and also the fossil assemblages in the underlying sediments of subtropical/tropical regions, the coccoliths of other LPZ species have been found, especially in cores taken from shallower waters (<1000 m; Okada and Matsuoka, 1996; Takahashi and Okada, 2000). Thus, future studies of cores taken from outer shelf and slope areas should add to the usefulness of the proxies developed for *F. profunda*.

9.4.3.2 Alkenone Studies

The presence of long-chain (C_{37}–C_{39}) unsaturated ketones in an *E. huxleyi* culture was first reported by Volkman *et al.* (1980a, b), although an earlier study had demonstrated their occurrence in marine sediments (Boon *et al.*, 1978). Further analysis showed that three non-coccolith-producing haptophyte species also possessed these long-chain alkenones, in particular C_{37} alkenones (Marlowe *et al.*, 1984). However, of these four taxa only *E. huxleyi* had a known fossil record. *E. huxleyi* first appeared about 294 ka ago (Wei and Peleo-Alampay, 1993), but the alkenone record extends back to at least the Eocene, with one report of alkenones in Cretaceous shales (Farrimond *et al.*, 1986). Marlowe *et al.* (1990) hypothesised that the ancestors of *E. huxleyi*, namely species of the genera *Gephyrocapsa* and *Reticulofenestra*, were largely responsible for the Cenozoic part of this record. The presence of alkenones in a culture of *Gephyrocapsa oceanica* (Fig. 9.4) has since been reported (Volkman *et al.*, 1995).

It is still not known for sure where these compounds are located in the cell or what their function is (Conte *et al.*, 1994), but they are not part of the lipid layer in the cell wall. It has been shown that the peak of alkenone and alkyl alkenoate production occurs between May and July, with lesser or equal concentrations during the autumn (Sawada *et al.*, 1998; Sicre *et al.*, 1999).

The production depth has been demonstrated or hypothesised to be in the shallow mixed layer (about 0–50 m). In spring, this layer coincides with the peak in *E. huxleyi* abundance. However, the highest *E. huxleyi* and alkenone concentrations were found in the chlorophyll maximum during the warm season (Prahl *et al.*, 1993), when *E. huxleyi* is known to sink out of the upper photic waters prior to the summer months (Okada and McIntyre, 1979; Jordan and Winter, 2000). In a separate study (Sawada *et al.*, 1998), $U^{K'}_{37}$-based temperatures (see explanation below) were 3–4 °C higher in the chlorophyll maximum (60–80 m), which resulted from alkenones derived from cells that had sunk one to three months previously (i.e. during late spring to midsummer).

From the studies mentioned above it was recognised that there were three C_{37} unsaturated compounds with 2, 3 or 4 double bonds, written as $C_{37:2}$, $C_{37:3}$ and $C_{37:4}$ respectively, which varied quantitatively. The alkenone unsaturation ratio, U^{K}_{37}, was first proposed by Brassell *et al.* (1986), whereby $U^{K}_{37} = ([C_{37:2}] - [C_{37:4}])/([C_{37:2}] + [C_{37:3}] + [C_{37:4}])$, but as $C_{37:4}$ is rarely detected

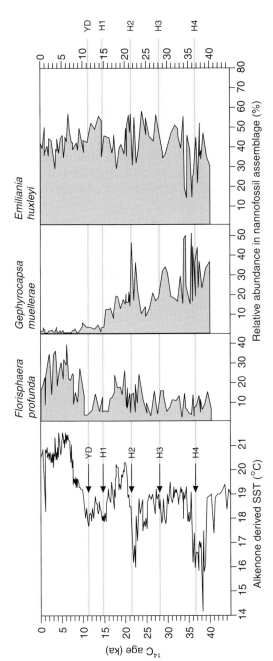

Figure 9.4 Example of palaeoceanographic data from coccolithophorids. The U^{K}_{37} sea-surface temperature and coccolith abundance records for ODP Hole 658C, off NW Africa at 20°N, are taken from Jordan *et al.* (1996). After the Last Glacial Maximum (at c.15 ka) the temperature record increases significantly, and *Florisphaera profunda* increases markedly in abundance; this is interpreted as being due to a reduction in the intensity of upwelling, allowing seasonal thermocline development. Within the glacial interval (pre-15 ka), a series of marked cooling events, with high abundances of *G. muellerae* and low abundances of *Florisphaera*, are interpreted as low-latitude expressions of 'Heinrich events', HI–4, i.e. ice-sheet break-up episodes.

in the sediments, the equation could be simplified to $U^K_{37} = ([C_{37:2}])/ ([C_{37:2}] + [C_{37:3}])$. This latter ratio became known as $U^{K'}_{37}$ (Prahl et al., 1988). Prahl and Wakeham (1987) grew E. huxleyi in culture at different temperatures (8 to 25 °C) in order to calibrate the ratio to temperature and to assess its use as a proxy palaeotemperature tool. They noticed that $C_{37:4}$ only became important when their cultures were grown at <15 °C. Volkman et al. (1995) later showed that the calibration for G. oceanica was different, suggesting that when both species are abundant in the fossil assemblage, adjustments to the SST calculations have to be made.

One advantage of using alkenones is that they are conserved during gut passage and so the feeding activities of zooplankton may not affect the production ratios (Harvey et al., 1987; Grice, 1998). Furthermore, they do not appear to be affected by freezing or carbonate dissolution, and the results are easily replicable and therefore deemed reliable (Sikes et al., 1991). However, some diagenesis in the water column and surface sediments does occur, but this does not appear to affect the SST calculations. Using the alkenone ratio as a proxy for SST in marine sediments has allowed scientists to confirm a number of points. The alkenone and faunal transfer-function temperature estimates are reasonably similar. This led Zhao et al. (1993) to state that alkenones can be used to cross-correlate the stratigraphy when hiatuses or gaps occur in nearby holes, and that the ratio produces a reproducible result. It has been suggested that alkenone productivity is independent of glacial–interglacial climatic changes, and that the alkenone peaks, which occurred every 23 ka, were controlled by precessional forcing, perhaps caused by increased trade-wind intensity and equatorial upwelling (Villanueva et al., 1997). Zhao et al. (1993) also suggested that long-term SST changes are controlled by Milankovitch-type insolation cycles, while the short-term changes may be influenced by oceanic circulation, meltwater events (e.g. Heinrich events) or upwelling strength. In fact, Eglinton et al. (1992) showed that rapid changes of up to 2.5 °C in periods of c.300 years can occur in the alkenone proxy SST data.

Some workers have compared the Holocene with the Eemian (the last interglacial) and have shown that the latter lasted approximately 3000 years and was 2–3 °C warmer (Ikehara et al., 1997; Villanueva et al., 1998). Others have compared the Holocene with the LGM (the last glacial maximum) and have shown that the LGM SST was 4 °C lower than that of the present (Ikehara et al., 1997; Tasman Plateau), while Villanueva et al. (1998) stated that the LGM was 4.5 °C cooler than the previous glacial period. Palaeotemperatures calculated for the surface sediments and LGM of subtropical/tropical sites have corroborated previous findings using oxygen isotopes in foraminiferal assemblages, i.e. that during the LGM the subtropical/tropical waters were only 1.5–4 °C cooler than at present (Brassell et al., 1986; Prahl and Wakeham, 1987; Lyle et al., 1992; Ohkouchi et al., 1994).

One of the applications of combined alkenone–coccolith studies has been to detect Heinrich events in geographical areas away from the iceberg melting zones (Fig. 9.5). Basically, Heinrich events (Broecker et al., 1992) are

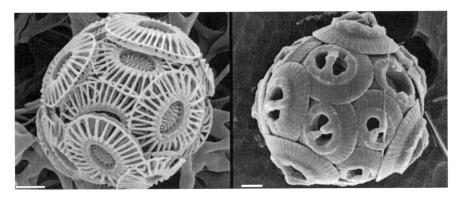

Figure 9.5 *Emiliania huxleyi* and *Gephyrocapsa oceanica*. Electron micrographs of the two main alkenone-producing coccolithophores. Scale bars – one micron. The alkenones produced by these species have very high preservation potential and provide an excellent record of past temperatures.

warming episodes that are recorded in marine sediments as a distinct series of ice-rafted debris (IRD) layers, as first documented by Heinrich (1988) in the North Atlantic. This debris was originally picked up by glaciers moving across the frozen continents, which then formed ice shelves at the coast. Icebergs that calved off these ice shelves during warm episodes carried this debris on their bases as they travelled towards warmer waters. As the icebergs melted they released the IRD, and thus the IRD became incorporated into the underlying sediments. In the case of the North Atlantic, this melting zone lies between 40 and 55 °N. However, it is assumed that the cold water derived from these melting icebergs may be detected as temperature anomalies in alkenone-derived SST records in areas outside of these melting zones. Jordan *et al.* (1996) showed that the timing of down-core U^K_{37} SST anomalies in a core off NW Africa corresponded closely with the dates of the IRD layers recorded further north. Furthermore, associated with these anomalies, the coccolith assemblages also changed dramatically, suggesting that these cooler waters either brought in a cold-water flora or that a cold-water flora developed due to the incoming cooler waters. The alkenone record and *F. profunda* abundance also suggest that upwelling was strongest during the deglaciation, and not at the LGM (Zhao *et al.*, 2000).

9.5 SUMMARY

In the first part of this chapter a general explanation is given of the taxonomy and morphology of modern coccolithophorids and thoracospheres. This is followed by a brief overview of the environmental factors affecting these marine organisms, in order to provide the reader with some background ecological information.

In the last part of the chapter, two key topics within current coccolithophorid research are discussed. The first of these topics covers the uses of an ecologically restricted species, *Florisphaera profunda*, for palaeoceanographic studies. Haptophyte-specific biomarkers (long-chain alkenones) are reviewed in the second topic and, as proxy indicators of past sea-surface temperatures, their usefulness in reconstructing palaeoclimates is demonstrated.

FURTHER READING

Bown, P.R. (ed.), 1998. *Calcareous nannofossil biostratigraphy*. Chapman and Hall, London, 315 pp.
This is the state-of-the-art book on coccolith stratigraphy. It is extensively illustrated and contains useful introductory chapters on coccolithophorid biology and methodology.

Green, J.C. and Leadbeater, B.S.C. (eds), 1994. *The haptophyte algae*. The Systematics Association Special Volume No. 51, Clarendon Press, Oxford, 446 pp.
This book contains 22 excellent mini-reviews on a range of topics, written by the leading experts in each field. Despite being sparsely illustrated, the chapters have extensive bibliographies.

Paasche, E., 2001. A review of the coccolithophorid *Emiliania huxleyi* (Prymnesiophyceae), with particular reference to growth, coccolith formation and calcification–photosynthesis interactions. *Phycologia*, in press.
This is an invaluable and very clear review of recent multidisciplinary research.

Winter, A. and Siesser, W.G. (eds), 1994. *Coccolithophores*. Cambridge University Press, Cambridge, 242 pp.
This book contains 11 key review articles, is well illustrated (including an atlas of many of the extant coccolithophorids) and is printed in a large format. A number of the chapters relate to the ecology and distribution of natural populations. There are also chapters on the distribution of fossil assemblages and the use of the stable isotopes found in coccoliths.

CHAPTER 10

ENVIRONMENTAL APPLICATIONS OF DINOFLAGELLATE CYSTS AND ACRITARCHS

Barrie Dale and Amy L. Dale

Micropalaeontology in its broader sense includes both mineralised microfossils and the organic-walled microfossils (palynomorphs) traditionally covered by the subject palynology. The groups of palynomorphs most used for environmental applications include the spores and pollen of plants, dinoflagellate cysts and acritarchs. As with other prominent groups of microfossils, their widespread abundance in both space and time was first used in biostratigraphy, mainly for providing routine age determinations and correlations for hydrocarbon exploration (Stover *et al.*, 1996). Subsequently, it was realised that their large concentrations are also particularly well suited to quantitative statistical treatments, and this combination has proved useful for documenting palaeoenvironmental change. To date, this has been applied mainly to investigations of climate change on time-scales of millions to thousands of years, where pollen is used to document vegetational responses on land, and dinoflagellate cysts and acritarchs indicate changes in aquatic environments. More recent work summarised here shows possibilities for a broad range of environmental applications for the aquatic palynomorphs which should help the earth sciences meet the new challenges posed by the environmental sciences.

One of the greatest challenges facing the earth sciences, as with other natural sciences, is how best to contribute to human understanding of the wide range of environmental problems that are the focus of much public concern at this time. These problems mostly involve the impact on the natural environment of a rapidly growing human population and industrialisation, including destruction of natural habitats and many different forms of pollution. The quality of future management of natural resources is reliant on critical input from the natural sciences, including the geosciences.

The environmental issues of most concern relating to this chapter include global warming due to the enhanced greenhouse effect, and marine pollution including eutrophication. Understanding these issues will require providing answers to such questions as:

1. What is the extent of the particular human impact?
2. When did it start? and
3. How does it compare with the natural variation to which the biota is subjected over time?

Answering these could be fairly straightforward if the relevant environmental parameters had been monitored over a long enough time. However, as indicated by the X in Fig. 10.1, the usual situation is far from ideal, and environmental monitoring often begins only after the impact is realised. This reveals a pressing need for developing retrospective methods, i.e. methods that allow an assessment of the issues after the fact. The aim is not just to be able to trace the development of environmental deterioration, but, more importantly, to develop robust methods for assessing environmental recovery resulting from attempts to reverse this.

It is becoming increasingly clear that micropalaeontology has a large potential for helping to investigate these issues. The scientific principles previously applied to interpret palaeoenvironmental change from changing assemblages of microfossils in sedimentary rocks offer a sound basis for tracing recent environmental change in the sediments accumulating in aquatic environments today. Within the past decade, efforts have been made to develop the use of microfossils in the sedimentary record as an archive of environmental change on the scale of tens to hundreds of years necessary for addressing the environmental issues. In some cases this has involved no more than scaling up already established methods (e.g. pollen as climatic indicators) to cover shorter time-spans. In other cases, microfossil groups have been investigated as possible indicators of new environmental signals (e.g. for eutrophication and marine pollution). These efforts have provided micropalaeontology with greater insights into the interactions between organisms and environments, contributing both to ecology and other environmental sciences. This opens up

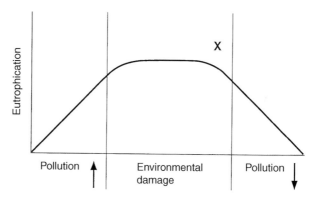

Figure 10.1 Diagram representing the development of cultural eutrophication with time, as pollution increases and causes environmental damage before diminishing following clean-up. Note that environmental monitoring often first begins after the environmental damage is realised (shown by X).

increasing possibilities for applications of environmental micropalaeontology in geology, oceanography, geography and archaeology. This exciting new challenge to micropalaeontology forms the basis of this chapter, particularly as it applies to one group – the dinoflagellates.

Dinoflagellates are a numerically important group of microplankton inhabiting all major freshwater, brackish and marine environments. They account for a substantial amount of the planktonic biomass, particularly in coastal and neritic waters, and in many regions they regularly produce extensive blooms that may colour the water green, brown or red. Species of dinoflagellates that produce toxins are responsible for many of the so-called 'red tides' (harmful algal blooms) that cause massive fish kills, and seriously threaten human health.

Dinoflagellates are single-celled organisms with characteristically biflagellated motile stages previously considered to be algae. Since many are photoautotrophs (i.e. primary producers employing photosynthesis), they are often referred to in the literature as phytoplankton. In fact, they are an extremely varied and complex group of organisms also including many species that are heterotrophs, feeding by predation on other microplankton, and some that combine both autotrophic and heterotrophic strategies. They are now considered to be protists, classified within their own Division, the Dinoflagellata (Fensome *et al.*, 1993). Their trophic complexity is reflected

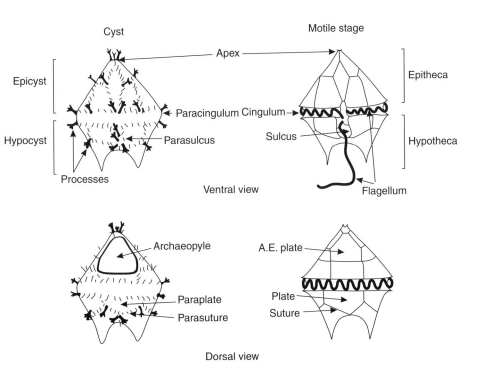

Figure 10.2 Comparison of the basic morphological features of dinoflagellate motiles and cysts, and the terminology used for these (A.E. plate = archaeopyle-equivalent thecal plate).

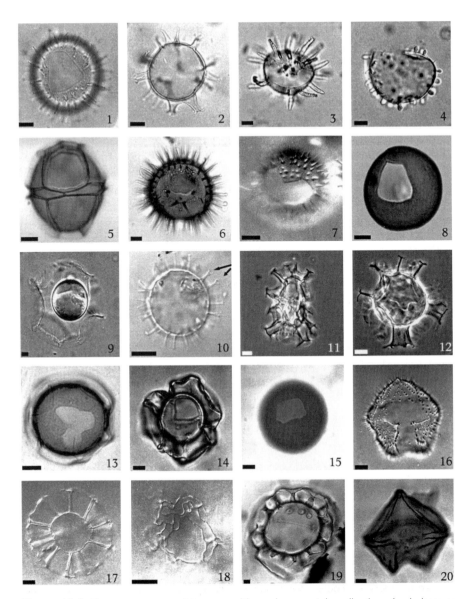

Figure 10.3 Some common cyst types used in environmental applications (scale bars = 10 μm). 1. *Operculodinium centrocarpum*, New Zealand, showing the five-sided intercalary archaeopyle; the most cosmopolitan living cyst (subpolar–equatorial, globally). 2. *Spiniferites bulloideus*, NE USA; cosmopolitan. 3. *Lingulodinium machaerophorum*, western USA, showing three-plate equivalent archaeopyle; warmer water indicator (equatorial–temperate, globally), and eutrophication indicator. 4. *Lingulodinium machaerophorum*, Baltic Sea, showing reduced, bulbous processes characteristic for low-salinity environments. 5. *Impagidinium patulum*, Mediterranean Sea; oceanic indicator. 6. *Protoperidinium conicum*, Norway, in apical view showing intercalary archaeopyle with dislodged operculum; cosmopolitan (subpolar–equatorial, globally) heterotroph, upwelling/eutrophication indicator. 7. *Multispinula minuta*,

also in the ecology of the group, and consequently has to be one of the main considerations in attempts to understand their palaeoecology.

In addition to the motile stage commonly found swimming in the plankton, many species produce a non-motile resting stage (a **cyst**), widely considered to represent a zygotic stage (hypnozygote) in the sexual cycle. It is the resistant walls of these cysts that accumulate in sediments, producing the fossil record of the group. Some species have calcareous cyst walls, and a few siliceous cysts are recorded from the fossil record, but most known living and fossil cysts have walls containing a sporopollenin-like material similar to that found in the spores and pollen of plants. This reveals a basic strategy where similar, heavy organic molecules are employed to protect cells that are vitally important for the successful completion of sexual cycles, in these protists and throughout the plant kingdom. Basic morphology and terminology of dinoflagellate cysts and motile stages is shown in Fig. 10.2.

Dinoflagellates have an unusual geological record; geochemically detected biomarkers suggest that they were already present in the Cambrian, but the fossil record shows only an isolated occurrence in the Silurian, followed by a consistent record from the Triassic to the present. This is best explained by considering the nature of the fossils (resting cysts that are only produced by some species, and then not always capable of fossilisation), and the broad classification of **acritarchs** that probably allows some earlier dinoflagellate cysts to be classified as acritarchs (as discussed below).

The group Acritarcha was erected in 1963 as an informal grouping of the palynomorphs of unknown affinities remaining after others identified as dinoflagellate cysts were formally placed within the dinoflagellates. Acritarchs are presumed to include fossils from a wide variety of organisms.

Newfoundland, Canada, showing two-plate equivalent archaeopyle; bipolar, cold-water indicator species characteristic of polar–subpolar coastal waters. 8. *Protoperidinium conicoides*, NE North Atlantic; bipolar cold water indicator. 9. *Impagidinium pallidum*, Fram Strait, northern North Atlantic, with cell contents; bipolar, cold (polar–subpolar) oceanic indicator. 10. *Peridinium faeroense* (=*Pentapharsodinium dalei*), Norway, showing some characteristic bifurcated processes (arrows); cooler water species characteristic of subpolar–cold-temperate coastal waters. 11. *Spiniferites elongatus*, Norway; cooler water species characteristic of subpolar–cold-temperate coastal waters. 12. *Spiniferites membranaceum*, Norway; cooler water species characteristic of subpolar–cold-temperate coastal waters. 13. *Protoperidinium americanum*, Peru; indicator for upwelling. 14. *Planinosphaeridium choanum*, Italy; cool–warm temperate. 15. *Protoperidinium avellana*, Newfoundland, Canada; a cooler water species characteristic of subpolar–temperate coastal waters. 16. *Protoperidinium pentagonum*, western North Atlantic; cosmopolitan (subpolar–equatorial, globally), coastal–neritic. 17. *Nematosphaeropsis labyrinthus*, Quaternary, North Sea; an outer neritic–oceanic indicator species. 18. *Nematosphaeropsis labyrinthus*, Eire; high focus, showing thread-like trabeculae connecting the process tips. 19. *Tuberculodinium vancampoae*, Crete; a warm-water indicator species (equatorial–warm temperate). 20. *Lejeunacysta oliva*, Eire, a cyst type with a pronounced protoperidinioid morphology.

Of particular interest here is the increasingly obvious probability that some dinoflagellate cysts remain unidentified within the acritarchs.

In order to be classified as a dinoflagellate cyst, an otherwise unknown palynomorph must show morphological affinity to the dinoflagellates (Fig. 10.2). Since the morphological features used to define this group are almost exclusively taken from the motile stages, and since motile stages do not fossilise, this usually involves identifying features on cysts that are presumed to be genetic reflections from the motile stage. Cysts may be identified as such in this way, occasionally based on overall body shape (e.g. Fig. 10.3, part 20), but more often based on the characteristic furrows housing the flagella (**cingulum** and **sulcus**), or details of the patterns of plates covering many motiles (**thecal tabulation**) (e.g. Fig. 10.3, part 5). The one distinctive feature common to all cysts is the excystment opening (**archaeopyle**), through which the emerging new motile stage exits. In many cases this reflects a recognisable part of the tabulation (one or more plates). However, it should be noted that one large group of dinoflagellates (athecate – or naked – dinoflagellates) do not have thecae, and therefore produce cysts lacking all forms of reflected tabulation.

While many dinoflagellate cysts have been identified using such criteria, several lines of evidence suggest that the taxonomic boundary between dinoflagellate cysts and acritarchs remains blurred. One striking feature of acritarchs is their stratigraphical distribution, since by far the most species are recorded from the Late Precambrian to the Carboniferous (i.e. long before the persistent dinoflagellate record begins). As a result, palynologists are reluctant to apply the term archaeopyle to the regular types of openings shown by some ancient groups of acritarchs, with the implication that these are related to dinoflagellates, even though exactly the same feature would be called an archaeopyle if present in younger rocks with an established dinoflagellate record. At the other end of the record, palynologists working with Recent and Quaternary cysts largely ignore the acritarchs, and regularly publish accounts of cysts that are not known to show definitive dinoflagellate cyst features, and are thus acritarchs by definition. This taxonomic and nomenclatural problem is further complicated by the fact that at least one indisputable living acritarch (illustrated in Fig. 10.3, part 10, showing none of the criteria used to identify cysts), in incubation experiments produced a motile thecate dinoflagellate identified as *Peridinium faeroense*. A full discussion of this issue is beyond the scope of this chapter, but it is important for the reader to realise that the Recent and Quaternary cysts referred to here and elsewhere are an ill-defined mixture of cysts *and* acritarchs.

This chapter concentrates on applications of organic-walled dinoflagellate cysts (from here on referred to as cysts in the interest of brevity), in marine to brackish-water environments. Mineralised cysts and freshwater cysts are not yet developed as environmental indicators. Acritarchs are included only to the extent that they co-occur with the cysts. The focus is on the Quaternary to Recent applications most relevant for the environmental sciences.

10.1   METHODS

The methodology used for this work is based on:

1. studies of the distribution of living/Recent cysts;
2. relating distribution patterns of these cysts to known environmental factors; and
3. using this information as a basis for interpreting environmental change from the cysts archived in the record preserved in bottom sediments.

10.1.1   Distribution of Living/Recent Cysts

The upper 1–2 cm of bottom sediment is collected, and the fossilisable cysts are extracted using palynological preparation methods (hydrochloric acid, HCl, followed by hydrofluoric acid, HF, to remove minerals, and sonication and sieving to concentrate cysts within the remaining organic fraction). The cysts are counted using ordinary light microscopy, and the results are expressed either qualitatively (per cent composition of species in assemblages) or quantitatively (cysts/unit of sediment – as dry weight or volume).

Quantitative measurement of cysts in surface sediments is of limited value, since this provides no unequivocal information regarding cyst productivity. The cysts account for only a very small amount of the sediment in most samples (far less than 1 per cent). The broad range of concentrations obtained (e.g. <20/g dry sediment in coarse sands to >150,000/g in some fine-grained fjord sediments) therefore may well reflect sedimentary factors rather than actual differences in cyst production. Measurement of cyst concentration in volume of wet sediment is particularly inadequate, since water content varies greatly according to sediment type and degree of consolidation. Quantitative methods based on cysts/g dry weight of cored sediment are recommended for environmental work (Dale, 2001). It should also be realised that sedimentation rates vary enormously, such that whereas a sample of the upper 1–2 cm of coastal sediment may represent several years, the equivalent amount in deep-sea sediments may represent several thousands of years (acknowledged as a major source of error in transfer functions based on mixed data from these different environments).

10.1.2   Relating Cyst Distribution to Environmental Factors

This may be attempted through a wide range of empirical and statistical methods. Arguably, this is best carried out from the fullest possible understanding of the ecology of dinoflagellates as expressed through their cysts. This approach therefore draws on the experience and statistical methods developed by ecologists aimed at investigating the complex web of environmental factors to find out their combined influences on the distribution of organisms (see discussion of ordination methods in the Appendix).

However, palaeontologists have developed their own methods which circumvent much of the ecological complexity (see references to the transfer-function method in sections 10.2.3, 10.3.1 and 10.4.1).

While there is broad agreement that cyst distribution does reflect environmental factors, there is disagreement as to how best to represent this statistically. Much of the effort so far has been devoted to the application of the transfer-function method to cysts, since this method has been widely adopted for other microfossil groups, such as foraminifera, for which it was developed. More ecologically based methods such as correspondence analysis, although not yet as extensively applied, suggest possible flaws in the use of the transfer-function method for cysts (discussed in section 10.4.1 and Appendix).

10.1.3 Interpreting Environmental Change

This is achieved by applying the above methods to the cyst record recovered from cored bottom sediments, within a time-frame established through dating the sediments. The precision assigned to the interpretations varies according to the methods, with transfer functions providing precise estimates of the environmental factors selected, e.g. sea-surface temperature, salinity and ice cover in the first two case studies in section 10.3.1, while more empirical methods provide less precise, but still useful information allowing interpretation of other types of environmental factors (e.g. movements of a major oceanographic front in the third case study in section 10.3.1).

A full description of the methods and equipment used in the various stages of this work is not possible in the restricted space available here. Fortunately, the standard palynological preparation methods are to be found in many books and articles dealing with palynology (e.g. Wood *et al.*, 1996). Similarly, descriptions of the standard equipment for sampling bottom sediments (dredges, grabs, box corers, and smaller gravity corers for surface and short cores, and larger piston corers for longer cores) are readily found in textbooks on oceanography and marine geology. However, usually these sampling devices can be used only from a boat with an appropriate winch, which may pose economic and logistical problems for those lacking access to oceanographic facilities. A simple, easily made alternative is illustrated (Fig. 10.4), suitable for collecting surface sediments from small boats in coastal waters. In addition, supplementary notes are added on *where* to sample, since the importance of this is neglected all too often. The main statistical treatments used for cyst data are described in the Appendix, since these are increasingly important for environmental applications, but the relevant information is sometimes hard to locate elsewhere, and may be even harder to understand for the non-specialist.

10.1.4 Site Selection

Work covered in this chapter is based on bottom-sediment samples, and one of the most important factors affecting the quality of data is site selection. Surface sediments should be targeted specifically for the job to be done (e.g.

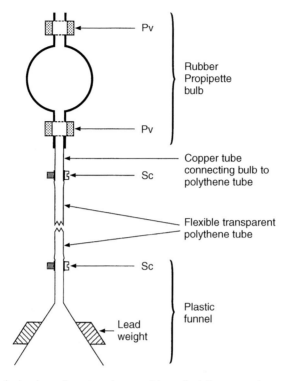

Figure 10.4   A simple sediment sucker used to collect the upper layer of bottom sediments. (Pv = pinch-valve, Sc = screw clamp). After carefully lowering the funnel on to the sea-bed, the sediment is sucked up into the funnel and the lower part of the polythene tube by repeated manipulation of the pinch-valves and bulb. After retrieval from the water, the retained sediment sample is pressed into a suitable container by reversing the manipulation procedure of pinch-valves and bulb. Designed, constructed, and used for over 30 years by B. Dale.

large-scale biogeographical surveys along north–south-trending coasts to document water-temperature signals should avoid local features such as rivers, human pollution, and so on, that may impose their own overprinting signals). It should be realised, too, that fishing (especially bottom trawling) regularly disturbs bottom sediments, and increasingly this has to be avoided if possible. Both this, and the natural bioturbation caused by benthic fauna, are usually circumvented by choosing sites with low enough oxygen levels in the bottom water. Thus optimal sediments for cores would be from anoxic basins with high sedimentation rates, such that a core of up to 1 m length could represent an undisturbed, several hundred year old record.

10.1.5   Cyst Transport

One major factor to be considered is transport. Most cysts fall within the size range of 25–75 μm, thus behaving as fine silt particles in the sedimentary regime. A basic understanding of the sedimentary system is therefore needed to select the most representative site. In practice, this often means sampling the deepest available part of a minor sedimentary basin, in the most sheltered location along a section of the coast (bays, fjords or even artificial boat basins). Whereas this usually allows locally representative sampling of coastal waters, the opposite is true for offshore and deep-sea sediments, where long-distance transport of cysts may well be the rule rather than the exception, since the observed assemblages often are dominated by more coastal rather than oceanic species. Samples from deep-sea sediment traps have shown transport to the deep sea of cysts from the continental margin of Norway, and the injection at depth into the North Atlantic of cysts from Barents Sea bottom water (Dale and Dale, 1992). That this type of sedimentary particle is transported long distances was dramatically illustrated by evidence of characteristic Palaeozoic palynomorphs washed down the St Lawrence River and sedimented over a long swath of ocean floor south of Bermuda (Needham *et al.*, 1969).

Cysts are thus well suited to applications in coastal environments, where there is a particular need for microfossil coverage, since this is where human impact on the marine environment is most concentrated, and where many of the other main groups (planktonic foraminifera, coccoliths and radiolaria) are largely absent due to their predominantly oceanic distributions. Cysts may prove unreliable in at least some deep-sea settings that are heavily influenced by long-distance transport.

10.2 HISTORY OF THE ENVIRONMENTAL APPLICATIONS OF CYSTS

A few palaeoenvironmental applications using cysts had already been developed in the late 1950s and 1960s, as cysts began to prove useful in biostratigraphy within the oil industry. Ratios between the marine cysts and various categories of spores and pollen transported in from land were used to estimate distance and direction to the palaeo-shoreline, and to help unravel the complex details of deltaic sediments. However, the presumption at that time was that fossils of benthic rather than planktonic organisms would provide the best palaeoenvironmental indicators. This was based on comparisons with known marine ecology, where the standard biogeographical zones are defined using the attached organisms (e.g. molluscs and macroalgae) or those otherwise capable of maintaining populations within recognisable environmental conditions (e.g. some fish). Planktonic organisms such as dinoflagellates, lacking this kind of control, were presumed to be more widely dispersed over the oceans – their distributions reflecting more large-scale movements of water-masses than local environmental conditions. Two main lines of research have changed this view: studies of living cyst ecology revealing plausible ecological signals, and attempts to apply the transfer-function methods to cysts.

10.2.1 Ecology of Living Cysts

Ecology provides the *credibility* for applications of organisms (living or fossil) as environmental indicators. Ecological studies usually begin by establishing which species live where, in order to explore how and why they do so, but this has proved too extensive so far for the more than 2000 species of marine dinoflagellates. This is because of the enormous number of plankton samples that would need to be counted worldwide, given the characteristically large variation in species composition often recorded from different water depths, sometimes at time intervals of up to just a few days. Fortunately for micropalaeontology, the cysts offer a much more viable alternative (Dale, 1983), at least for the more than 200 cyst-forming species, based on the following:

- Cyst assemblages in one sample collected from the upper 1–2 cm of bottom sediment represent an integrated record from the water column of up to tens of years in coastal systems, and hundreds or even thousands of years offshore.
- The cysts are generally easier to identify than motiles.
- Cysts rather than motiles offer the relevant ecological information for palaeoecology, since only the cysts fossilise.
- Cysts offer the *only* relevant ecological information to date for many living species where the cyst/motile relationships are not yet known.

However, several features of the life-cycle impose obvious restrictions on the use of cysts as environmental indicators. The cyst-forming dinoflagellates appear to be mostly coastal/neritic species, in which the cyst acts as a **benthic resting stage**. A higher proportion of such species is found at higher latitudes, where the cyst functions as an over-wintering stage. These cysts have an obligatory resting period of up to several months; they are capable of surviving at least 10 years if necessary in conditions of temperature and water chemistry prohibitive to survival of the motile stages. In a Norwegian fjord, for example, colder water species form plankton blooms in the spring, before encysting for the remainder of the year, while other, warmer water species bloom only in the few weeks of summer and remain in the sediments for at least nine months of the year. These cysts certainly may reflect a definite temperature window of potential use for interpreting palaeotemperatures, but it is imperative to realise that this is the spring *or* summer temperatures – the encysted stage is extremely resistant and able to survive a wide range of winter conditions including temperature. Similarly, it is important to realise that the benthic resting stage imposes limits (probably less than 100 m) on the water depth from which the newly emerged motile may be expected to swim up and re-establish plankton blooms. Cysts of well-known coastal forms recovered from deep-sea sediments are most likely transported in from more coastal waters, and therefore not representative of overlying waters. Similarly, the heterotrophic forms are from species that do not require light, and therefore may live at other depths in the water column.

However, a few truly oceanic, photo-autotrophic, cyst-forming species are documented; these are presumed to follow an alternative strategy to the benthic resting stages, and to provide the basis for oceanic environmental signals.

The extensive distribution of abundant living cysts in coastal/neritic bottom sediments was first discovered in the 1960s by micropalaeontologists seeking living representatives of the fossils (e.g. Wall and Dale, 1968). Prior to this, biological studies of dinoflagellates were almost exclusively restricted to motiles found in the plankton. From the 1970s to the present, several micropalaeontologists have investigated the distribution of cysts in Recent sediments as a basis for developing palaeoecological applications. These are mainly concentrated on the North Atlantic, but other regions have been added (summarised by Mudie and Harland, 1996), and at least one ongoing study has targeted samples from some of the remaining gaps to attempt global models of cyst distribution (Dale and Dale, in press). The first distribution studies in the 1970s showed that cyst assemblages were related to water-mass characteristics, suggesting an obvious potential for use as environmental indicators. This was taken one step further by the realisation that at least some cyst distributions reflect the basic standard biogeographical zones established for other marine organisms. In other words, the cysts offer comparable environmental signals to those shown by molluscs, macroalgae, and so on. This probably reflects somewhat comparable life strategies to these other organisms, involving a benthic stage (cyst) 'anchoring' the otherwise planktonic dinoflagellates to their biogeography.

10.2.2 Ecological Signals from Cysts

The term **ecological signal** is used here as elsewhere to denote consistent ecological features of organisms (in this case cysts) that allow recognition of at least some aspects of specified environmental factors. Although not necessarily quantitative, such empirical signals may represent the extent of current knowledge regarding the ecological relationship between an organism and a particular environmental factor. As such, they may provide useful applications even without being able to give precise estimates of environmental factors. For example, a eutrophication signal may be used to demonstrate how far environmental recovery has progressed, following clean-up of a polluted aquatic system, by tracing the extent of decreasing eutrophication relative to pre-eutrophication levels.

Recent cyst distributions suggested ecological signals that could be used for environmental applications, but this is a relatively new field of research where the limits to many of these signals are not yet fully understood, and the potential for routine applications is still being developed. In considering these signals, it is necessary to distinguish between coastal/neritic assemblages (less likely to be affected by transported cysts, and therefore providing more reliable signals) and the less reliable offshore/deep-sea assemblages. The main ecological signals shown by coastal cysts (summarised by Dale, 1996) are as follows.

10.2.2.1 Surface-water Temperature Signals

These are expressed by some cysts with distributions restricted to specific temperature windows. Many cyst assemblages are dominated by one or more of several **cosmopolitan species** with broad tolerances for temperature (subpolar to equatorial). However, a few species show restricted distributions with respect to temperature, including several that occupy standard biogeographical zones, and several that occupy the transition regions straddling the boundaries between zones. Many of the colder water species characteristic for polar and subpolar zones are bipolar, and thus may be used in both hemispheres. This allows applications in Quaternary sequences where the migrating **biogeographical zones** (polar, subpolar, temperate and equatorial) can be distinguished as indicators of shifting temperatures with climatic change. However, at best this suggests estimates of the temperature limits defining the zones (e.g. *Lingulodinium machaerophorum* may denote summer temperatures of at least 12 °C), rather than the precise average seasonal temperatures desired by climate modellers.

10.2.2.2 Coastal/Oceanic Signal

This signal reflects the distinctly different water-mass characteristics of most coastal versus oceanic waters, with the boundary usually located just seawards of the shelf break. Oceanic assemblages usually contain some of the around ten truly oceanic cyst types, mainly species of *Impagidinium* or the outer neritic to oceanic species of *Nematosphaeropsis* (Fig. 10.3, parts 5, 9, 17, and 18). These cysts are almost never found in shelf or coastal sediments, and the few recorded exceptions from the east coast of the USA may be explained by transport in over the shelf by unusual landwards sediment transport, or possibly by warm-water rings separated off from the Gulf Stream. This signal therefore comprises:

1. oceanic assemblages denoted by the presence of oceanic cysts (although numerically these may be outnumbered by coastal cysts transported in from the more productive coastal/neritic zone);
2. coastal/neritic assemblages lacking oceanic cysts;
3. the boundary zone between these two may be characterised by a marked increase in the cosmopolitan *Operculodinium centrocarpum* (considered to reflect the relatively unstable conditions where these very different water-masses converge).

10.2.2.3 Salinity Signals

These have mainly been described from the Baltic Sea (results summarised by Dale, 1996). Not surprisingly, the coastal/neritic cyst types show broad tolerances within the salinity range from normal marine (around 35) to about 20, since this degree of variation may well be encountered in coastal waters affected by freshwater runoff. However, while many of these species seem not to tolerate salinities much below 20, the few (more cosmopolitan) species that do so provide useful signals of brackish waters. Cysts of the most cosmopolitan species (*O. centrocarpum*, *L. machaerophorum* and *Spiniferites bulloideus*, Fig. 10.3, parts 1–3) all consist of more or less spherical central bodies with numerous, relatively long spines (processes) projecting out from the surface. These spines are progressively reduced in length with lower salinities, until at salinities around 5–0 they may be almost unrecognisable (see Fig. 10.3, part 4). This reduction seems to follow a linear relationship to salinity, such that it may be possible to use *average* process length as a reasonable estimate of palaeosalinities within the 20–0 range. However, it should be noted that not all individuals are affected, and a few cysts with atypical process lengths may be found, especially where large populations are examined.

10.2.2.4 Nutrient Signals

These have been demonstrated both from upwelling regions (e.g. off Peru, California, northwest Africa and South Africa) and associated with coastal eutrophication (e.g. the case study on the Oslofjord, in section 10.3.2.1). All the examples of strong upwelling systems show similar signals involving dominance by cysts of heterotrophic species, mainly within the genus *Protoperidinium*, although a particular species may vary in amounts from region to region. This is thought to reflect the basic predominance of diatoms and small flagellates in the plankton affected by upwelling, therefore favouring the heterotrophic dinoflagellates feeding on them. Both off California and NW Africa, the water-masses over the shelf are also somewhat enriched by nutrients from the upwelling offshore, and this too produces a characteristic cyst signal with a high abundance of *Lingulodinium machaerophorum*. It is interesting to note that these two main cyst signals associated with more oceanic upwelling are also repeated in the few cases of cultural eutrophication (i.e. from extra nutrients resulting from human pollution) so far studied from coastal sites. More heavily polluted sites (Tokyo Bay, Japan, and several Norwegian fjords) are characterised by proportional increases in cysts of heterotrophic species, while large increases in *L. machaerophorum* are associated with cultural eutrophication in the less polluted Oslofjord (discussed in section 10.3.2.2).

10.2.3 The Development of Quantitative Methods

Palaeoclimatic modelling began to incorporate increasingly quantitative methods in the early 1970s, with the introduction of the Imbrie and Kipp (1971) transfer-function method. This method allowed, for the first time, estimation of precise sea-surface temperatures from statistical analysis of microfossil distributions. Although first developed for planktonic foraminifera, transfer functions also quickly became the most widely applied modelling method for a variety of other microfossil groups, including dinoflagellate cysts. During the 1980s and 1990s, cyst-based transfer functions were developed to give precise estimates of a variety of oceanographic parameters, including: sea-surface temperature (August and February), salinity (August and February), seasonality (difference between summer and winter), ice-cover, and, more recently, nutrients. The vast majority of these studies were carried out on cyst distributions in the North Atlantic.

The method combined multivariate analyses of principal component analysis (PCA) and multiple regression, (1) to calibrate Recent cyst data to measured oceanographic parameters, and (2) to obtain equations to relate these to fossil cyst data as a basis for interpreting palaeoenvironments (see Appendix). The Imbrie and Kipp (1971) method has been modified for use within palynology, and since the 1990s the 'best analogue' method of Guiot (1990) has been applied instead of regression to calculate the transfer functions, but the cyst groupings were still obtained using PCA.

The application of transfer functions to obtain precise estimates of past sea-surface conditions requires a number of basic assumptions, including:

- The microfossil assemblages upon which the transfer functions are based do indeed reflect the selected hydrographic conditions in the overlying waters.
- The pelagic system being sampled today has acted in a consistent manner throughout the time period being studied, and the species have responded in a consistent way to the physical and chemical parameters of the ocean.
- The ecological variables being predicted and/or the species present do not interact or combine in ways that make the transfer-function extrapolation unreliable.

Alternative modelling methods have also been applied that rely more on empirical approaches and do not yet attempt to provide palaeoecological information as precise as that given by transfer functions. Examples of these include comparisons of ratios of known indicator species to infer sea-surface temperatures based on their documented ranges (summarised by Edwards *et al.*, 1991), and, more recently, applications of multivariate statistical techniques of correspondence analysis (CA) and canonical correspondence analysis (CCA) (see case study on Recent and fossil cysts from the Angola Basin in section 10.3.1.3 and Appendix).

CA had become the preferred ordination method over PCA for modern-day ecological studies since its introduction (as 'reciprocal averaging') by Hill (1973), and has been applied to cyst studies since the early 1990s. Although CA provides more plausible ecological subdivisions than PCA in comparative studies of Recent data (see Appendix), it has not been incorporated in the development of transfer functions. Indeed, some proponents of CA question that such precise data should be estimated using the transfer-function method, due to the inevitable oversimplification of ecology imposed by the critical assumptions listed above. Yet, the development of palaeoclimatic models is reliant on precise estimates of palaeoceanographic parameters. In effect, this has spurred the development of new transfer-function techniques, while at the same time hindering the development of more empirical approaches such as CA.

10.3 Case Studies

The following brief summaries of a few case studies are meant to illustrate the main areas of application of cysts rather than the total volume of applications so far. The conclusions reported here are those of the original authors. Possible alternatives to some of these are discussed in section 10.4.1, which also mentions further potential for other kinds of future applications.

10.3.1 Case Studies in Palaeoclimatology and Palaeoceanography

To date, the vast majority of applications of cysts have been concentrated in this important area of research, reflecting the current dominance of climate studies within the environmental sciences. These applications have almost exclusively employed transfer-function methods in response to the pressing need to produce the precise estimates of values for palaeoenvironmental factors that are critical ingredients for models of past climate change. Two case studies are included from the many published reports of this work, while a third illustrates an alternative approach.

10.3.1.1 Transfer-function Derived Surface-water Conditions in a Norwegian Fjord over the Past 11,300 Years

This study by Grøsfjeld *et al.* (1999) is included here as an example of the use of current transfer-function methods to reconstruct precise estimates of past sea-surface conditions. It is one of the first attempts to use such methods in coastal waters, where problems of long-distance transport of cysts are presumably minimised. It therefore serves also as a good example to illustrate the widening credibility gap between these methods and known cyst ecology (discussed in section 10.4.1).

A total of 77 samples were analysed from an 8.5-m long core taken in Voldafjord, western Norway, with age control provided from AMS radiocarbon dates. The transfer functions used were based on a database of cyst assemblages in surface sediments from 439 sites from middle to high latitudes of the North Atlantic and adjacent basins. These included mainly oceanic and shelf sites, with no directly comparable sites for a Norwegian fjord, with waters which often reflect a complex mixture of local hydrography and various influences from more offshore waters.

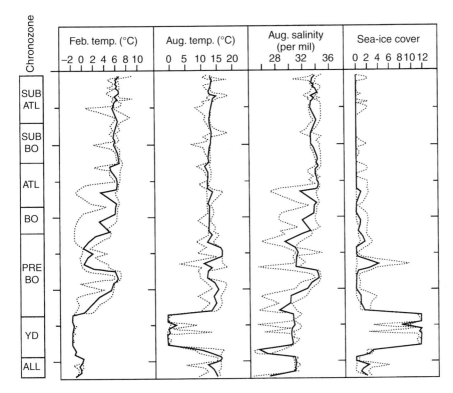

Figure 10.5   The estimated past water conditions in the core from the Voldafjord. Redrawn from Grøsfjeld *et al.* (1999).

The study shows, as have many others, that the transfer methods used for other microfossils can be applied in the same way for cysts. The authors provide estimates for sea-surface temperatures in February and August, salinity in August, and the number of months with more than 50 per cent of sea-ice cover, for the past 11,300 years (redrawn here in Fig. 10.5). The general interpretations of climatic change inferred from the cyst record using transfer functions are similar to those previously suggested from empirical studies of cyst distribution. The details allow closer comparisons with proxy data from lakes in western Norway, and from further offshore records. The main conclusions are:

- *c.* 11,300–10,800 BP – cool-temperate surface waters, with high interannual variation in sea-surface water temperatures which possibly resulted in strong stratification of the water column;
- *c.* 10,800–10,000 BP – sea-surface waters close to freezing, with long-lasting seasonal sea-ice. Estimated changes in temperature within just a few decades at the beginning and end of the Younger Dryas are larger than those estimated from lakes or offshore;

- *c.* 10,000 BP – surface-water warming, followed by established interglacial conditions between 10,000 BP and 9500 BP;
- *c.* 7000 BP – probable onset of the modern Norwegian Coastal Current;
- the past 7000 years – increased exchange of water-masses with coastal regions; relatively stable conditions with possible interruptions for periods of cooling.

10.3.1.2 Checking the 'Meltwater Spike' Hypothesis for Triggering the Younger Dryas

De Vernal *et al.* (1996) provide an example where the transfer-function method and cysts were applied to check the feasibility of an important hypothesis concerning the cause of the Younger Dryas cold event (between 10,800 and 10,300 BP).

Broecker and co-workers (1989) proposed that this abrupt climatic change was driven by a decrease in the rate of North Atlantic Deep Water production, triggered by a sudden dilution of North Atlantic surface water in response to the diversion of Laurentide ice-sheet meltwater from the Mississippi drainage system to the St Lawrence River.

De Vernal *et al.* (1996) studied the cyst record covering this time interval in three sediment cores from a transect along the present-day outflow from the St Lawrence River across the shelf on to the slope. Oxygen-isotope stratigraphies from foraminifera provided age control, and the sampling intervals, estimated to be from between 50 and 100 ^{14}C years, were judged to be suitable for determining hydrographic changes on the scale needed. Transfer functions used were derived from a similar database of cyst assemblages in surface sediments to that used in the previous case study. This reportedly included many neritic samples that allow reconstructions of salinities ranging from 25 to 36.

Cyst assemblages in surface sediments from the region that includes the cored sites are dominated by the cosmopolitan *O. centrocarpum* and spherical brown protoperidinioid cysts. The authors recorded similar assemblages from the cored intervals including the Younger Dryas, and were thus able to conclude that salinities at that time were about the same as at present, with even a suggestion of higher salinities nearer the mouth of the river. This was interpreted to suggest no appreciable increase, but rather a significant decrease in meltwater runoff, suggesting that the hypothesis of Broecker *et al.* (1989) is incorrect.

10.3.1.3 Tracing Past Migrations of the Angola–Benguela Front Using Correspondence Analysis of Cysts

In a study of Recent and fossil cysts from the Angola Basin (Dale *et al.*, in press), correspondence analysis was applied to a set of 49 Recent surface-sediment samples in order to analyse the main ecological gradients influencing cyst distributions at the present day. The Recent samples covered a wide range of oceanic environments, including the Congo River fan, an area of strong upwelling off the coast of southwest Central Africa and a nearshore–offshore gradient spanning from coastal waters to open ocean. The present-day hydrography is heavily influenced by the Congo River plume, and the Benguela Current to the south (Fig. 10.6). A front situated between 14° and 16°S – the Angola–Benguela Front (ABF) – separates the colder waters of the Benguela Current from the warmer waters of the Angola Current (labelled ABF, BC and AC, respectively, in Fig. 10.6). The strength of the upwelling system at present is strongly related to the position of the Angola–Benguela Front, and this has important implications for (palaeo)productivity and (palaeo)climate in the region. The objective of the study was to construct a quantitative portrayal of the present-day oceanic environment in the Angola Basin, which could be used to trace the movements of the Angola–Benguela Front (ABF) back in time.

The results from the correspondence analysis indicated five main groupings (see Fig. A.8) that could be related to the main environments known to be present in the study area (compare Figs 10.6 and 10.7). Axis 1 of the CA suggested a nearshore–offshore gradient, and Axis 2 a nutrient gradient, and these were confirmed by a subsequent canonical correspondence analysis constraining the species data to reflect depth (Fig. A.12), and a classified CA reflecting total percentage heterotrophs and autotrophs at each site (see Fig. A.13).

Having established the main gradients affecting the study area at present and the main cyst types associated with them, CA was applied down-core to trace changes in the oceanographic system over the last 200,000 years (Dale *et al.*, in press). A 20.7-m long piston core (T89–32) was collected from 3330 m water depth near the outer sample of the coastal upwelling zone shown in Fig. 10.7. The core location was chosen due to its strategic position relative to the Angola–Benguela Front (ABF). Some 41 subsamples were taken at approximately 30–100 cm intervals, and these were dated isotopically and analysed for dinoflagellate cysts.

The main variation, interpreted according to the CA of Recent samples, was characterised as a contrast between upwelling conditions such as today along the Angola–Namibia coast (dominance by spherical brown protoperidinioid cysts, SBPr) and non-upwelling conditions resembling those in the outer river plume (dominance by *Spiniferites* spp.; see Figs A.7 and A.8). The occurrence of the upwelling assemblage (SBPr) at the core site was presumed to reflect northward hydrographic displacement via a strengthened Benguela Current, and the occurrence of the river plume assemblage (*Spiniferites* spp.) at the core site was interpreted to reflect southward penetration of the Southern Equatorial Counter Current (SECC, see Fig. 10.6), comparable to that observed today. Thus, 'northward' and 'southward' movements of the front were substituted for 'negative' and 'positive' directions along Axis 1 of the CA.

A down-core plot reflecting the past migrations of the ABF was obtained by weighting the CA sample scores of the main axes by their respective

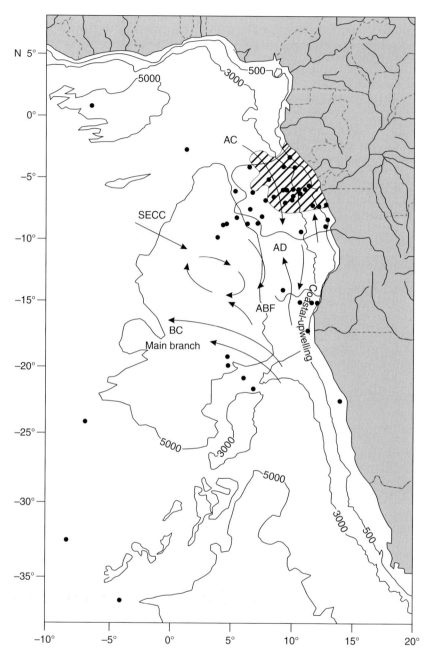

Figure 10.6 Map of the Angola Basin, showing the main hydrological features and sites for surface-sediment samples (dots). (AC = Angola Current, SECC = Southern Equatorial Counter Current, and BC = Benguela Current.) Wavy line denoted by 'ABF' marks the Angola–Benguela Front. Hatched area out from the Congo River shows the area of the inner river plume, and the white outer area shows the extent of the deep-sea fan and what is here referred to as the outer river plume.

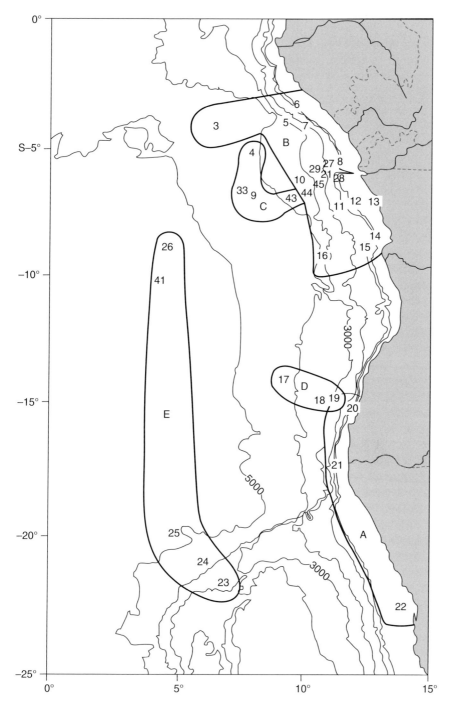

Figure 10.7 Map of the Angola Basin, showing sample groupings (A–E) from correspondence analysis of recent samples shown in Fig. A.8. Note the close similarity to the hydrographic features shown in Fig. 10.6.

eigenvalues. The results from the composite curve of the first two axes of the CA analysis of Core T89-32 are shown alongside a foraminiferal curve which incorporates changes observed in four cores ranging from 6.5°S to 20°S lati-

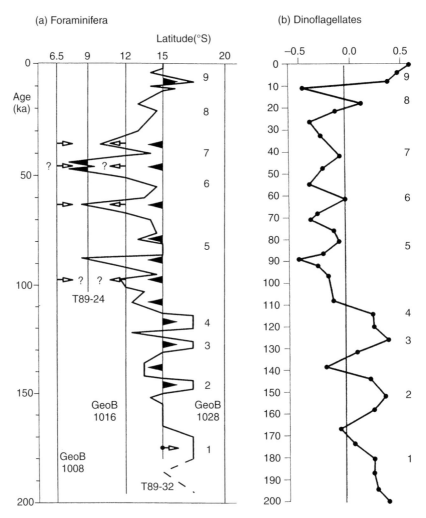

Figure 10.8 Oscillations in the Angola–Benguela Front (ABF) over the past 200,000 years, as reflected by (a) foraminifera (using four cores), and (b) dinoflagellate cysts (using one core near the outer sample from the coastal upwelling zone shown in Fig. 10.6). The foraminifera curve is based on the documented presence of indicator species at the respective latitudes represented by the cores, whereas the cyst curve is based on weighted CA scores (see section A.3.2.2) from a single core. Positive oscillations reflect dominance by *Spiniferites* spp., reflecting southward movements of the ABF. The solid line in (B) is the mean weighted CA score, reflecting the mean position of ABF during the last 200,000 years. Southward oscillations are numbered 1–9 to aid comparison.

tude (Fig. 10.8). The ability of the CA ordination, based on one core, to identify all of the major southward excursions of the ABF that were identified based on foraminiferal analysis using four cores, illustrates the potential of correspondence analysis for providing reliable, ecologically based palaeo-environmental analysis from cysts.

10.3.2 Case Studies in Eutrophication and Marine Pollution

This is an important field where other groups of microfossils, notably benthic foraminifera and diatoms, have proved to be useful indicators of various forms of marine pollution (see Chapter 4). It is a relatively new area for research using cysts, but the first few studies reported so far reveal strong signals that should allow the development of robust applications. Since the dinoflagellates account for a large amount of primary production in coastal waters, their cyst record is particularly well suited for tracing the development of cultural eutrophication in the coastal zone. Furthermore, since the group also includes many heterotrophic species feeding mainly on other groups of phytoplankton, the relative proportions of heterotrophs versus photo-autotrophs seems to provide signals for the pollution-induced changes affecting the balance between different groups within the phytoplankton. As with other environmental studies, the main difficulty to be overcome concerns the identification of the various elements from the complex factors involved in human impact, in order to assess possible effects on the biota. Here, we include the first attempts to document eutrophication signals (from the Oslofjord, with relatively little industrial pollution), and discuss the current status of a second probable eutrophication signal – so far only suggested from sites that are also affected by heavy industrial pollution.

10.3.2.1 *The Eutrophication Signal from the Oslofjord, Norway*

Cultural eutrophication is an increase in the rate of supply to an ecosystem of organic matter resulting from human activities. It has been associated with harmful effects on marine ecosystems in different parts of the world, including anoxia that kills most organisms. Identifying cultural eutrophication has proved difficult, and there is a need for methods suitable for tracing its development over time. Dale *et al.* (1999) reported the first attempts to develop such methods from the cyst record, in the bottom sediments of the Oslofjord. The hydrography of the inner basins of this and many such fjords is characterised by restricted circulation, resulting from narrow channels and shallow sills. Such systems are particularly vulnerable to cultural eutrophication.

The inner Oslofjord, adjacent to the city of Oslo, is well suited for the studies reported here. One of the earliest documented examples of cultural eutrophication had already been recognised there in the 1960s by marine scientists responding to public concern over deteriorating water quality and the loss of a local fishery due to increasing anoxia. These problems were associated with nutrient loading due to sewage from the growing human population of the region, and therefore water quality improved as better sewage treatment was developed, starting in the 1970s. This allowed the option of searching for signals of eutrophication in the sedimentary record. It also suggested possibilities for testing such signals as potential indicators of environmental recovery, from the more than twenty-year record of improving conditions since the 1970s.

Four 20–60 cm long sediment cores from water depths of 100–200 m were analysed from the two innermost basins and the narrow channel leading to these. The ^{210}Pb dates showed records spanning several centuries in all cores, thereby including the time period for development of eutrophication. Comparisons between the cyst records and the known history of eutrophication strongly suggested consistent cyst signals tracing the development of eutrophication. These are considered to be particularly robust, since they reverse following improved sewage treatment, in some cases back to pre-eutrophication levels.

The eutrophication signal documented from the Oslofjord, illustrated in Fig. 10.9, comprises two main elements:

- increased total cyst concentrations (approximately a doubling of the amounts of cysts/g dry sediment);
- a massive increase of one cyst type in particular, *Lingulodinium machaerophorum* (=the cyst of the photo-autotrophic *Gonyaulax polyedra*).

The increase in total cysts/g is interpreted as a reflection of increased phyto-plankton production, including the cyst-forming dinoflagellates. The preferential increase of *L. machaerophorum* probably reflects a combination of the fact that this species thrives on high nutrient levels in the water (shown by its association with nutrient-enriched shelf waters elsewhere, as described in section 10.2.2.4), and represents a warmer water species that blooms in summer at these latitudes. The Oslofjord is a nutrient-limited system where the often-strong spring bloom utilises much of the dissolved nutrients available, and phytoplankton growth in summer is limited to what is left over. Nutrients added throughout the year from sewage effluent would presumably not have increased spring productivity, since this was not previously limited, but may well have increased summer plankton, including *L. machaerophorum*. This is supported by the fact that a cold-water species, *Peridinium faroense*, which blooms in spring, showed no appreciable increase with eutrophication (Fig. 10.9).

The documentation of a eutrophication signal featuring large amounts of *L. machaerophorum* suggests the possibility of identifying eutrophication in other regions, based on high proportions of this species in surface sediments. River influence may increase nutrient levels and therefore possibly lead to elevated amounts of *L. machaerophorum* anywhere in its range. However, whole regions characterised by such assemblages include: parts of the

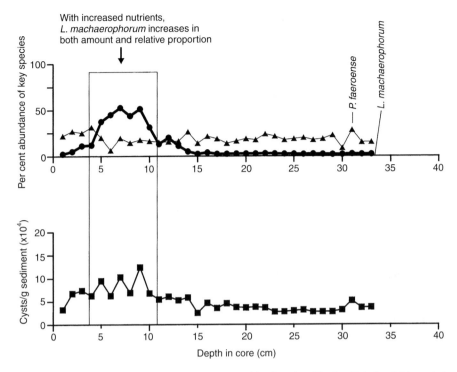

Figure 10.9 X–Y plot showing the cyst eutrophication signal in the Oslofjord. Note that levels of *Lingulodinium machaerophorum* increase both in proportion of the assemblage (%) and in total concentration (cysts/gram) at times of documented increased eutrophication (shown by the box). Note that amounts of the cooler water species *Peridinium faeroense* show relatively little change throughout.

Swedish coast, the Irish Sea and parts of the California coast (including Los Angeles harbour). These would be worth investigating further to see if they do indeed represent similar examples of cultural eutrophication.

10.3.2.2 A Probable Eutrophication Signal Based on the Heterotrophic–Autotrophic Ratio in Cyst Assemblages

The Oslofjord was chosen by Dale (2001) to study possible eutrophication signals, partly due to the relatively low amounts of industrial pollution that otherwise might obscure these. Another site, the Frierfjord in southwestern Norway with relatively low amounts of sewage but large amounts of industrial waste, was chosen to study the possible effects of industrial pollution, from the cyst record in two sediment cores. Sætre *et al.* (1997) described cyst signals from this site that were in some ways opposite to those from the Oslofjord, with increased pollution associated with an overall decrease in total cyst concentration and a proportional increase of cysts of hetero-trophic species (mostly belonging to the genus *Protoperidinium*). These were tentatively interpreted to represent possible adverse effects from pollution (chemical toxicity, reduced light penetration, etc.) that affected the autotrophs more than the heterotrophs in planktonic dinoflagellates, including the cyst-formers. Subsequent work from a fjord in western Norway affected mainly by sewage also showed similar cyst signals to those from the Frierfjord, suggesting that this may represent an alternative eutrophication signal (Thorsen and Dale, 1997). These authors discussed the possibility of increased diatom production as prey for the heterotrophs. Matsuoka (1999) also found similar signals in the heavily polluted Tokyo Bay, Japan, and interpreted the shift from more autotrophic to more hetero-trophic dominance of cyst assemblages as evidence for eutrophication causing increased diatom production, and thereby increased prey for heterotrophic dinoflagellates.

Dale (2001) pointed out that as yet there is no unequivocal evidence sup-porting this interpretation, since the reported shift is a *proportional* shift, with no appreciable increase in total cyst concentrations, suggesting more adverse effects on the autotrophs, rather than increased production of heterotrophs. Moreover, sites such as that in Tokyo Bay are so heavily affected by *both* mas-sive eutrophication and industrial pollution that it is not possible to separate these factors and identify the possible eutrophication signal.

Nevertheless, since elsewhere the ratios between autotrophs and het-erotrophs seem to shift more towards the latter in the extremes of nutrient enrichment represented by upwelling systems, this strongly implies a con-nection with nutrients. In any case, these examples of probable pollution-induced shifts within the planktonic groups have important implications for other members of the food web reliant on the phytoplankton for a supply of suitable prey (e.g. in co-operation with fisheries biologists, we are currently investigating the possiblity that this could have caused collapse of local fisheries at various times along the southern coast of Norway). Ongoing studies of eutrophicated systems lacking industrial pollution should soon allow us to confirm this proportional shift to more heterotroph species as a useful indicator of the state of the environment with respect to eutrophication.

10.3.3 Case Studies in Marine Archaeology

While this type of work is not yet reported in the literature, it is mentioned here as another field of research where the practical application of cysts is proving useful. Pollen analysis has well-established applications in archae-ology – mainly in recording terrestrial vegetational changes associated with human settlement and climatic changes which can provide a basis for stratigraphy and dating (see Chapter 11). With a background understanding of their ecological distributions, cysts may be used similarly in the marine realm.

10.3.3.1 Identifying Viking-Age and Middle Ages Sediments in the Oslofjord

Within the past ten years, major new road projects have exposed valuable archaeological sites near the central railway station in Oslo. Included were waterfront installations from the old Viking-Age settlement, and several ships ranging in age from this time to the early Middle Ages. These marine sediments are now raised above present-day sea-level due to isostatic uplift.

Plans to extend the road works across part of the inner fjord require prior assessments from marine archaeologists to protect against the possible destruction of further remains. To facilitate this, B. Dale and A.L. Dale (unpublished reports) were asked to investigate the possibility of using the cyst record to identify sediments of Viking Age to early Middle Ages across the inner basins of the fjord. Several factors serve to complicate the sedimentary record concerned, including several small rivers that produce greatly increased sedimentation rates locally, and many centuries of human activity such as dredging and the dumping of large quantities of sawdust.

The cyst record was first described from a sequence of archaeologically dated marine sediments exposed by the excavations, allowing recognition of the Medieval Warm Period, from approximately AD 1000 to 1300. A series of 15 sediment cores (up to 17 m in length) were analysed for cysts, which showed systematic changes in assemblages (example shown in Fig. 10.10) that could be correlated within the inner fjord. Based on climatic signals recognised from a global database of recent cyst distributions (B. Dale and A.L. Dale, in press) and local records from the reference section on land, this was interpreted as a short-term climatic record. The record of spruce pollen from the same palynological preparations (shown on Fig. 10.10) was also used as a convenient time marker, since its characteristic increase marking the establishment of the species in local forests has been dated by other pollen studies.

Combining the spruce and cyst records allowed the identification of a distinctive warm period, interpreted as the Medieval Warm Period, bracketed by the first cold periods immediately below and immediately above the establishment of spruce. This was informally called the C–D interval (see Fig. 10.10), and corresponds with the time period to be identified. For reasons mentioned above, this was absent in some parts known to have been dredged in historical times, and varied between depths within the sediment of 1 m and deeper than recorded in the 17 m sampled, with a thickness of between 1.9 and 3.2 m. Identifying these sediments helped free some parts of the proposed road project from further restrictions, eliminated the need

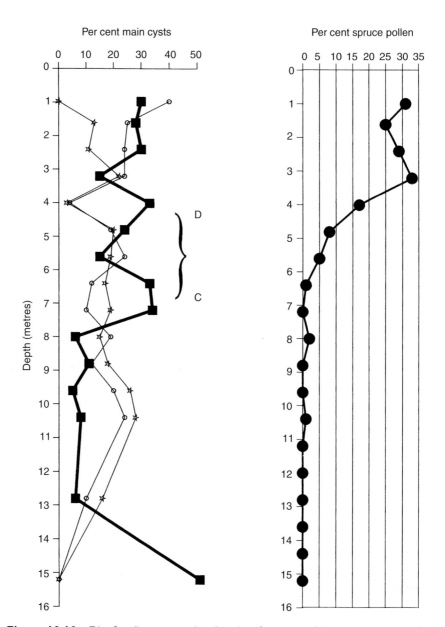

Figure 10.10 Dinoflagellate cyst and pollen data from one of the sediment cores from the inner Oslofjord, studied for archaeology. The cooler water species *Peridinium faeroense* (black squares), and warmer water species *Lingulodinium machaerophorum* (open circles) and total *Spiniferites* species (open stars) were used to define two periods of cooler climate (C and D) bracketing the establishment of spruce locally (from % spruce pollen in the total pollen record). The CD interval, including sediments from the Viking Age and Middle Ages, was traced in cores along the projected site of a major road project to aid the environmental impact assessment by archaeologists.

for more comprehensive and costly (e.g. seismic) surveys, and allowed archaeologists to concentrate their work on other parts that might contain valuable remains.

10.4   FUTURE WORK

Much of the work reported here is relatively new, and many of the observations are based on few examples so far. Future work should be concentrated on testing these applications in new examples. This is particularly the case in fields such as marine pollution and eutrophication, but even the more established applications such as transfer functions should be reassessed. There is a widening credibility gap developing between these palaeontological methods and known ecology of dinoflagellates, discussed below, and this needs to be addressed. However, there are also many possibilities for developing new applications for cysts, and some of these are mentioned in closing.

10.4.1   Transfer Functions: the Widening Credibility Gap

The main research on dinoflagellate ecology has been largely ignored by Quaternary palynologists developing the use of dinoflagellate cysts as palaeoenvironmental indicators based on the transfer-function method. The method itself is not reliant on ecological interpretations, although some relationship between the organisms producing the microfossils and the water conditions that are estimated has to be assumed. This raises several fundamental questions:

- How much seeming deviation from ecology can the method tolerate while still maintaining scientific credibility?
- What alternative processes may be invoked to explain the apparently close relationship between the organisms and the selected water parameters which allows such precise palaeo-estimates?
- Are there basic differences between the cysts and other microfossil groups used in transfer functions that make the cysts less credible in such applications?

The pertinent ecological factors to be considered include:

- Almost all of the known cysts are coastal species with relatively broad tolerance for water temperature and salinity, while a few are restricted to oceanic waters.
- At higher latitudes, such as those used in the first two case studies (see section 10.3.1), the cysts are produced in the warmer half of the year, and they often have a mandatory resting period which enables them to function as over-wintering stages.

- Dormant resting cysts are much less sensitive than planktonic cells to environmental parameters such as temperature and salinity.
- Most of the cysts that are often frequent in assemblages represent species with broad tolerance levels for water temperature and salinity.
- A few are characteristically warmer water species or colder water species, allowing us to recognise limits to the water temperatures represented.
- Salinity probably seriously affects assemblages only at the brackish level (<20).
- Nutrition and other environmental factors are often more important than temperature and salinity for determining the composition of cyst assemblages.

Quaternary palynologists have followed the methodology previously applied for other fossil groups (planktonic foraminifera, diatoms, etc.) to develop transfer functions relating cyst distribution to water temperature, salinity and ice-cover. The database from which transfer functions are developed in the North Atlantic still contains many oceanic sites. Whereas the other main fossil groups used are known to live in the waters overlying these sites, available evidence suggests that most of the cysts are transported in from more coastal regions. There is no support from 'living ecology' for the use of these mainly tolerant coastal species for interpreting relatively small differences of temperature and salinity across the deep ocean, even if one assumes them to be *in situ*. Similarly, there is no known ecological link between ice-cover and these species – most of the yearly ice occurs in surface waters while the cyst is dormant in bottom sediments, and the emerging motile stages seemingly thrive regardless of ice-cover. Thus, there is no sound ecological basis for assuming a direct link between the cysts in deep-sea surface sediments and temperature, salinity and ice-cover in the overlying surface waters. Nevertheless, many Quaternary palynologists assume a statistically based link between the cyst assemblages and surface-water parameters which is used to interpret water temperature, salinity, and ice-cover from the fossil record with a precision rivalling that of contemporary oceanographic measurements.

The first case study in section 10.3.1 largely avoids problems with long-distance transport of cysts to the site in a fjord, but relies on a database of mixed shelf and deep-sea sites for its proxy data. Since the fjord is a radically different environment from the more open-ocean North Atlantic, the assumption made has to be that the dinoflagellate assemblages from these very different systems simply reflect factors of temperature, salinity and ice-cover that are of interest to Quaternary geology. This opens an obvious credibility gap between so-called 'quantitative' Quaternary palynology and plankton ecology. The assemblages from the deep sea and shelves of the North Atlantic do not offer an ecologically sound basis for interpreting assemblages in a fjord; the interpreted factors such as winter temperature and sea-ice occurring during dormancy are not generally of a magnitude likely to be critical for determining variance in assemblages; and the total range of salinities interpreted from the whole core is no more than these coastal species may experience within one day's tidal cycle in estuarine environments.

The differences between these two views of cyst applications may be illustrated by reconsidering the second case study (checking the meltwater spike hypothesis). The more ecologically based view presented here (sections 10.2.2, 10.2.3 and the third case study) suggests several reasons why the observations of de Vernal *et al.* (1996) should not be taken as unequivocal evidence against the meltwater hypothesis of Broecker *et al.* (1989). These authors did not consider environmental factors such as water stability and nutrients, considered to be important in the third case study, on the second largest freshwater plume in the world. Their data-set of cysts related to present-day conditions in the North Atlantic therefore may not provide a good enough proxy for recognising the effects of the proposed meltwater plume.

The main cyst signal described from the Congo River plume is produced by assemblages dominated by the cosmopolitan *O. centrocarpum* and heterotrophic protoperidinioids (mainly spherical brown cysts). Cyst assemblages in present-day surface sediments off the St Lawrence shelf edge are also characterised by just such assemblages. In our opinion, this is more likely to be due to factors like nutrients and water stability, as discussed here, than temperature and salinity – not least, since both cyst groups show broad tolerance for both temperature and salinity at the present day. Based on observations from the Congo plume, it is not unreasonable to suppose that a massive plume of fresh water from the St Lawrence around 11,000 BP could also have produced cyst assemblages dominated by *O. centrocarpum* and spherical brown protoperidinioids. However, such assemblages would be interpreted erroneously as representing salinities close to modern values based on the transfer functions of de Vernal *et al.* (1996).

In the absence of a plausible ecological basis for transfer-function based interpretations, other as-yet unknown factors must be invoked for producing the statistically established links between cysts in surface sediments of the North Atlantic and environmental parameters in the overlying surface waters. At best, this uncertainty seriously limits the possibilities for interpreting environmental changes through time within the North Atlantic system, and it offers no sound scientific basis for interpreting palaeoenvironments in a Norwegian fjord. There is an obvious need for alternative methods for quantitative statistical treatments based on more modern ecological principles (see Appendix).

10.4.2   New Applications in Harmful Algal Bloom Research

Harmful algal blooms (HABs) pose a severe threat to human health and commercial fisheries worldwide. Dinoflagellates cause many of these, and many of the HAB species also produce cysts, although some do not fossilise. Understanding the role of cysts in the life-cycle, and as seed-beds for initiating blooms, is important for understanding the dynamics of HABs, and separate studies are under way covering these biological aspects. However, there is a new and rapidly growing field of HAB research that overlaps with the environmental sciences and opens up new possibilities for cyst applications. Basically, this has to do with two central issues:

- The spreading of HABs through transport of the offending organisms in ships' ballast tanks; and
- The enhancement of HABs through other human activities.

There is reasonable evidence suggesting transport of a variety of marine organisms by the large volumes of ballast water taken in one part of the world, after delivery of cargo, and discharged into another prior to loading the next cargo. Of interest here is the observations of cysts of toxic dinoflagellates in the sediments of ships' ballast tanks, and accusations between nations of introduction of HABs internationally by these means (Australia has already imposed laws requiring ships to exchange ballast water in the open ocean). The methods described in this chapter have clear applications regarding this issue, in that the cyst record in bottom sediments may be used as evidence for when a given species was or was not introduced. Increasingly, this kind of application is also providing useful information on the periodicity and varying intensity with which HABs occur.

There is also concern about other possible ways in which human activities may be contributing to the frequency and intensity of such blooms. This is a complex issue, since there is obviously much more scientific effort applied to these phenomena at present than was previously the case. Therefore, the extent to which HABs are being enhanced or simply more frequently recognised is far from certain. Nevertheless, there are plausible arguments suggesting such effects (e.g. from cultural eutrophication and intensive fish-farming), and cyst applications may help to resolve such issues. Especially where the offending HAB species has a fossilisable cyst, the amounts of those cysts may be traced through time from the sedimentary record, and compared directly with the cyst signals for eutrophication.

10.5 SUMMARY

Organic-walled microfossils of dinoflagellate cysts and acritarchs show possibilities for a broad range of environmental applications in marine and brackish waters. As with other microfossils, their assemblages archived in bottom sediments are particularly useful for providing retrospective evidence of environmental change through time. With possible resolution of just a few years, on time-scales of tens to hundreds of years, this allows documentation of natural levels of change as background for assessment of human impact on the environment. The methodology used involves studies of the distribution patterns of living/Recent cysts, relating these patterns to known environmental factors to identify environmental signals from cyst assemblages, and applying these signals to interpret environmental change from the cysts archived in the sedimentary record. Global cyst-distribution patterns reveal ecological signals reflecting surface-water temperatures, and gradients for coastal to oceanic waters, salinity, and dissolved nutrients. While there is broad agreement that cyst assemblages in bottom sediments do reflect environmental factors, there is a basic disagreement on how best to apply these to environmental studies. Much effort has been invested so far in developing transfer functions for cysts, based on methods first used for foraminifera in the deep sea. This certainly provides the detailed estimates of palaeotemperature, salinity and ice-cover needed for modelling climate. However, the assumptions on which the transfer functions are based lack ecological credibility, and alternative statistical treatments based on correspondence analysis are presented here and in the Appendix which have provided more plausible ecological signals in comparative studies of Recent cyst data. Case studies are summarised illustrating the use of both methods in palaeoceanography and palaeoclimate studies. Other case studies cover the use of cysts as indicators of eutrophication and marine pollution, in developing applications in marine archaeology, and studies of harmful algal blooms.

FURTHER READING

Dale, B., 1983. Dinoflagellate resting cysts: 'benthic plankton'. In G.A. Fryxel (ed.), *Survival strategies of the algae.* Cambridge University Press, Cambridge, pp. 66–136. This reference provides a comprehensive review of the biological and geological background to the study of dinoflagellate cysts: the history of cyst studies; terminology and morphology; cyst and dinoflagellate life-cycles; cyst classification; cyst ecology; and application of cyst work to living dinoflagellate studies.

Dale, B., 1996. Dinoflagellate cyst ecology: modelling and geological applications. In J. Jansonius and D.C. McGregor (eds), *Palynology: principles and applications.* American Association of Stratigraphical Palynologists Foundation, 3, 1249–1275. This reference includes fuller descriptions of the ecological cyst signals reported here, and their specific use in Quaternary studies.

Mudie, P.J. and Harland, R., 1996. Aquatic Quaternary, Chapter 21. In J. Jansonius and D.C. McGregor (eds), *Palynology: principles and applications.* American Association of Stratigraphical Palynologists Foundation, 2, pp. 843–877. This provides the most comprehensive review of published Recent cyst distribution studies, and the application of cysts and acritarchs in biostratigraphy and palaeoenvironmental interpretation of marine Quaternary sequences.

Rochon, A., de Vernal, A., Turon, J.-L., Matthiessen, J. and Head, M.J., 1999. Distribution of recent dinoflagellate cysts in surface sediments from the North Atlantic Ocean and adjacent seas in relation to sea-surface parameters. *AASP Contributions Series,* **35**, American Association of Stratigraphical Palynologists Foundation, Dallas, Texas, 146 pp.

This reference provides a comprehensive summary of cyst distributions, and sea-surface temperatures, salinities and ice-cover, mainly from the northern North Atlantic, on which palaeoenvironmental transfer functions are developed. It includes useful, well-illustrated descriptions of the cyst taxa.

CHAPTER

THE ENVIRONMENTAL APPLICATIONS OF POLLEN ANALYSIS

Frank M. Chambers

Pollen analysis has the distinction of being the most widely used micropalaeontological technique – at least for the Quaternary (Birks and Birks, 1980; Williams *et al.*, 1998). In northwest Europe, pollen analysis commenced with the work of von Post (1916, 1946), and was developed further in Scandinavia by Iversen (1941, 1944) and Erdtman (1943, 1969). It was introduced to Britain by Godwin (1934a, b, 1940); developed from Denmark for Ireland by Jessen (1949) and explored further by Mitchell (1951); then evaluated from The Netherlands by Janssen (1970); applied to Pleistocene deposits in Europe by Zagwijn (1960), West (1956, 1980, 2000) and Woillard (1978), *inter alia*, and to Holocene mire and lake sediments in Europe by a raft of workers. In the United States, where investigation of lake sediments has predominated, use of absolute pollen analysis (see below) was pursued by Davis (1967, 1976), and extended by many others. Environmental applications of pollen analysis have expanded to all continents and to many islands, including the most isolated habited oceanic landmass – Easter Island (see section 11.6.3) – as well as its use in some marine environments.

Pollen analysis, or palynology (from the Greek *palynos* for pollen), is not solely concerned with identifying and counting the pollen (literally, 'fine flour') of higher plants, but may also involve counting fossil spores, particularly of pteridophytes (ferns) and of some bryophytes (notably *Sphagnum* moss, especially in the northern hemisphere); for pre-Carboniferous material, spores may be all that are recorded. Its methods are long established but are continually being refined and improved (e.g. Bryant and Wrenn, 1998; AASP, 2000) to separate, concentrate, recognise and identify the polleniferous material in a sediment from the remainder (colloquially: 'crud'). Pollen analytical methods are labour-intensive; attempts to automate preparation procedures or pollen identification have hitherto been found to be neither cost-effective nor particularly successful (but see France *et al.*, 2000).

Pollen analysts require lengthy training; pollen analysis is not a technique to be picked up overnight or to embark on lightly. There are six principal stages:

1. Sampling/subsampling
2. Preparation
3. Identification/counting
4. Numerical manipulation/analysis
5. Drafting (or tabulation)
6. Interpretation.

Stages 4 and 5 can be reversed, or passed through several iterations. All stages require training and practice, but at present the third (identification) and final (interpretation) stages are the most skilled. Some experienced analysts might delegate stages one to five to others, but will reserve Stage 6 to themselves; others would never relinquish Stage 3, as this is the basis for all interpretation.

Pollen identification is a hard-won skill. Some of the more common pollen types can be recognised relatively quickly, but identification of some of the more 'difficult' (but nevertheless common) or the more obscure types can take years to master. Three years is a good minimum apprenticeship for pollen counting, which means that towards the end of a doctoral study in anglophone countries, or after completion of a Masters thesis in continental Europe, a pollen analyst might be considered reasonably competent at identification – albeit only really in the pollen flora of the region and/or of the time period in which he or she has worked. This is partly why few *practising* pollen analysts dominate the world stage: their published work is too defined and confined by their regional and geological-time expertise. It is also why most Quaternary palynologists are at a complete loss in pre-Cretaceous material, because this has no angiosperm pollen – the pollen types with which they are likely to be most familiar.

Pollen can be found, and may be abundant, in a wide range of sedimentary environments such as lakes, mires, soils, colluvium, alluvium and estuarine sediments (Moore *et al.*, 1991) and in some archaeological contexts (see Dimbleby, 1985; Davis, 1994). In acid anaerobic environments, such as in bog peat and in some lake sediments, most types are remarkably resistant to decay. Pollen grains are tiny (10 μm to 120 μm in diameter, with the vast majority of pollen types between 20 μm and 60 μm) and are abundant in suitable sediments, so small subsamples of *c.* 0.5 cm^3 can be analysed. Subsampling is undertaken initially at broad regular intervals, with samples interdigitated as appropriate (a sample every 32, 16, 8, 4, 2, 1 cm, etc.), or at very close intervals using fine-resolution pollen analysis (FRPA: see Green and Dolman, 1988; Simmons, 1993). Sampling advice is beyond the scope of this volume; for guidance, see Dimbleby (1985), Berglund (1986) and Jones and Rowe (1999).

11.1 PREPARATION METHODS

Preparation methods will differ according to sample material, to local practice and custom, and to the differing perceptions of the health and safety hazards of chemical reagents. For pre-Quaternary samples, it is unlikely that

Table 11.1   Principal stages used in pollen extraction

Stage	Treatment
1	10% hydrochloric acid, HCl (include spore tablets for absolute pollen frequency (APF) technique, if required)
2	Wash using distilled water
3	Hot sodium hydroxide, NaOH (10%) for 10–15 min
4	180 μm sieve to remove coarse debris
5	Wash using distilled water, H_2O
6	10% HCl
7	Zinc chloride, $ZnCl_2$ (density > 2.0) – then centrifuge, discard pellet and decant the supernatant for next stage
8	Wash using dilute HCl and H_2O
9	Wash using H_2O
10	10 μm sieve to remove fine particles. Residue > 10 μm retained for next stage
11	Wash using acetic acid, CH_3COOH
12	Acetylation, using acetic anhydride and conc. sulphuric acid, H_2SO_4
13	Wash using CH_3COOH
14	Wash using H_2O
15	Stain (optional; make alkaline first if using Safranin)
16	Dehydrate, first in 95% alcohol, then in 100% ethanol; and extract in isopropanol
17	Store in vial with silicone oil of 2000 cs viscosity

The method above avoids use of hydrofluoric acid, HF, and is effective for Holocene clay-rich sediments (Bjork *et al.*, 1978). For other mineral-rich sediments, stages 6–8 can be replaced as follows:

6	Add very small amount of 2 M HCl
7	Simmer in 40% HF for 20 min, centrifuge and decant for safe disposal
8	Add 2 M HCl and heat for 5 min, avoiding boiling; centrifuge while still hot, and decant.

Note that stages 6–10 are unnecessary for highly organic, pollen-rich samples. If only a small amount of (especially coarse) mineral matter is present, the method of swirling the pollen suspension between beakers (leaving the sand residue) can also be effective.

For further details of the range of methods, see Barber (1976), Berglund and Ralska-Jaisewiczowa (1986), Faegri and Iversen (1989), MacDonald (1990), Moore *et al.* (1991), Wood *et al.* (1996), Jones (1998), Nakagawa *et al.* (1998), Batten (1991) and Bruch and Pross (1999). For simultaneous extraction with phytoliths (plant opal), see Lentfer and Boyd (2000).

the use of hydrofluoric acid (HF) can be avoided (see Batten, 1999). For organic-rich Pleistocene samples, for many Holocene sediments and for modern surface samples, the steps can be used successively, each separated by centrifugation at 3000 r.p.m., to disaggregate the sample (use dilute sodium hydroxide, NaOH), to remove the larger fragments (sieve at 180 μm), to wash (distilled water, H_2O), acidify (acetic acid, CH_3COOH), remove cellulose by acetylation (*sic*.: more commonly termed 'acetolysis'), acid-wash (CH_3COOH), wash (H_2O), dehydrate (95 per cent spirit, then 100 per cent ethanol), extract (using a permitted extractant, such as isopropanol) and mount in silicone oil (cf. Table 11.1). Non-dehydrated mounts in glycerine jelly (so avoiding the use of alcohols) are also feasible.

Pollen preparation is part art and part science; it takes practice and skill to make suspensions that are neither clumped nor contaminated with fine mineral material.

11.2 Identifying and Counting Pollen and Spores

Pollen and spores are usually identified by light microscopy (see Coxon and Clayton, 1999) of preparations mounted on microslides in a suitable medium of appropriate optical properties and viscosity (e.g. silicone oil of 2000 cs viscosity). Counting takes place under magnification of 400× or 600× along traverses (transects) at least a field-width apart, with critical identifications attempted at 1000× or 1500×. Owing to the non-random distribution of pollen types (Brookes and Thomas, 1967), traverses are usually spaced evenly across the slide. Identification hinges on discriminating between various structural types (including, for pollen, the number of pores (pori) and furrows (colpi), and, for spores, the existence of trilete or monolete scars), sculpturing types (the surface pattern of the outer layers or 'exine') and less reliable but sometimes diagnostic features such as shape and size (see Table 11.2; Fig. 11.1). For reliable results, a research-quality microscope is required, preferably with a good (e.g. Apochromat or better) 40× objective, and also with the option of phase contrast to reveal the detail of the exine.

11.2.1 Aids to Pollen Identification

For some regions of the world, published 'pollen floras' list the pollen types that may be identifiable from well-preserved material (e.g. for the British Late Quaternary: cf. Andrew, 1970, 1984; Bennett *et al.*, 1994). Published keys

Table 11.2 A simplified list of structural and sculpturing types for pollen grains

Structural types[a]	Sculpturing types (for any structure)
Tetrad: four grains fused together	Psilate: smooth or finely dotted
Dyad: two grains fused together	Scabrate: roughly dotted
Inaperturate: single grain, no pores or furrows	Verrucate: warty
Monoporate: one pore	Striate: lineations
Diporate: two pores	Rugulate: thick striations
Triporate: three equatorial pores	Cerebrate: brain-like
Stephanoporate: four or more equatorial pores	Clavate: club-shaped projections
Periporate: four or more pores (not in one plane)	Echinate: spine-shaped projections
Monocolpate: one furrow	Reticulate: net-like
Tricolpate: three meridional furrows	Foveolate: like the surface of an orange
Stephanocolpate: four or more meridional furrows	
Pericolpate: four or more furrows	
Syncolpate: furrows fused to a ring	
Tricolporate: three meridional furrows, each with pore	
Stephanocolporate: four or more meridional furrows, each with pore	
Fenestrate: large, window-like openings	

[a]Note: This structural scheme is used by Faegri and Iversen (1989); for an alternative structural nomenclature see Moore *et al.* (1991).

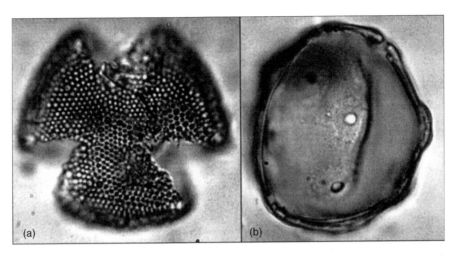

Figure 11.1 Subfossil pollen grains of cultivated flax – *Linum utissitassimum* (left specimen) and walnut – *Juglans* (in this instance, *J. regia*, right specimen). Subfossil pollen grains of (a) can be identified to species level. Diameters of pollen grains (in silicone oil) are c. 45 µm (left) and c. 40 µm (right).

(e.g. Faegri and Iversen, 1989; Moore *et al.*, 1991) and photomicrographs (e.g. for Europe: Erdtman *et al.*, 1961, 1963; Punt and Clarke, 1976–; Reille, 1992) can aid in their identification, and there are promising attempts to automate the process (France *et al.*, 2000); but despite various Web-based catalogues of images (see Anon., 2000) for Quaternary material, there is presently no adequate substitute for a comprehensive collection of comparative reference material, prepared like the fossil material and mounted on pollen microslides, to show the range of morphological and size variation.

Pollen analysis demands the confident identification of pollen 'types'. A competent analyst will attempt to identify pollen and spores to the lowest taxonomic level possible: in some instances this will be to species, more often to genus or to family, and sometimes to a higher taxonomic level. Because identification is conducted to various taxonomic levels, and relatively infrequently to species, each pollen (and spore) type is best referred to as a taxon (plural: taxa). These taxa should conform as far as possible with the botanical nomenclature used in international (rather than national) floras (see Frodin, 2001).

11.3 CALCULATING AND PLOTTING THE DATA

The pollen data are usually presented graphically in a pollen diagram. The various proportions of each taxon in a pollen count are designated a pollen spectrum, and (for sediment cores) each count (or pollen spectrum) represents a horizon (of whatever thickness) of the sampled material. The pollen diagram will not correspond with the original pollen counts exactly; the data will have been 'cleaned' to show only pollen types confidently identified. This may mean amalgamation of tentatively identified taxa to produce a curve for a taxon at a higher taxonomic level. For example, the analyst of Holocene material may be cautious of tentative separate identification of various different cereals, and these may later be amalgamated to give a combined 'Cerealia' or cereal-type curve.

Until recently, pollen diagrams were drawn manually, which required considerable skill in drafting, with local or national conventions as to appearance. In the last decade, various computer pollen-plotting programs (e.g. TILIAGRAPH – Grimm, 1991; SIMPOLL – Bennett, 1994a) were devised to make the appearance rather more uniform and the task quicker (or at least, replotting in slightly different formats is quicker), but each counting and plotting program has its advocates and its critics! Data for sediment cores may be shown plotted against depth, preferably with the sample thickness clearly shown, but for Holocene sediments – provided that the sediment accumulation rate can be calculated sufficiently reliably (see Pilcher, 1993; Bennett, 1994b) – then data can be plotted against either radiocarbon age (BP, where 0 BP is AD 1950) or calibrated calendar age (as cal. BC/ cal. AD, or cal. BP).

11.3.1 Pollen Sum

Percentage pollen data are plotted with reference to a pollen 'sum'; that is, data are plotted as percentages of a total. For example, the pollen sum used can be total land pollen – TLP – excluding spores and pollen types of obligate aquatics; or 'tree' pollen; or dry-land taxa, which might exclude not only aquatics but also mire taxa that have local over-representation. The viewer of a pollen diagram needs to be aware of the sum used. Representation of spores is either as a percentage of the pollen sum, or more usually (correctly: Faegri and Iversen, 1989) as a percentage of the sum-plus-spores. It aids visual interpretation if taxa are plotted on the same linear scale, but in diagrams with many taxa, in which only a few dominate, it may be appropriate to halve the scale of those dominants and to extend the linear scale of the minor taxa to make them legible.

Data for individual stratigraphically unrelated samples can be shown on the same plot, provided that the range of taxa originally recorded was not too dissimilar and that the same pollen sum is used.

11.3.2 Ordering of Taxa

The convention for Quaternary pollen diagrams from temperate regions is to plot the tree taxa in order of immigration, or in cooler regions to adopt a similar strategy for tundra taxa. However, this immigration order is not always apparent, so individual analysts have tended to adopt an order for their first study and then to apply it in all their diagrams (cf. Figs 11.2, 11.4) For pre-Quaternary palynology, other strategies can be adopted. There is still

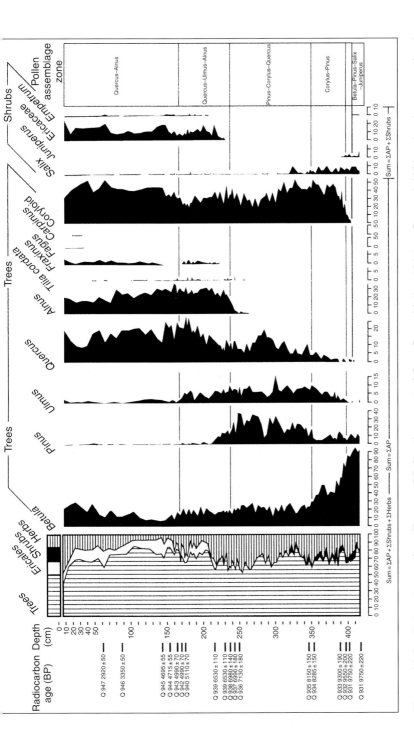

Figure 11.2 Pollen diagram from Tregaron S.E. Bog (now part of Cors Caron National Nature Reserve), Wales, showing conventional ordering of arboreal taxa, supposedly in order of immigration. Note the pollen assemblage biozones, implying significant changes to forest types through time. The impact of human activity on the landscape in recent millennia can be inferred from the rise in non-arboreal pollen. (After Hibbert and Switsur (1976); also in Berglund et al. (1996: p. 82).)

no consensus on the ordering of non-arboreal taxa; some analysts attempt to present them in broad ecological groups, but other criteria can be used. For further details on the plotting of pollen diagrams, see Berglund and Ralska-Jasiewiczowa (1986).

11.3.3   Other Data

In addition to pollen grains, pteridophyte and some bryophyte spores, various non-pollen microfossils (NPMs) are routinely counted by some analysts. These include testate amoebae (especially in peats), coprophilous fungi and a range of other NPM types (see van Geel, 1986), which can also be plotted on the pollen diagram. For Holocene material, charcoal counts are commonly made from pollen slides (see Clark, 1982), and are plotted. It is also commonplace for various summary curves to be presented, such as the ratio between arboreal pollen (AP, comprising tree and shrub) and non-arboreal pollen (NAP), which gives an immediate visual impression (albeit potentially misleading) of the degree of 'openness' of the landscape (see section 11.6.1). Plant macrofossils may also be recorded, but these are usually plotted in a separate diagram (e.g. Chambers *et al.*, 1999).

11.4   NUMERICAL MANIPULATION AND ANALYSIS

11.4.1   Correction Factors

Owing to differential production and dispersal, some workers have devised 'correction' factors for the principal arboreal pollen taxa (e.g. Andersen, 1970; Prentice, 1986b). However, these have not been universally applied, and it is more usual nowadays to plot the original pollen percentage data, and to let the viewer perform corrections mentally (see section 11.6).

11.4.2   Pollen Concentration and Pollen Influx

During the pollen preparation process it is possible to add a known quantity of exotic 'marker' grains – either as an aliquot or as 'tablets' (Stockmarr, 1971) – to a specific volume of sample, which, when counted alongside the sample's pollen, will enable the calculation of pollen concentration values in grains/cm^3. If the sediment core from which the samples were taken has been dated satisfactorily – for example by radiocarbon dating – then pollen influx values can be calculated. Absolute pollen frequency (APF) data are pollen influx values (in grains/cm^2 per year), based on sediment-accumulation rates (see Birks and Birks, 1980; Berglund and Ralska-Jasiewiczowa, 1986); when compared with percentage data these can give a markedly different impression of the environment, especially for the Pleistocene Late Glacial to Holocene transition (see Fig. 11.3).

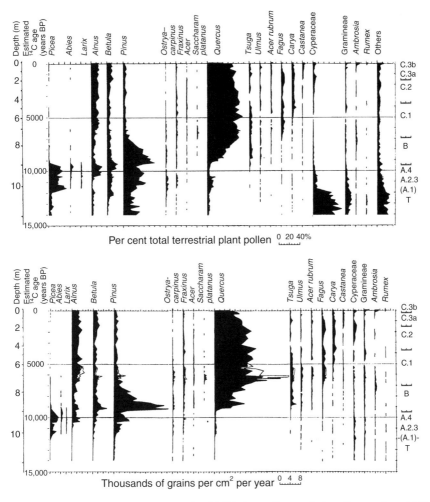

Figure 11.3 A classic study by Davis (1967) from Rogers Lake, Connecticut, USA, showing the differences between pollen percentage data (top) and pollen-influx data (bottom) from the same counts. Note the major differences pre-10,000 years BP.

11.4.3 Zonation

Pollen diagrams are usually zoned, principally to aid description, analysis and interpretation, although zones are also used as a method of relative dating, especially within Pleistocene temperate stages (see Turner and West, 1968; West, 1980, 2000). They have also been used to characterise and (in the absence of adequate radiometric dating) to correlate Pleistocene temperate stages (e.g. West, 1980, 2000) – a practice that in Britain has given rise to a good deal of critical debate, particularly after other correlating and dating methods have been applied (cf. West, 1956, 1980; Turner, 1970; Bowen *et al.*, 1989; Rowe *et al.*, 1999; see also Jones and Keen, 1993).

Holocene pollen zones (pollen assemblage (bio)zones: PAZ) were previously assumed to be contemporaneous across a wide area (e.g. Godwin, 1940), but radiocarbon dating showed this is seldom the case (e.g. Smith and Pilcher, 1973), so analysts now usually devise local (i.e. locality- or profile-specific) pollen assemblage zones (LPAZ). Analysts who are uncomfortable with either the spatial or the internal homogeneous connotations of the word 'zone' have sometimes opted instead for 'phase'. Zonation (or phasing) can be performed visually by the analyst (by eyeballing) or numerically using computer programs (Birks, 1986).

11.4.4 Numerical Analysis

Various other numerical manipulations of the data can be performed, including the use of principal component analysis, (detrended) correspondence analysis (Hill and Gauch, 1980), cluster analysis, and so on (see Birks and Birks, 1980; Birks and Gordon, 1985; Prentice, 1986a, b; Maddy and Brew, 1995; Appendix). Some of these can be helpful in aiding interpretation (see below), although they are no substitute for ecological knowledge. For analysis of the range of taxa present in a pollen diagram, see Birks and Line (1992).

11.5 INTERPRETATION

11.5.1 Taphonomy and Taxonomy

For interpretation of any pollen spectrum, or series of spectra, its taphonomy (namely, the processes leading to the fossil assemblage) and its taxonomy (i.e. the pollen types identified, and what species/taxa they refer to) are both critical. Each will be influenced by pollen production and dispersal. Some plant species produce hundreds of times more pollen than others: wind-pollinated species (anemophilous types) produce a superabundance of pollen; other species produce fewer, rather sticky pollen grains, designed particularly for transfer by insects and, in some cases, by birds. The dispersal distance of anemophilous pollen can be prodigious; some gymnosperms (e.g. boreal pines) produce pollen complete with air-sacs, which are designed to buoy up the grain during atmospheric transport, resulting in transfer of some grains for hundreds of kilometres – even to treeless islands! So, in interpreting pollen diagrams, the inherent differential production and dispersal of pollen types are important considerations. However, these are only two (albeit major) factors that may influence the taphonomy and taxonomy of the fossil assemblage. Light-demanding plants when growing in shade will normally produce far fewer pollen grains, which are also likely to be less well-dispersed, than will an individual of the same species growing in the open. Moreover, pollen production of herbaceous species may be suppressed severely by heavy grazing. So a range of factors, plus those that affect the eventual types seen in the pollen diagram, needs to be considered (see Table 11.3).

Table 11.3 Factors affecting the pollen spectra in pollen diagrams

Factor	Cause
Natural factors	
Differential production	Interspecific genetic (also intraspecific phenotypic) differences; climate; microclimate; soil; grazing
Differential dispersal	Dispersal agent; predisposition for mechanism of dispersal; filtration
Selective transport	Ease of movement or buoyancy in air or water; down-profile movement
Differential destruction	Intermittent oxidation of sediment; bacterial and microbial attack; differing resistance to decay
Incorporation of derived fossils	Reworking of sediment
Analytical factors	
Taxon gross overrepresentation	Incorporation of anther into subsample
Loss during preparation	Differential loss on or through sieves; in discarded supernatant; differential chemical destruction
Operator bias	An 'eye' for some types; inadvertent 'blind' eye to others; or failure to take account of non-random distribution and so to space traverses during counting of slide
Misidentifications	Inexperienced analyst; overoptimistic 'splitting' of taxa; faults in, misuse of, or 'sinks' in pollen keys (especially binary keys); contaminated type-slides
Recording/plotting errors	Errors in recording, in calculations, or in plotting of diagram

There is plentiful advice and information published on the interpretation of pollen diagrams, some of it explicit (as a good starting point, see Birks and Birks, 1980; Faegri and Iversen, 1989) and much of it implicit in the numerous publications that report the results of palynological work. Interpretation is an acquired skill; it requires a good knowledge of plant ecology, an awareness of the possible rate, magnitude and frequency of natural and human-induced environmental changes, and it is helped considerably if the age and duration of accumulation of the samples can be ascertained with some precision and accuracy (Pilcher, 1993). However, it is not immune from fashion. For example, the classical 'elm decline' horizon in north European pollen diagrams has been attributed variously to differential pollen transport (filtration), climatic change, soil depletion, human influence, elm disease, or a combination of the last two. The suggestion of elm disease was made decades ago (Heybroek, 1963), but was then overshadowed by the ascendant viewpoint of those workers who saw in it the handiwork of Neolithic colonists (see Smith, 1981); it took the ravages of the 1970s to 1980s Dutch elm disease to resurrect its profile, and later some skilled hypothesis-testing by Peglar and Birks (1993) to re-establish disease as a credible causal agent.

11.5.2 Indicator Types

In interpreting pollen data, analysts will place considerably more emphasis on relatively few taxa. Particular 'indicator' types may be endowed with special meaning. For example, in Holocene material, early studies by Iversen

(1944) placed emphasis on the pollen records of *Hedera* (ivy), *Viscum* (mistletoe) and *Ilex* (holly), from which inferences were made as to the changing Holocene climate in Scandinavia. The climate response of arboreal pollen taxa was taken further by Huntley (1992a, b; but see Chambers, 1995). Aalbersberg and Litt (1998) used multiproxy measures (pollen, beetles, periglacial data) for the last interglacial in Europe – the Eemian.

In studies that attempt to chronicle the human impact on the landscape (for example, Birks *et al.*, 1988; Berglund, 1991), emphasis may be placed on pollen and spores from those taxa that are believed to indicate human presence or activity. In some continental European literature, a distinction is made between 'apophytes' (shrubs, herbs and graminids whose abundance has increased as a result of human activities) and 'anthropochores' (herbs and graminids, including cultivars, introduced by humans – see Fig. 11.5 later). Elsewhere, 'cultural' or so-called 'anthropogenic' indicators in pollen diagrams have been used by many to infer human influence (see Behre, 1986). Of particular interest is the pollen of crop plants (cultivars: see Fig. 11.1), of segetal species (plants that grow almost exclusively in cultivated ground, e.g. *Spergula arvensis* – corn spurrey, *Centaurea cyanus* – cornflower), and of 'ruderal' taxa (plants that grow in 'waste' places, or in disturbed ground). Some workers have calculated ratios between the proportions of so-called 'arable' and of 'pastoral' (grazed grassland) indicators; others try to track the introduction and subsequent naturalisation of anthropochores.

11.5.3   Spatial Representation

Holocene pollen data from dated sediment profiles can be used to plot the spread of taxa through time and space (see Davis, 1976; Huntley and Birks, 1983). In Europe, Berglund *et al.* (1996) developed the concepts of 'reference sites' and 'reference areas' to chronicle palaeoenvironmental change, while for North America the results of a wealth of Late Quaternary site-data can now be viewed on-screen using the North American Pollen Database (WDC, 2000).

The appearance of the landscape through time can be shown using summary curves (e.g. of AP/NAP ratios). These can aid interpretation, but may also be misleading: for example, if in reality the tree and shrub taxa that produce abundant well-dispersed pollen are succeeded by ones that are typically palynologically under-represented, the pollen diagram might show this by a fall in AP relative to NAP, even though the landscape has remained just as wooded or has become even more so. Sugita *et al.* (1999) show how simulation studies can help to clarify the degree of 'openness' of the landscape.

11.6   CASE STUDIES

Throughout the 1980s and 1990s, thousands of papers were published that reported the results of pollen analysis; so, in order to search literature databases (such as the Web of Science) meaningfully, students and researchers must use additional keywords. Below, published case studies illustrate the wide range of environmental applications of pollen analysis in Quaternary studies.

11.6.1 A Pleistocene Interglacial Sequence from Les Echets

Pollen analysis can be used to reconstruct environmental changes over tens of thousands of years. Here, part of a pollen diagram from Les Echets, east of the Massif Central, France (Fig. 11.4) implies the successive spread of taxa in the last interglacial – the Eemian (oxygen-isotope substage 5e) – *c.* 125–120 ka. Notable is the remarkable early expansion of elm (*Ulmus*), and the successive dominance of oak (*Quercus*, later with hazel – *Corylus*), hornbeam (*Carpinus*) and silver fir (*Abies*).

11.6.2 A Holocene Pollen Diagram from Rotsee, Central Swiss Plateau

The mountain ranges of Central Europe potentially form a barrier to plant migration. Rotsee, near Lucerne, exhibits a typical pollen sequence for a site just north of the Alps for the Early–Mid Holocene, showing successive expansion of forest trees, with the rather late expansion of silver fir (*Abies*) and beech (*Fagus*). The influence of human activity on the landscape in the latter part of the Holocene is clearly recorded (see Fig. 11.5), with the introduction of cereals (Cerealia), walnut (*Juglans*), chestnut (*Castanea*), rye (*Secale*) and hop and/or hemp (*Humulus/Cannabis* type).

11.6.3 Late Holocene Pollen Data from Easter Island

The application of pollen analysis to deposits on Easter Island provides one of the most compelling environmental applications of pollen analysis. The story is recounted for a lay audience by Bahn and Flenley (1992); the scientific results are reported in Flenley and King (1984), Dransfield *et al.* (1984) and in full in Flenley *et al.* (1991). In brief, sediment cores from crater lakes on Easter Island were analysed for their pollen content, and the sediments were dated by radiocarbon assay (Fig. 11.6) . Evidence suggested that Easter Island, which is now devoid of native trees, used to be forested up to the Late Holocene – or, at least, formerly contained significant areas of forest, including an endemic but now extinct species of palm, whose subfossil fruits have been found on the island. The data then show evidence for forest decline, perhaps from before AD 800 at one site (Rano Kau) and *c.* AD 950 at two others. By *c.* AD 950 the main episode of forest decline was largely complete at Rano Kau, and the last forest remnants were gone by *c.* AD 1400. This accords with accounts by European explorers of the late eighteenth century, who reported the island devoid of trees. Bahn and Flenley (1992) speculated

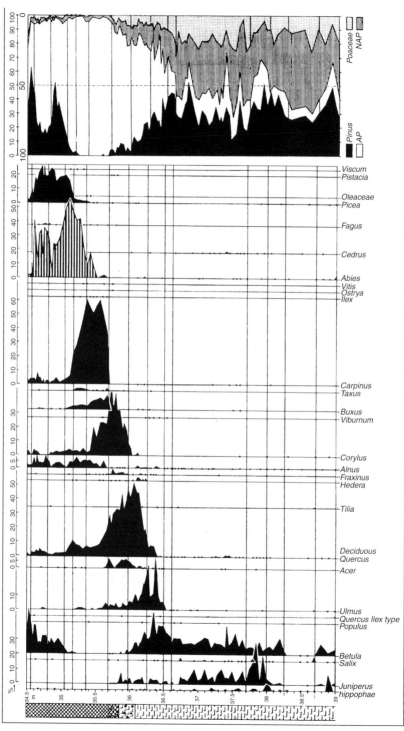

Figure 11.4 Part of a pollen profile from Les Echets, near Lyon, France, showing a remarkable immigration sequence for the last interglacial – the Eemian. (After de Beaulieu and Reille (1984).)

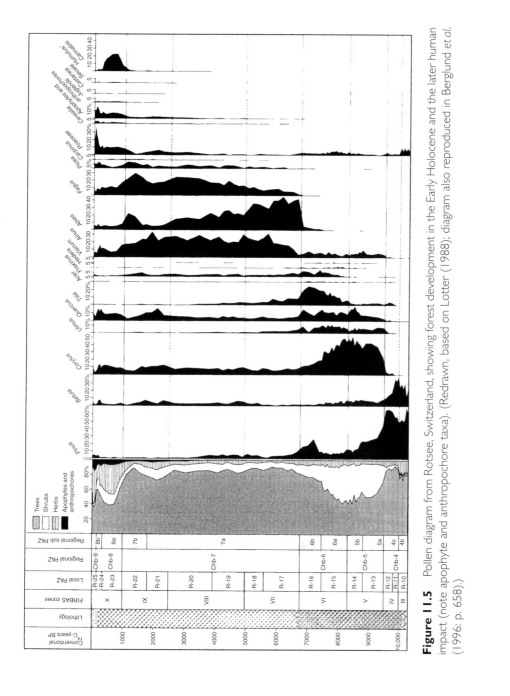

Figure 11.5 Pollen diagram from Rotsee, Switzerland, showing forest development in the Early Holocene and the later human impact (note apophyte and anthropochore taxa). (Redrawn, based on Lotter (1988); diagram also reproduced in Berglund *et al.* (1996: p. 658)).

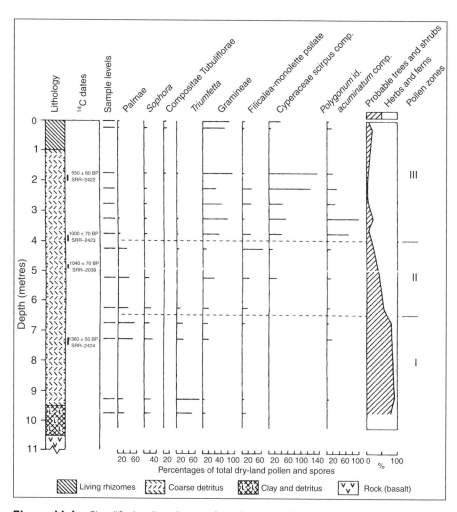

Figure 11.6 Simplified pollen diagram from Rano Kau, Easter Island, showing decline of palm (Palmae) in Zone II and rise of grasses (Gramineae). (Adapted from Bahn and Flenley (1992). See also Flenley *et al.* (1991).)

that clearance of trees was carried out by the island's Polynesian population; that forest regeneration was inhibited by the ravages of introduced Polynesian rats; and that the island's human population, which had expanded exponentially from 'a boatload of [Polynesian] settlers in the first centuries AD' to perhaps six to eight thousand, then seemed to have experienced a major resource crisis – intimately linked with the disappearance of trees – and that it crashed (well before contact with white slave-traders), to perhaps fewer than 2000 by the late eighteenth century. As an object lesson in the folly of overexploitation of natural resources, the example of Easter Island is hard to beat (Fig. 11.7).

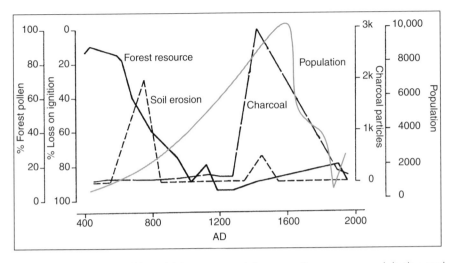

Figure 11.7 Model of Late Holocene population growth, resource exploitation and population crash on Easter Island, inferred from the Rano Kau pollen record. (After Bahn and Flenley (1992).)

This example demonstrates that even a brief pollen diagram (here containing fewer than 15 analysed horizons) can provide astonishing results.

11.7 SUMMARY

Pollen analysis has made remarkable progress since its early applications in the first half of the twentieth century. It is now applied on all continents, on many islands, and even in shallow-water marine sediments. The careful use of detailed identification keys and use of reference type-slides can produce more precise pollen identifications, with the result that pollen diagrams contain many more taxa than used to be the case; nevertheless, the accuracy of the pollen identifications depends upon the experience of the individual analyst. Pollen analysis remains the most widely applied micropalaeontological technique in the Quaternary, and is now often used in combination with other techniques in so-called multiproxy studies (of climate, of human influence, etc.). It has become increasingly more sophisticated in its applications, presentation and interpretation. Despite attempts at automation, it is still reliant on the technical skills of sampling, preparation and identification; while its value to science remains dependent on the ecological knowledge of the individual analyst.

FURTHER READING

For concise, informed introductory-level descriptions of pollen analysis, see Lowe and Walker (1997, chapter 4) or Williams *et al.* (1998, chapter 10). For more detail, see Moore *et al.* (1991). For researchers, the classic work is Faegri and Iversen (1989). Although over 20 years old, the exemplary accounts in Birks and Birks (1980: chapters 8–12) remain essential reading – for both newcomers and established researchers. Journals that frequently publish articles containing pollen diagrams include *Geoscience and Man, Grana, Journal of Quaternary Science, Palynology, Review of Palaeobotany and Palynology, The Holocene* and *Vegetation History and Archaeobotany*; a previous repository was *New Phytologist* (from 1940 to 1995).

Consult the glossary of palynology at http://www.bio.uu.nl/~palaeo/glossary/glos-int.htm revised by Hoen (1999); for other links, consult the compendium of palaeobotanical sites at http://www.unimuenster.de/GeoPalaeontologie/Palaeo/Palbot/links.html (accessed 1/12/01).

APPLICATION OF ECOLOGICALLY BASED STATISTICAL TREATMENTS TO MICROPALAEONTOLOGY

Amy L. Dale and Barrie Dale

Most of the statistical applications commonly applied in micropalaeontology today are based largely on assumed relationships between organisms and selected environmental parameters, rather than on the known living ecology of the studied group. This is because the ecology of most of the microfossil groups concerned is still poorly understood, and the objective of most applications is to produce estimates of a few selected environmental parameters needed for modelling palaeoclimate, rather than developing palaeoecology as such. Transfer functions, for example, which have been the most widely used statistical method since their introduction three decades ago (Imbrie and Kipp, 1971), were developed to produce detailed estimates of past sea-surface temperature and salinity from assemblages of planktonic foraminifera in the deep sea. The ecological relationships between the living organisms today and such environmental parameters were not understood sufficiently well to provide a direct ecological basis for the palaeo-estimates, and transfer functions were developed as a more statistically based alternative.

The transfer-function methods applied in micropalaeontology differ in fundamental ways from the statistical methods commonly utilised in ecology. Ecologists usually assume that organisms have evolved in response to a complex web of biological and environmental factors, such that it is often difficult to assess the extent to which their biogeographical distributions reflect any one given environmental parameter. Statistical treatments in ecology therefore often follow an approach that seeks to find out which environmental factors are influencing the distribution of organisms, and then to quantify their relative importance. In contrast, transfer-function methods assume that the Recent distribution of the organisms reflects a few selected environmental parameters (e.g. those of importance to palaeoclimate), and use statistical methods to produce a best-fit between these and the assemblages of organisms. It is further assumed that this best-fit relationship is sufficiently robust through time that it may be used to generate reliable estimates of past environmental parameters.

The basic assumptions used in transfer functions may be considered to be too simplistic, given the complexity of real-life ecology. In fact, the method is not reliant on the statistically established relationship between organisms and environmental parameters being ecologically sound, if the assumption that it is in any case consistent through time holds true. However, if the relationship to environmental parameters (basic to transfer functions) is not ecological (i.e. reflecting the organisms' ecological response to temperature or salinity, etc.), then other as-yet unknown factors have to be invoked to explain the relationship. It is not difficult to imagine other possible factors (e.g. sedimentary and hydrological) that could influence the composition of the Recent assemblages used, but until the nature of the relationship between organisms and environmental parameters is understood, the assumed consistency through time lacks credibility.

As more ecological information becomes available for the groups of organisms producing the microfossil record, it is possible to begin to assess the extent to which the relationships developed through transfer functions may be truly ecological (e.g. Murray, 2000, has raised such questions regarding the foraminifera). For dinoflagellate cysts, their relatively well-known present-day ecology suggests that attempts to use transfer functions so far lack ecological credibility (see section 10.4.1). To some extent, this is due to factors that possibly affect this group more than most others (e.g. the long-distance transport of many coastal species into the deep sea). The distribution of more oceanic groups such as planktonic foraminifera may therefore prove to be more ecologically viable for the transfer-function approach. Nevertheless, the authors maintain that palaeoenvironmental modelling in micropalaeontology will be improved to the extent that it can incorporate quantitative methods based as much as possible on known ecological responses characteristic for the particular group. This Appendix outlines an approach that considers the species' ecological response to the environment in which it is normally found. Although developed for work on dinoflagellate cysts (see Chapter 10), the basic considerations and statistical methods presented are applicable to a wide variety of microfossil groups. In the examples presented, dinoflagellate cysts are referred to simply as 'cysts', in the interests of brevity.

A.1 STATISTICAL METHODS

A.1.1 Choice of Method

The most suitable method for a statistical analysis will depend on a number of factors, including the inherent nature and complexity of the data, the objective at hand, and the basic ecology of the microfossil group being studied. For the analysis of a single gradient, for example, a simple X/Y plot may be sufficient, whereas more complex data may require a cluster analysis to reduce the data into its main elements.

Palaeoenvironmental modelling functions best when basic assumptions of the species' response to its environment are incorporated into the mathematical algorithm. For this, multivariate methods belonging to a category known as *ordination* are most common. Once environmental variability is established, time-series and spectral analysis may be applied to assess the extent to which the observed changes are cyclical, directional, or both, and if cyclical, on what time-scales.

This discussion mentions some of the main methods of cluster analysis, because it is a useful tool for elucidating the main trends in micropalaeontological data, but focuses mainly on the methods of ordination commonly used in palaeoenvironmental modelling. A number of these ordination methods are described, and background information is supplied to help evaluate which techniques are best suited to which types of environmental analysis. Due to space limitations, time-series analysis is not discussed here, since the main focus is on the analysis of spatial distributions, which require basic assumptions of living ecology that vary from group to group. Since most of the published literature applies the ordination methods of principal component analysis (PCA), correspondence analysis (CA), detrended correspondence analysis (DCA) or (detrended) canonical correspondence analysis ((D)CCA), these methods will be illustrated and compared using the same Recent database. As will be seen, different types of analyses produce different results with the same data-set, and therefore it may be advisable to try several different methods and use scientific judgement as to which is the best suited. Some guidelines are provided below, but ultimately, the analysts' own familiarity with the innate ecology of the specific microfossil group will be his or her best guide.

A.1.2 Options within Various Methods

Within the broader method that is ultimately chosen, there will be additional options that can influence the effectiveness of a statistical analysis, concerning, for example, data transformation, scaling of axes, and so on. Frequency distributions that are highly skewed are often best analysed after log transformation, for example, to prevent a few high values from unduly influencing the data. Data-sets containing many zero values are in general best left untransformed. When possible, it is advisable to count a fixed number of individuals (e.g. 400), such that the margin of error is comparable for all samples and as small as practically possible. Once the broader approach is decided upon, using the guidelines suggested below, a textbook or user manual should be consulted for details of the different scaling and standardising options that are specific to the chosen method (e.g. Jongman *et al.*, 1995; Ter Braak and Smilauer, 1998).

A.2 CLUSTER ANALYSIS

There is a variety of clustering methods available but only a few are commonly used in micropalaeontology. One main distinction is between divisive and agglomerative clustering methods. In divisive methods, the larger group

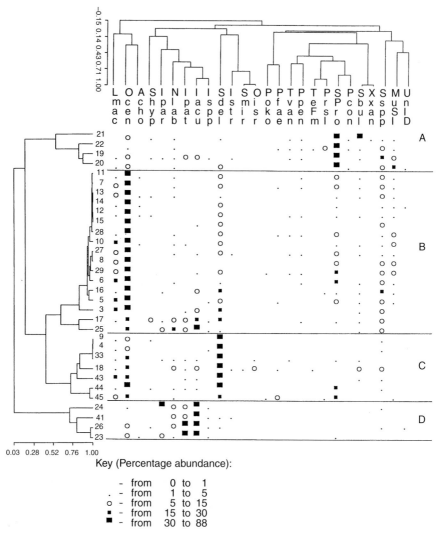

Key (Percentage abundance):

-	-	from	0	to	1
.	-	from	1	to	5
o	-	from	5	to	15
▪	-	from	15	to	30
■	-	from	30	to	88

Figure A.1 Cluster diagram of Recent cysts and samples from the Angola Basin, plotted together with a schematic representation of the raw data. Cyst abbreviations (top dendrogram) consist of the first letter of the genus name, followed by three letters of the species name, according to the names listed for the correspondence analysis in Table A.1. Locations of samples (side dendrogram) are listed in the geographical plot of CA groupings shown in Fig. A.9. The cluster analysis provides an effective data summary, useful for interpreting the ordination analyses shown in Figs A.5 to A.13. Note the clear assemblage groupings underlying the clusters A–D. The scale indicates the Pearson's product moment correlation coefficient (1.00 = strongest possible correlation).

is first divided into smaller groups, and then these are divided into subsequent branching groups, until some sort of stopping rule is satisfied. The assumption of this method of clustering is that large differences should prevail over less important, smaller differences within the data. Alternatively, agglomerative methods start with individual objects (e.g. microfossil species), which are combined into larger groups by collecting species with similar distributions into larger groups. In this case, 'local' similarity is considered more important than the larger differences. Most agglomerative methods require the construction of a matrix of similarity or dissimilarity, in which each object (e.g. microfossil species) is compared with every other. The choice of similarity or dissimilarity index (e.g. geometrical distance, correlation, or covariance) is dependent on the inherent nature of the data (e.g. continuous or presence/absence, degree of species turnover, etc.). Among the different types of agglomerative methods, there are a variety of alternatives that vary in the manner in which the similarity between clusters is calculated. Details of the available options will generally be included in the program manual.

Clustering methods may also be 'hierarchical' or 'non-hierarchical'. Hierarchical methods assume that certain differences are more important than other differences, and arrange the groups into a hierarchical system. Non-hierarchical methods do not impose a hierarchical structure on the data.

The example shown below from the Angola Basin (data used in Section 10.3.1.3) shows a type of cluster analysis that is very useful for elucidating the main trends, while at the same time giving a succinct summary of the raw data (Fig. A.1). It was carried out using the hierarchical, agglomerative UPGMA (unweighted pair-group method using arithmetic averages) method, with Pearson's product moment correlation coefficient as the similarity measure. The program MVSP (acronym for **M**ulti-**V**ariate **S**tatistical **P**ackage) was used for the cluster analysis (Kovach, 1998), and the SORTDATA subroutine within version 2.2 (Kovach, 1995) was used to schematically plot the raw data.

A.3   ORDINATION

Most approaches to palaeoenvironmental modelling, including transfer functions, are based on groupings of species or samples that are obtained through a method of ordination. 'Ordination' is a collective term for multivariate statistical techniques that incorporate species abundance information to produce graphical representations of relationships between cyst types and samples in multidimensional space. The assumption is that species distribute themselves in nature in such a way as to exploit fully the ecological factors shaping their environment, and that the distribution thus carries information about these factors (i.e. temperature, nutrient levels or other hydrographic parameters). It is also assumed that the majority of species are responding to a few main ecological gradients, such that complex sets of data may be reduced to a few main elements, which can be represented by axes in a geometrical plot. Species and site scores along the geometrical axes are calculated in such a way that the relative species' positions in the ordination plot reflect their optimal niche dispersion along the main ecological gradients.

One important way in which the various ordination methods differ is in the underlying species-response model upon which the mathematical algorithm is based: e.g. linear, unimodal or non-parametric. Some methods incorporate both linear and unimodal methods, e.g. canonical correspondence analysis (CCA), and these are especially useful for testing hypotheses when there is a suspected relationship between ecological variable(s) and species distribution (see below). Non-parametric multidimensional scaling (NMDS) focuses only on rank order abundance and ignores proportional relationships between the species. The information loss caused by converting to rank order is considered undesirable in palaeoenvironmental modelling, and thus non-parametric methods are not discussed further here.

A.3.1 Linear Methods

In methods based on a linear species-response model (e.g. factor analysis, principal-component analysis (PCA)), an incremental increase/decrease in a given environmental variable (X) results in an incremental increase/decrease in the abundance (Y) of that species. The species response is incorporated into the mathematical algorithm as a linear relationship:

$$Y = a(X) + b$$

where: Y = species abundance; a = the slope parameter of the regression coefficient of the straight line; X = the value of the environmental variable; and b = the Y-intercept (the value at which the line crosses the vertical axis) (Fig. A.2(a)).

The linear method PCA has been the primary ordination technique used in palaeoenvironmental modelling, where its main application has been in the calculation of transfer functions. (The method was first referred to as a 'Q-mode factor analysis' by Imbrie and Kipp, 1971, but was later noted actually to be a Q-mode PCA.) Briefly, it consists of the following steps:

- PCA is carried out to quantify variation among Recent assemblages in the study area and group them into 'factor assemblages'. The similarity among samples is quantified according to the similarity of the species they contain and the contribution of the species to the overall variation between samples. Species scores (also called 'factor loadings') are then assigned to each statistically coherent geographical group.
- Multiple regression is then used to correlate the species scores (factor loadings) to modern oceanographic parameters and to obtain palaeoecological transfer-function equations relating the species variation to measured ecological parameters.
- PCA is then carried out down-core to obtain factor scores for fossil assemblages, and these are related using core-top regression coefficients to palaeoecological parameters back in time.

A.3.2 Unimodal Methods

In methods based on a unimodal species response (e.g. correspondence analysis (CA), detrended correspondence analysis (DCA)), the assumption is that a species increases as its limiting resource increases, but then *decreases* as its limiting resource reaches a concentration beyond which that particular

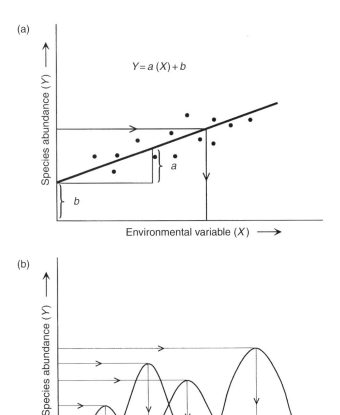

Figure A.2 Species-response curves underlying the main ordination methods: (a) linear, and (b) unimodal. In linear methods (a), $Y = a(X) + b$, where y = species abundance, a = the slope parameter, X = an environmental variable, and b = the y-intercept. In unimodal methods (b), the species-response curve is a parabola, with species abundance decreasing to either side of a maximum value of Y. In (b), u is the optimum – the value of x that gives a maximum abundance value for Y. The u is calculated as a weighted average, weighted proportionally to the species abundance: $u = (Y_1X_1 + Y_2X_2 + \cdots Y_nX_n)/(Y_1 + Y_2 + \cdots Y_n)$.

species thrives. The unimodal relationship may be represented by a parabola (Figs A.2(b) and A.3), expressed by the quadratic equation:

$$Y = a(X) + b(X^2) + c$$

where Y and X are as above, a and b are constants related to y, and c is the Y-intercept. However, the important feature of the unimodal curve for ordination is that it allows the estimation of an indicator value (termed 'optimum' (u) in Figs A.2(b) and A.3(a)), and the ecological tolerance (t in Fig. A.3(a)) of the species. The indicator value (u) is obtained by taking the weighted average of the values for x over the sites where the species is present:

$$u = (Y_1X_1 + Y_2X_2 + \cdots Y_nX_n)/ (Y_1 + Y_2 + \cdots Y_n),$$

where $Y_1, Y_2, \ldots Y_n$ are the abundances of the species, and $X_1, X_2, \ldots X_n$ the values of the environmental variables at sites $1, 2 \ldots n$.

A.3.2.1 Unimodal Methods and the Normal Probability Distribution

The interpretation of the ordination resulting from unimodal methods requires a small digression into some basic principles of population statistics. The objective of statistics in general is to make inferences about a population based on information contained in a sample or set of samples. As background for interpreting unimodal models, a brief description of two basic tenets of population statistics is provided: (1) the 'Empirical Rule', and (2) the 'Central Limit Theorem', which relates the 'Empirical Rule' to probability theory. Familiarity with these is necessary because the curve upon which unimodal methods are based – the parabola – approximates the familiar bell-shaped distribution, also called 'normal' or 'Gaussian' distribution, upon which population statistics is based (compare (a) and (b) in Fig. A.3), allowing the application of these theorems in the interpretation of unimodally-based ordination models.

The Empirical Rule is derived from repeated sampling and states that: given a distribution of measurements that is approximately bell-shaped (Fig. A.3(b)), the interval:

1. The mean +1 standard deviation (=s.d.) will contain approximately 68 per cent of the measurements.
2. The mean +2 s.d. will contain approximately 95 per cent of the measurements.
3. The mean +3 s.d. will contain all or almost all of the measurements.

The Central Limit Theorem, which is basic to probability theory, states that: 'The sums and means of samples of random measurements drawn from a population tend to possess, approximately, a bell-shaped distribution in repeated sampling.' This means, in other words, that repeated sampling from a population will result in a great majority of measurements falling near the sample mean (=average), with relatively few being extremely high or low. Thus, since repeated sampling will tend to produce a distribution

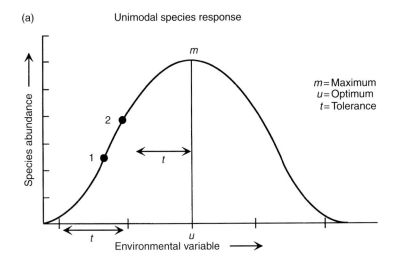

(a)   Unimodal species response

(b)   Normal probability distribution

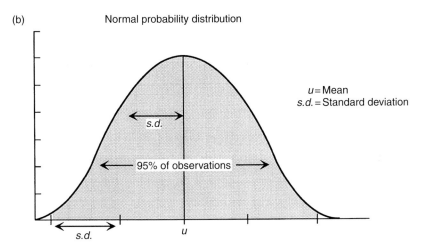

Figure A.3 Comparison of the unimodal species-response curve (a) and the normal probability distribution (b) upon which population statistics are based. In (a), m is the species' maximum abundance; u is the optimum value of X that gives maximum abundance for Y, and t is the tolerance (a measure of ecological amplitude). According to the unimodal species-response model, using Hill's scaling, a species is considered to appear, rise to its maximum, and disappear within $4(t)$ of its optimum (u). From 1 to 2 shows a linear species response – typical for a shorter ecological gradient. (b) illustrates the 'Empirical Rule' of the normal probability distribution, which states that $u \pm 1$ standard deviation (=s.d.) will contain approximately 68 per cent of the measurements; $u \pm 2$ s.d. will contain approximately 95 per cent of the measurements; and $u \pm 3$ s.d. will contain all or almost all of the measurements. The similarity between these two curves allows the application of probability theory in the interpretation of ordination models based on a unimodal species response.

curve that is approximately bell shaped, the Central Limit Theorem and the Empirical Rule can be applied in interpreting unimodal methods. Since the algorithms of unimodal models assign ordination scores on the basis of weighted averages (=means) of the species occurrences, 95 per cent or more of the individuals can be expected to occur within two standard deviations to either side of that mean (see Fig. A.3).

In order for the term 'standard deviation' to have meaning in the ecological as well as the statistical sense, however, a method of scaling the species' response on the ordination axes was needed that would express the species range in equivalents of standard deviations. This method of rescaling was introduced with the algorithm of detrended correspondence analysis (DCA, see below), and has since become a scaling option also within CA, where it is called 'Hill's scaling', after its inventor. The differences between CA and DCA are discussed below, but first it is necessary to give some general background information about the main elements of ordination which are common to all these methods.

A.3.2.2 Important Traits of the Method: PCA, CA and DCA

1. The initial input for these ordination methods is a data table (called a 'matrix') of species by site occurrences. The sites are plotted ('ordinated') along geometrical axes on the basis of their species composition, such that species/sites that are similar to each other plot near each other whereas species/sites that are dissimilar plot far apart on the axes.
2. Each species and site is associated with a score (called 'factor loading' or 'principal component loading' in PCA) for each of the main axes derived from the ordination, and this is a quantitative representation of the preferred ecological optimum of the species along the environmental gradient represented by the axes (e.g. Table A.1)
3. The species and sites are dispersed along axes in such a way that the first axis always represents the gradient of greatest species variation; the second axis the greatest source of variation remaining after the first axis is accounted for; the third axis the majority of variation remaining after the previous axes are accounted for, and so on, until all of the species variation is accounted for.
4. Each axis is associated with a coefficient reflecting the degree of species dispersion along that axis, which is its 'eigenvalue'. The eigenvalue thus gives a measure of the relative importance of the various axes in explaining the overall species variation. If the dispersion (reflected by the eigenvalue) is large, then a given variable (reflected by the axis) separates the species well, but if the dispersion is small, it is considered ineffective (unrelated to the species variance). The first axis will always have the highest eigenvalue, the second axis the next highest, and so on (except in a hybrid method or a (D)CCA when the number of environmental variables is small; see below).

Table A.1 Correspondence analysis of surface-sediment samples from the Angola Basin: eigenvalues and species scores for the first three axes. Axes 1–2 are plotted in Fig. A.7. Cyst type codes representing end-member species on Axes 1 to 3 (which had eigenvalues higher than 3.0) and/or having statistical weights in the highest quartile are marked in bold, as are their corresponding weights and scores for the axes on which they score highly. This is to emphasise their relatively stronger influence on the ordination results

Cyst name	Code (eigenvalue):	Axis 1 0.61	Axis 2 0.43	Axis 3 0.34	Stat.wt.
Impagidinium aculeatum	I. acu	**3.78**	0.03	−0.05	**291**
Impagidinium paradoxum	I. par	**3.34**	0.12	−0.13	53
Impagidinium patulum	I. pat	**3.83**	0.24	−0.12	**136**
Impagidinium strialatum	I. str	1.91	−0.14	0.94	9
Impagidinium spp. indeterminate	I. spp	**3.55**	−0.03	0.10	15
Lingulodinium machaerophorum	**L. mac**	−0.97	−0.52	−0.66	**186**
Multispinula spp indeterminate	MuSI	−0.96	1.40	−0.35	88
Nematosphaeropsis labyrinthus	N. lab	2.06	0.37	0.12	77
Operculodinium centrocarpum	**O. cen**	−0.83	−0.24	**−1.21**	**1191**
Operculodinium israelianum	O. isr	0.30	−1.43	2.51	6
Peridinium faeroense	P. fae	−1.02	0.85	1.16	13
Ataxiodinium choanum	A. cho	0.64	−0.17	−1.72	7
Polykrikos kofoidii	Po. ko	−1.13	0.47	0.20	2
Protoperidinium conicum	P. con	−1.23	4.53	2.86	6
Protoperidinium pentagonum	P. pen	−0.87	1.93	0.28	13
Protoperidinium spp. indeterminate	Pro. SI	−1.23	5.21	3.27	15
Spherical brown protoperidinioid spp.	**SBPro**	−0.93	**2.48**	1.11	**320**
Spiniferites bulloideus	S. bul	−0.93	2.75	2.01	73
Spiniferites delicatus	**S. del**	−0.67	**−1.99**	1.96	**519**
Spiniferites hyperacanthus	S. hyp	0.83	−0.22	0.74	19
Spiniferites mirabilis	S. mir	0.47	−1.09	2.26	1
Spiniferites spp. indeterminate	**Sp. SI**	−0.39	0.37	−0.20	**178**
Tectatodinium spp. indeterminate	Tec. frn	−0.70	3.66	3.23	3
Tuberculodinium vancampoae	T. van	−0.91	1.44	1.22	10
Xandarodinium xanthum	X. xan	−1.32	5.00	3.38	1

5. The axes are constructed in such a way that the variation represented on a given axis is uncorrelated with the variation explained on any other axis.

6. CA and DCA are methods of 'reciprocal averaging', in which the species and site scores are calculated simultaneously. Less directly but somewhat true also for PCA, the placement of one category (species or sites) can be directly interpreted from the other, and vice versa.

7. The following main terms are therefore essential for interpreting ordination diagrams and tables:

- The *eigenvalues* give a measure of reliability of the respective ordination axes, and of their relative contributions in explaining the overall species variance.
- The species and site *scores* ('loadings' in PCA) provide a measure of the ecological optimum along the ecological gradient represented by the axis.

- Because CA and DCA are based on weighted averages, the *statistical weight* of a species is an additional criterion important for assessing its significance in unimodally based ordination methods. Total percentages presumably equal 100 for sites; however, statistical weights commonly are relevant only for evaluating relative species' contributions. A species' weight is a rough measure of its overall influence on the resulting plot. Species with both high weights and high scores (on axes having high eigenvalues), for example, will be indicator species: they are influential in both defining the main directions of variation (=axes, representing ecological gradients), and for representing end-member positions along the important gradients. A species with a high weight and a low score on all the important axes is likely to be a cosmopolitan species, whereas a species with both a low weight and a low score is considered to be on the outer fringe of its ecological tolerance, or a random occurrence (= 'statistical noise').

8. PCA and (D)CA differ in the ways they assign mathematical weights to the species. In PCA, the sample loadings are a *sum* of a set of species loadings ('scores' in (D)CA) for those species that occur in the samples. In (D)CA, an *average* of a set of species scores is used. This means that, in effect, PCA emphasises the dominant species and typically gives rarer species very small loadings (scores in (D)CA), whereas in a CA even rare species may have high scores. Thus, the main gradients in a PCA reflect the proportions of the most dominant species, whereas relative distribution plays a larger role in a (D)CA.

A.3.2.3   CA vs. DCA

DCA is a variant of CA that was introduced to correct two perceived 'faults' with CA. The first of these is that, despite the mathematical requirement that the second axis be uncorrelated to the first (see above), it is often related to it, producing an arched pattern in the ordination plot known as the 'horse-shoe effect'. To correct this, a procedure known as 'detrending' was devised, in which the first axis was divided into a number of segments and within each segment the site scores on the second axis were adjusted by subtracting their means, so that the axes were made to be independent from each other. Subsequent axes were derived by similarly 'detrending' with respect to each of the existing axes.

The second perceived fault with CA was that the distance at the ends of the ordination axis tended to be compressed with respect to the middle, so that the ordinational space did not have consistent meaning. This was corrected by rescaling the axes and standardising the site and species scores such that the within-site variance was equal to 1. The tolerances of the site and species curves would both therefore approach 1, and the distances on the respective axes would have consistent meaning.

As mentioned above, this method of rescaling ('Hill's scaling') is now an option also within CA, and it gives the ordination plot a definite ecological meaning. By rescaling such that the within-site variance is equal to 1, the lengths of the ordination axes reflect the range of the site scores, and this length can be expressed in multiples of the standard deviation (s.d.). The sample scores are thus interpretable in standard-deviation units of species turnover, and the spacing of the site scores represents distances of species turnover. According to the Gaussian response curve (Fig. A.3(a) and discussed above), a species with a tolerance of 1 can be interpreted to appear, rise to its optimum and disappear again over an interval of about 4 s.d. Species which appear more than 4 s.d. from each other on an ordination axis therefore do not appear in the same samples, and samples which appear more than 4 s.d. from each other do not contain species in common.

In nature, it is observed that a unimodal species response is typically observed in cases where longer ecological gradients are sampled, whereas a linear species response is more typical in cases where shorter ecological gradients are sampled (e.g. from 1 to 2 in Fig. A.3(a)). Thus, a DCA or a CA using Hill's scaling may be a useful first step for determining whether a linear or unimodal method would be best suited to a particular data-set: if the gradient represented by the first axis is short (e.g <3 standard deviations), a linear method would be best suited, whereas if it is long (e.g. >4 standard deviations), a unimodal method would be indicated.

A.4   CANONICAL CORRESPONDENCE ANALYSIS

The last category of ordination method to be discussed before examples are presented is a variant of CA known as canonical correspondence analysis, or CCA. In CCA, the species distributions are *constrained* to be linear combinations of pre-assigned environmental variables, so that the assemblage variation is related directly to measured environmental variation. Detrended canonical correspondence analysis (DCCA) may also be carried out, and alternative constrained ordination methods exist for exploring assemblage–environment relationships directly (e.g. redundancy analysis, Gaussian canonical ordination) and for removing the effects of particular environmental influences or of covariance (e.g. partial constrained ordination). However, a full treatment of all these methods of constrained ordination is beyond the scope of this chapter, so, for the sake of simplicity, discussion will be limited to CCA.

A CCA can be used to assess the strength of the influence of the measured ecological variables for explaining the observed species variance (reflected by the eigenvalues and the species–environment correlations). The statistical significance of the relationship may be evaluated by a Monte Carlo permutation test included in the CCA analysis, which supplies probabilities (p-values) that the observed variation is attributable to chance. Ideally, p-values should be 0.05 or lower – corresponding to a 5 per cent or lower probability that the relationship is attributable to chance.

Additional criteria for assessing the strength and extent ('power') of the relationship include: the *angle* of the environmental axes (arrows) with the species axes, indicating the degree of correlation between the environmental and species axes, and the *relative length* of the arrows, indicating their explanatory power (e.g. does the observed relationship apply to all, or only part, of the total data-set?).

A final assessment of the contribution of environmental variables should be made by comparing the CA with the CCA ordinations. The unconstrained analysis will always convey the species variation maximally (e.g. a CCA will never be associated with higher eigenvalues than a CA of the same data). This is because the species distributions *themselves* carry more information about the ecology than any external forcing of the data can impose on them. If the ordinations resulting from a CA and a CCA are similar, then the measured factors expressed in the CCA can be considered likely to be influencing the data. If the unconstrained CA and constrained CCA are *not* similar (even if the statistical results of the CCA suggest a strong relationship), then it is likely that additional unmeasured factors are also playing an important role or that several factors are interacting to affect the species variation.

If significant and strong relationships are indicated by a CCA, and end-member species or samples can be associated with specific values of environmental parameters, the axes may be calibrated accordingly and incremental values inferred for the remaining species and/or sites. This gives the potential for direct values for palaeoceanographical conditions to be obtained from a CCA for inclusion in quantitative palaeoenvironmental models.

A.5 Examples of Various Methods

The theoretical considerations discussed above are best illustrated by comparing the different methods, here applied to the same data-set of Recent dinoflagellate cysts from the Angola Basin. The hydrographic setting of the study region includes a number of different ecological subregions, as discussed in detail by Dale *et al.* (in press) and shown in Fig. A.4, including: a major tropical river plume (inner plume stippled; outer plume demarcated by solid line), an area of coastal upwelling (indicated), a major oceanographic front (wavy line labelled ABF), and samples ranging from nearshore to offshore. Further details of the hydrography can be found in section 10.3.1.3. This comparison focuses on the choice of ordination method employed, since any subsequent environmental modelling – e.g. calculating surface-water parameters if the transfer-function approach is applied, or

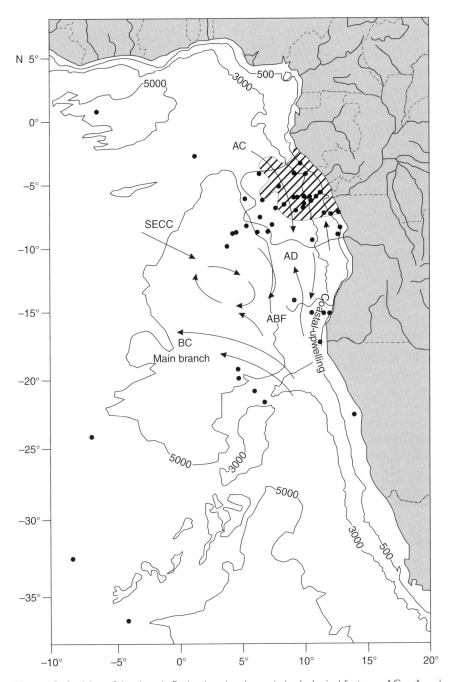

Figure A.4 Map of the Angola Basin, showing the main hydrological features. AC = Angola Current, SECC = Southern Equatorial Counter Current, and BC = Benguela Current. Wavy line denoted by 'ABF' marks the Angola–Benguela Front. The hatched area out from the Congo River basin shows the area of the inner river plume, and the white outer area shows the extent of the deep-sea fan and what is here referred to as the outer river plume.

extrapolating the identified main gradient values if a (D)CCA method is applied – can only be as reliable as the results of the original ordination upon which the modelling is based.

As mentioned above, the cluster analysis was carried out using MVSP (acronym for **M**ultivariate **S**tatistical **P**ackage). The principal component analysis, correspondence analysis, detrended correspondence analysis and canonical correspondence analysis were carried out using the program CANOCO™ (**CANO**nical **C**ommunity **O**rdination; Ter Braak, 1990; Ter Braak and Smilauer, 1998). In the ordination plots shown for unimodal methods, the species with statistical weights in the highest quartile are plotted in bold, in order to emphasise their relatively greater influence on the ordination results. No data transformations were carried out in the examples shown here.

A.5.1   Cluster Analysis

First, a cluster analysis was carried out, in which the sample and species cluster dendrograms were arranged around a schematic representation of the percentage abundances (Fig. A.1). This serves as a simplified summary of the main assemblages at each site, and provides a useful reference for interpreting subsequent analyses. The cluster analysis (Fig. A.1) revealed four main groupings based on dominance by (A) spherical brown protoperidinioid cysts; (B) *Operculodinium centrocarpum*; (C) *Spiniferites delicatus*; and (D) *Impagidinium* species.

A.5.2   DCA

Next, a DCA using Hill's scaling was carried out as a preliminary step in choosing the best-suited ordination method. Of the various detrending options available (see below), the method 'detrending by segments' was used in the example shown here, in order to be most comparable with the DCAs used in published cyst studies. It was determined that the primary gradient was greater than 4 s.d. (Figs A.5 and A.6), suggesting that unimodal methods would be better suited for these data than linear methods.

Axis 1 suggested a nearshore–offshore gradient (note the occurrence of nearshore cyst type *Lingulodinium machaerophorum* (L. mac) as a negative end-member, and offshore *Impagidinium* cyst types (e.g. *I. acu* and *I. pat*) as positive end-members in Fig. A.5), but the sample ordination showed too little dispersion to be interpretable (Fig. A.6). Options use alternative methods of detrending: detrending by 2nd-, 3rd- and 4th-order polynomials showed improved dispersion, and of the various detrending options, detrending by 2nd-order polynomials provided the closest correspondence to the known hydrography of the region. However, none matched the ecological conditions known in the area today as well as did the CA, so additional examples of DCA are not shown.

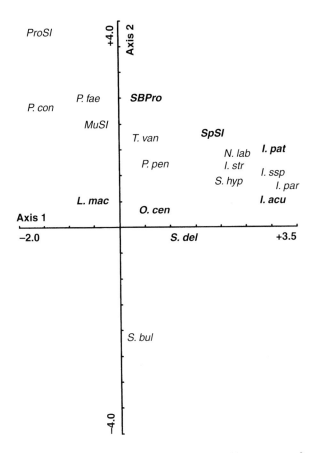

Figure A.5 Detrended correspondence analysis (DCA) of Recent cysts from the Angola Basin: species ordination. Detrending is by segments in order to be most comparable to the method most commonly used in published cyst studies applying DCA. A nearshore–offshore gradient is suggested for Axis 1 by the separation of nearshore species (e.g. *L. mac, SBPro, O. cen*) and offshore species (e.g. *I. pat, I. acu, I. par* and *N. lab*), but the hydrographic features known to be present in the region (Fig. A.4) are not clearly reflected by the species ordination. Cyst types with weights in the highest quartile are plotted larger in bold in order to emphasise their relatively greater influence on the ordination results, and the cyst types in the lowest quartile are omitted but included in Table A.1. Species abbreviations consist of the first letter of the genus name followed by three letters of the species name, according to the names listed for the correspondence analysis in Table A.1.

A.5.3 CA

Figs A.7 and A.8 show the species and sample ordinations, respectively, which were obtained using CA. Axes 1 to 3 were associated with high eigenvalues (0.61, 0.43 and 0.34, respectively; Tables A.1 and A.2). Five groupings could be identified from the ordination plots (A–E, Fig. A.8). These are shown geographically in Fig. A.9 and are interpreted to represent the five ecological subregions known to exist in the region:

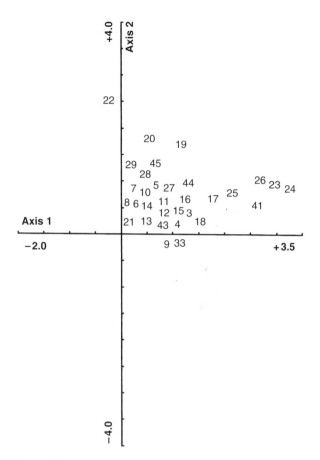

Figure A.6 Detrended correspondence analysis (DCA) of Recent samples from the Angola Basin: site ordination, using detrending by segments in order to be the most comparable to the method most commonly used in published cyst studies applying DCA. Sample locations are listed in the geographical plot of CA grouping shown in Fig. A.9. Note the crowding of nearly all the samples into the upper right quadrant.

(A) a zone of coastal upwelling along the Angola–Namibia Coast, characterised by dominance by spherical brown protoperidinioids (*SBPro*);

(B) the inner plume from the Congo River Delta, characterised by dominance by *O. centrocarpum* (*O. cen*);

(C) the outer river plume, characterised by dominance by *S. delicatus* (*S. del*);

(D) the Angola–Benguela Front, characterised mainly by dominance by *O. centrocarpum*, but with accessory high proportions of either *S. delicatus* or *Nematosphaeropsis labyrinthus* (*N. lab*); and

(E) the oceanic realm offshore, characterised by dominance by *Impagidinium* species (e.g. *I. acu* and *I. pat*) (see Figs A.7 to A.9 and discussed in section 10.3.1).

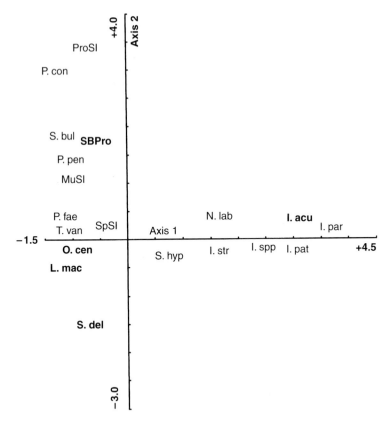

Figure A.7   Correspondence analysis (CA) of recent cysts from the Angola Basin, showing the main cyst types corresponding to the sample groupings shown in Fig. A.8 and plotted geographically in Fig. A.9. Cyst types with weights in the highest quartile are plotted larger in bold in order to emphasise their relatively greater influence on the ordination results, and cyst types in the lowest quartile are omitted but included in Table A.1. Cyst abbreviations consist of the first letter of the genus name followed by three letters of the species name, according to the names listed in Table A.1.

Axis 1 clearly suggested that the strongest ecological gradient affecting species distribution in the study area was relationship to shore (i.e. a coastal–oceanic gradient). On Axis 2, the remaining variation to be accounted for involved a gradient represented by Group A (upwelling) plotting most positively and Group C (outer Congo River plume) most negatively, suggesting a nutrient gradient.

Axis 3 suggested a gradient involving a transition from unstable, more nutrient-rich waters in the inner river plume to more stable, probably somewhat less but still nutrient-enriched waters in the outer river plume. The cosmopolitan species *O. centrocarpum* dominates in the most unstable waters nearest the river mouth, but *S. delicatus* outcompetes *O. centrocarpum* in the more stable waters of the outer river plume.

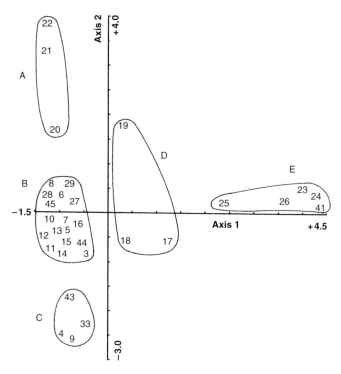

Figure A.8 Correspondence analysis (CA) of Recent cysts from the Angola Basin: site ordination, showing the groupings that are plotted geographically in Fig. A.9. Group A samples reflect coastal upwelling, Group B the inner river plume, Group C the outer river plume, Group D the Angola–Benguela Front, and Group E oceanic conditions.

A.5.4 PCA

Even though the gradient expressed by the primary axis was sufficiently long (e.g. >4 s.d.) to suggest that linear methods would be unsuitable, a PCA ordination was included in this comparison to provide an illustration of the ordination method upon which transfer functions are based. Figs A.10 and A.11 show a PCA of the Angola Basin Recent cyst data-set. The options available in PCA involve whether or how the data are transformed as a precursor to carrying out the PCA (e.g. uncentred/centred, standardised/ unstandardised, and which methods of rescaling are used if a standardised PCA is performed). The options in the PCA presented here are chosen so as to be most comparable to that performed in most published cyst-based transfer-function studies, and are as follows:

- The scaling was focused on the inter-species correlations
- The species data were divided by their standard deviations in order to reduce overdominance by the most abundant species
- The data were centred by species so that the PCA would be carried out on a matrix of covariance between species.

Table A.2 Correspondence analysis of surface-sediment samples from the Angola Basin: eigenvalues and samples scores for the first three axes. Axes 1 and 2 are plotted in Fig. A.8 and shown geographically in Fig. A.9

Station (eigenvalue):	Axis 1 0.61	Axis 2 0.43	Axis 3 0.34
3	−0.40	−0.54	−0.15
4	−0.65	−1.61	1.35
5	−0.74	−0.16	−0.42
6	−0.80	0.34	−0.44
7	−0.77	−0.09	−0.63
8	−0.85	0.28	−0.68
9	−0.65	−1.67	1.47
10	−0.77	−0.06	−0.76
11	−0.72	−0.22	−0.52
12	−0.79	−0.17	−0.73
13	−0.75	−0.15	−0.85
14	−0.78	− 0.26	−0.93
15	−0.75	−0.33	−0.93
16	−0.39	−0.14	−0.23
17	1.02	−0.26	0.05
18	0.31	−0.46	0.76
19	0.29	1.14	0.25
20	−0.71	1.08	0.50
21	−0.86	2.12	1.14
22	−0.84	2.55	1.25
23	3.37	0.17	−0.13
24	3.56	0.11	−0.12
25	2.51	0.03	−0.34
26	2.96	0.07	−0.06
27	−0.68	0.07	−0.55
28	−0.78	0.11	−0.64
29	−0.77	0.37	−0.37
33	−0.35	−1.33	1.26
41	3.48	0.03	0.04
43	−0.56	−1.02	0.41
44	−0.50	−0.38	0.54
45	−0.71	0.03	0.50

From the species ordination, it can be seen that the nearshore/offshore gradient so prominently displayed by Axis 1 of the CA is not clearly apparent in the PCA. The PCA identified *O. centrocarpum* and *L. machaerophorum* as end-members plotting negatively on Axis 1, with *S. delicatus* and offshore species of *Impagidinium* and *N. labyrinthus* plotting positively. Furthermore, the heterotroph-dominated upwelling assemblage (dominated by spherical brown protoperidinioid cysts, *Pro. sp*) occurs in the same quadrant as the offshore assemblage, despite the fact that these two groups are associated with markedly contrasting environments. No obvious gradient was suggested by Axis 2 of the PCA.

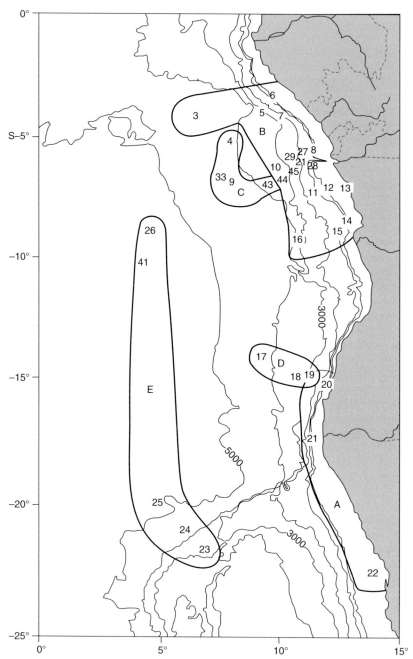

Figure A.9 Map of the Angola Basin, showing sample groupings (A–E) from correspondence analysis of Recent samples shown in Fig. A.8. Note the close similarity to the
hydrographic features shown in Fig. A.4. Group A samples reflect coastal upwelling, Group
B the inner river plume, Group C the outer river plume, Group D the Angola–Benguela
Front, and Group E oceanic conditions.

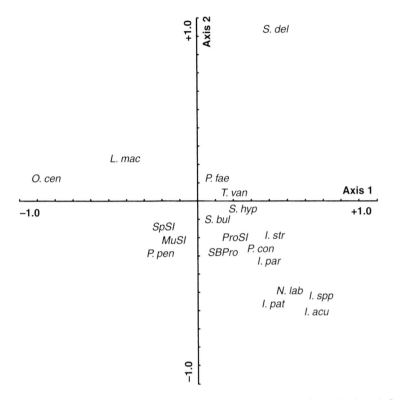

Figure A.10 Principal component analysis (PCA) of Recent cysts from the Angola Basin: species ordination. A nearshore–offshore gradient is suggested for Axis 1 by the separation of nearshore species (e.g. *L. mac*, *O. cen*) and offshore species (e.g. *I. pat*, *I. acu* and *N. lab*). However, the coastal upwelling assemblage (dominated by *SBPro*) is an exception to this, and there is no clear gradient reflected by Axis 2 (discussed in the text). Cyst types with a percentage abundance in the lowest quartile are omitted. Cyst abbreviations consist of the first letter of the genus name followed by three letters of the species name, according to the names listed for the correspondence analysis in Table A.1.

The site groupings are also less informative in the PCA than in the CA: whereas the CA and DCA using detrending by 2nd-order polynomials identified five main ecologically significant groupings (A–E in Figs A.8 and A.9), the PCA identified four groups (F–I in Fig. A.11), and these contained a mixture of sites from the different geographical subregions in the study area. Samples from the upwelling region were grouped together with samples from offshore (labelled Groups A and E in the CA of Fig. A.8, and Group I in Fig. A.11), for example, and the positions of the groupings in relation to each other is not consistent with the known hydrography of the region. Although the species and sample ordination plots are inevitably subjected to some degree of distortion through the collapsing of many dimensions into two, it is clear that the PCA groupings are more reflective of the absolute abundances

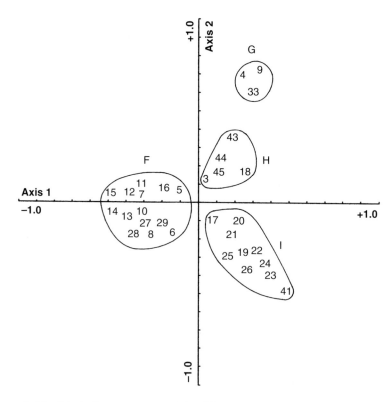

Figure A.11 Principal component analysis of Recent samples from the Angola Basin: site ordination. Groupings F, G, H and I are for comparison with correspondence analysis groupings A–E in Fig. A.8, which are plotted on the map shown in Fig. A.9. Sample locations are listed in the geographical plot of the CA grouping shown in Fig. A.9. The PCA sample groupings do not correspond with the known hydrography as closely as those produced by the CA (compare with Figs A.8 and A.9) or DCA using detrending by polynomials (discussed in text).

of the dominant species (indicated by the statistical weights for the CA, see Table A.1), than of their ecologically related variance.

A.5.5 CCA Test of Coastal–Oceanic Gradient for Axis I

Once the environmental gradients suspected of influencing the species variation are identified, a canonical CA (CCA) may be a useful method for evaluating the suspected relationship. Since both the species and site ordinations for the Angola Basin study suggested that the first ordination axis reflected a coastal–oceanic gradient, this was tested by constraining the data to reflect sample depth (Fig. A.12). The CCA revealed a strong correlation ($r = 0.81$) between depth and the primary ordination axis. Furthermore, it can be seen that all of the higher weight species, with the exception of one (*S. delicatus*),

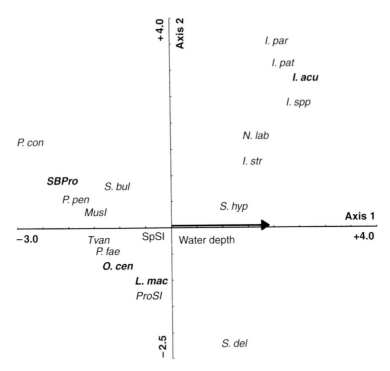

Figure A.12   Cyst-type ordination of canonical correspondence analysis (CCA) of Recent cysts from the Angola Basin, in which the data are constrained to reflect the variable *water depth*. Cyst types are classified as heterotrophic or autotrophic in order to elucidate the possible relationship between the trophic preference and the species dispersion along the axes. Note that all the end-member cyst types plot similarly when constrained to reflect water depth, as in the unconstrained CA, with the exception of S. *delicatus* which plots in the positive domain of Axis 1 in the CCA but negatively in the CA (see Fig. A.7). The CCA supports the interpretations of Axis 1 and Axis 3 discussed in the text. Cyst types with statistical weights in the highest quartile are plotted larger in bold in order to emphasise their relatively greater influence on the ordination results. Cyst abbreviations consist of the first letter of the genus name followed by three letters of the species name, according to the names listed in Table A.1.

plot in the same quadrant in the CCA as in the CA. The placement of all of the *Impagidinium* species and *N. labyrinthus*, for example, which are known oceanic indicator species, far to the positive extreme also on the CCA, strongly supports a gradient reflecting the relationship to shore.

S. *delicatus* plotted negatively on the CA (Fig. A.7), but positively on the CCA (Fig. A.12), indicating that this species showed higher abundance in the further offshore samples in the database, despite its pattern of abundance causing it to be grouped on the 'nearer shore' side of Axis 1 in the CA. This is consistent with an interpretation that the extensive penetration of the massive freshwater plume from the Congo River effectively displaces the coastal/oceanic boundary far out into the ocean in this region, so that, although S. *delicatus* is generally associated with nearshore cyst types, it dominates in samples from far offshore in the study region, as shown by the CCA (Fig. A.12). Thus, the CCA corroborates the interpretation of a depth gradient for Axis 1 of the CA, and supports the interpretation of a displaced neritic/oceanic boundary as an element of the 'stability' gradient represented by Axis 3.

A.5.6 Classified CA to Test Nutrient Gradient for Axis 2

The suggestion by the placement of cyst types and samples along Axis 2 of the CA, that the second-most important source of variance was related to nutrient regime, was investigated by classifying total proportions of autotrophic versus heterotrophic abundance at each site (Fig. A.13). It can be

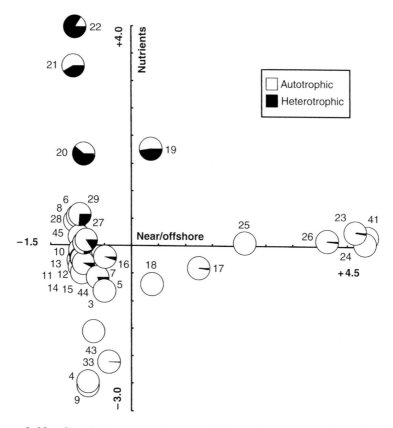

Figure A.13 Classified correspondence analysis of Recent samples from the Angola Basin, in which sites are shown as pie diagrams indicating the total proportions of heterotrophic and autotrophic cyst types. The clustering of sites with high proportions of heterotrophic species in the upper left quadrant supports the interpretation of a nutrient gradient for Axis 2 of the CA (discussed in the text).

seen that the stations with the highest dominance by heterotrophs plot negatively on Axis 1 and positively on Axis 2, thus supporting the influence of the nutrient regime on the cyst distribution in this area.

The combination of methods shown here laid a strong foundation for subsequent interpretation of climatic change back in time in this area, based on the results of the Recent cyst distribution in the region. The down-core study is summarised in section 10.3.1.

A.5.7   Summary of Examples

Cluster analysis is a useful method for displaying the main trends in complex data, but as a non-parametric technique (that is, not based on any assumption of species response) it is of limited use in palaeoenvironmental modelling. However, particularly when the cluster dendrograms are arranged around a schematic representation of the raw data as shown in Fig. A.1, cluster analysis provides a useful tool for helping to interpret all subsequent analyses.

Of the ordination methods DCA, CA and PCA used here to analyse Recent cysts from the Angola Basin and in comparisons from various regions of our Recent database (Dale and Dale, in press), CA consistently provided results which most closely expressed the ecological parameters known to be influencing the region. PCA and DCA tended to show distortion and/or a lower degree of species and site dispersion. In comparing, for example, the groupings from the DCA shown in Fig. A.6, the CA shown in Fig. A.8, and the PCA shown in Fig. A.11, with the hydrography of the region shown in Fig. A.4, it can be seen that the CA ordination (which is plotted geographically in Fig. A.9) clearly shows the best match.

It was expected from the lengths of the gradient on the main DCA axes that unimodal methods would perform better on this data-set than linear methods (e.g. PCA), but the reason for the closer match with known ecology by the CA over the DCA is less clear. It may be that distortion results from the mathematical correction inherent in the detrending algorithm – forcing independence of the axes, since often in nature the main gradients are in fact related (e.g. sea-surface temperature and salinity).

There is a great need to apply the various methods described here to new data-sets, particularly of Recent microfossil groups where the environmental parameters are known. Whereas the examples here illustrated the suitability of unimodal methods for analysing dinoflagellate cysts, alternative methods may have proved more useful in the analysis of other microfossil groups. Linear methods such as PCA have proved useful for analysing planktonic foraminiferal distributions, for example, since these typically consist of relatively fewer species inhabiting the more stable realm of the deep sea. Each environmental situation and data-set should be evaluated individually, and it is recommended that several different methods be attempted in order to compare their effectiveness in uncovering the underlying data structure. The main criterion for effective data analysis is that the chosen method accommodates as much as possible the real ecology of the group, and that the known ecology of living species provides the ultimate reference for evaluating the method.

Further Reading

Imbrie, J. and Kipp, N.G., 1971. A new micropaleontological method for quantitative paleoclimatology: application to a Late Pleistocene Caribbean core. In K. Turekian (ed.), *The Late Cenozoic glacial ages*, Yale University Press, New Haven, USA, pp. 71–181.
This paper is a pioneer study in the use of transfer functions to study climate change. Although some aspects of the transfer-function method have been modified and others called into question since this classical work was published, the underlying assumptions are presented clearly and the quantitative approach reflects a comprehensive understanding of palaeoceanography, which makes it valuable for all students of micropalaeontology. The study also includes an example of time-series analysis.

Jongman, R.H.G., Ter Braak, C.J.F. and van Tongeren, O.F.R. (eds), 1995. *Data analysis in community and landscape ecology*, reprinted edition, Cambridge University Press, Cambridge.
This book provides a useful synthesis of methods for the analysis of environmental data. It contains chapters on data collection, regression, calibration, and cluster analysis, with a comprehensive chapter on ordination that includes sections on correspondence analysis, detrended correspondence analysis, canonical correspondence analysis, principal component analysis and other methods commonly encountered in microfossil research. Explanations of the mathematics are clear and simple.

Ter Braak, C.J.F., 1996. *Unimodal models to relate species to environment*. DLO-Agricultural Mathematics Group, Wageningen, 266 pp.
The main author of this book, which includes a compendium of articles, is also the inventor of the CANOCO program, a leading program for carrying out multivariate statistical methods for analysing relationships between species and their environment. This book focuses on the importance of the underlying species response in methods for quantifying ecological data. Chapters 4 and 5, addressing correspondence analysis and canonical correspondence analysis, respectively, are particularly recommended.

REFERENCES

Aalbersberg, G. and Litt, T., 1998. Multiproxy climate reconstructions for the Eemian and Early Weichselian. *Journal of Quaternary Science*, 13, 367–390.

AASP, 2000. *American Association of Stratigraphic Palynologists Homepage*. Accessed 18/9/2000 at *http://www.bc.edu/bc_org/associations/aasp/*

Abelmann, A. and Gowing, M.M., 1997. Spatial distribution pattern of living Polycystine radiolarian taxa – baseline study for palaeoenvironmental reconstructions in the Southern Ocean (Atlantic sector). *Marine Micropaleontology*, 30, 3–28.

Abrantes, F., 1988. Diatom productivity peak and increased circulation during the latest Quaternary: Alboran Basin (western Mediterranean). *Marine Micropalaeontology*, 13, 79–96.

Absolon, A., 1973. Ostracoden aus einigen Profilen spät- und postglazialer Karbonatablagerungen in Mitteleuropa, Mitteilungen der Bayerischen Staatssammlung für Paläontologie und historische Geologie, 13, 47–94.

Agbeti, M.D. and Dickman, M., 1989. Use of fossil diatom assemblages to determine historical changes in trophic status. *Canadian Journal of Fisheries and Aquatic Sciences*, 46, 1013–1021.

Ahagon, N., Tanaka, Y. and Ujiié, H., 1993. *Florisphaera profunda*, a possible nannoplankton indicator of late Quaternary changes in sea-water turbidity at the northwestern margin of the Pacific. *Marine Micropaleontology*, 22, 255–273.

Alavi, S.N., 1988. Late Holocene deep-sea benthic foraminifera from the Sea of Marmara. *Marine Micropaleontology*, 13, 213–237.

Almogi-Labin, A., Perelis-Grossovicz, L. and Raab, M., 1992. *Ammonia* from a hypersaline inland pool, Dead Sea area, Israel. *Journal of Foraminiferal Research*, 22, 257–266.

Almogi-Labin, A., Siman-Tov, R., Rosenfeld, A. and Debard, E., 1995. Occurrence and distribution of the foraminifer *Ammonia beccarii tepida* (Cushman) in water bodies, Recent and Quaternary, of the Dead Sea rift, Israel. *Marine Micropaleontology*, 26, 153–159.

Almogi-Labin, A., Hemleben, C., Meisschner, D. and Erlenkeuser, H., 1996. Response of Red Sea deep-water agglutinated foraminifera to water-mass changes during the Late Quaternary. *Marine Micropaleontology*, 28, 283–297.

Altenbach, A.V., 1988. Deep sea benthic foraminifera and flux rates of organic carbon. *Revue Paléobiologie, Special Volume*, 2, 719–720.

Altenbach, A.V., 1992. Short term processes and patterns in the foraminiferal response to organic fluxes. *Marine Micropaleontology*, 19, 119–129.

Altenbach, A.V. and Sarnthein, M., 1989. Productivity record in benthic foraminifera. In W.H. Berger, V.S. Smetacek and G. Wefer (eds), *Productivity in the oceans: present and past*. Wiley, New York, pp. 255–269.

Altenbach, A.V., Pflaumann, U., Schiebel, R., Thies, A., Timm, S. and Trauth, M., 1999. Scaling percentages and distributional patterns of benthic foraminifera with flux rates of organic carbon. *Journal of Foraminiferal Research*, 29, 173–185.

Alve, E., 1991. Foraminifera, climatic change and pollution: a study of Late Holocene sediments in Drammensfjord, SE Norway. *The Holocene*, 1, 243–261.

Alve, E., 1994. Opportunistic features of the foraminifer *Stainforthia fusiformis* (Williamson): evidence from Frierfjord, Norway. *Journal of Micropalaeontology*, 13, 24.

Alve, E., 1995a. Benthic foraminiferal responses to estuarine pollution: a review. *Journal of Foraminiferal Research*, 25, 190–203.

Alve, E., 1995b. Benthic foraminiferal distribution and recolonization of formerly anoxic environments in Drammensfjord, southern Norway. *Marine Micropaleontology*, 25, 169–186.

Alve, E., 2000. Environmental stratigraphy. A case study reconstructing bottom water oxygen conditions in Frierfjord, Norway, over the past five centuries. In R. Martin (ed.), *Environmental micropaleontology*. Kluwer Academic/Plenum Publishers, New York, pp. 323–350.

Alve, E. and Bernhard, J.M., 1995. Vertical migratory response of benthic foraminifera to controlled oxygen concentrations in an experimental mesocosm. *Marine Ecology Progress Series*, 116, 137–151.

Alve, E. and Olsgard, F., 1999. Benthic foraminiferal colonization in experiments with copper-contaminated sediments. *Journal of Foraminiferal Research*, 29, 186–195.

Andersen, S.T., 1970. The relative pollen productivity and representation of north European trees, and correction factors for tree pollen spectra. *Danmarks Geologiske undersøgelse Series II*, 96, 1–99.

Anderson, N.J., Rippley, B. and Stevenson, A.C., 1990. Change to a diatom assemblage in a eutrophic lake following point source nutrient re-direction: a paleolimnological approach. *Freshwater Biology*, 23, 205–217.

Anderson, O.R., 1983. *Radiolaria*. Springer-Verlag, New York, 355 pp.

Anderson, O.R., 1996. The physiological ecology of planktonic sarcodines with applications to paleoecology: patterns in space and time. *Journal of Eukaryotic Microbiology*, 43, 261–274.

Anderson, O.R., Bennett, P. and Bryan, M., 1989a. Experimental and observational studies of radiolarian physiological ecology: 1. Growth, abundance and opal productivity of the spongiose radiolarian *Spongaster tetras tetras*. *Marine Micropaleontology*, 14, 257–265.

Anderson, O.R., Bennett, P., Angel, D. and Bryan, M., 1989b. Experimental and observational studies of radiolarian physiological ecology: 2. Trophic activity and symbiont primary productivity of *Spongaster tetras tetras* with comparative data on predatory activity of some Nassellarida. *Marine Micropaleontology*, 14, 267–273.

Anderson, O.R., Bennett, P. and Bryan, M., 1989c. Experimental and observational studies of radiolarian physiological ecology: 3. Effects of temperature, salinity and light intensity on growth and survival of *Spongaster tetras tetras* maintained in laboratory culture. *Marine Micropaleontology*, 14, 275–282.

Anderson, O.R., Bryan, M. and Bennett, P., 1990. Experimental and observational studies of radiolarian physiological ecology: 4. Factors determining the distribution and survival of *Didymocyrtis tetrathalamus tetrathalamus* with implications for palaeoecological interpretations. *Marine Micropaleontology*, 16, 155–167.

Andrew, R., 1970. The Cambridge pollen reference collection. In D. Walker and R.G. West (eds), *Studies in the Vegetational History of the British Isles*. Cambridge University Press, Cambridge, pp. 225–231.

Andrew, R., 1984. *A practical pollen guide to the British flora*. Technical Guide 1, Quaternary Research Association, Cambridge.

Andrews, J.E., Boomer, I., Bailiff, I., *et al.*, 2000. Sedimentary evolution of the North Norfolk barrier coastline in the context of Holocene sea-level change. In I. Shennan and J. Andrews (eds), *Holocene land–ocean interaction and environmental change around the North Sea*, Geological Society of London, Special Publication, pp. 219–251.

Andruleit, H., Rad, U.V., Bruns, A. and Ittekkot, V., 2000. Coccolithophore fluxes from sediment traps in the northeastern Arabian Sea off Pakistan. *Marine Micropaleontology*, 38, 285–308.

Angel, D.L., Verghese, S., Lee, J.J. *et al.*, 2000. Impact of a net cage fish farm on the distribution of benthic foraminifera in the northern Gulf of Eilat (Aqba, Red Sea). *Journal of Foraminiferal Research*, 30, 54–65.

Anon., 1975. Proposals for the standardisation of diatom terminology and diagnosis. *Nova Hedwigia Beiheft*, 53, 323–354.

Anon., 2000. *Annotated bibliography of WWW sites relevant to automated pollen identification, and collections of pollen images.* Accessed 18/9/2000 at *http://saturn.sees.bangor.ac.uk/~ian/pdbase/pollen_dbase.html*

Archer, D.E., 1991a. Equatorial Pacific calcite preservation cycles: production or dissolution. *Paleoceanography*, 6, 561–571.

Archer, D.E., 1991b. Modelling the calcite lysocline. *Journal of Geophysical Research*, 96, 17,037–17,050.

Athersuch, J., Horne, D.J. and Whittaker, J.E., 1989. *Marine and Brackish Water Ostracods. Synopsis of the British Fauna (New Series)*, No. 43, E.J. Brill for the Linnean Society of London, Leiden, 343 pp.

Austin, W.E.N. and Scourse, J.D., 1997. Evolution of seasonal stratification in the Celtic Sea during the Holocene. *Journal of the Geological Society*, 154, 249–256.

Bahn, P. and Flenley, J.R., 1992. *Easter Island, Earth Island*. Thames & Hudson, London.

Baltanás, A., Otero, M., Arqueros, L., Rossetti, G. and Rossi, V., 2000. Ontogenetic changes in the carapace shape of the non-marine ostracod *Eucypris virens* (Jurine). *Hydrobiologia*, 419, 65–72.

Barber, H.G. and Haworth, E.Y., 1981. A guide to the morphology of the diatom frustule. *Freshwater Biological Association Scientific Publication*, 44, 112 pp.

Barber, K.E., 1976. History of Vegetation. In S.B. Chapman (ed.), *Methods in plant ecology*. Blackwell Scientific, Oxford, pp. 5–83.

Barber, K.E., Battarbee, R.W., Brooks, S.J., *et al.*, 1999. Proxy records of climate change in the UK over the last two millennia: documented change and sedimentary records from lakes and bogs. *Journal of the Geological Society, London*, 156, 369–380.

Barmawidjaja, D.M., Jorissen, F.J., Puskaric, S. and Van der Zwaan, G.J., 1992. Microhabitat selection by benthic foraminifera in the northern Adriatic Sea. *Journal of Foraminiferal Research*, 22, 297–317.

Barmawidjaja, D.M., van der Zwaan, G.J., Jorissen, F.J. and Puskaric, S., 1995. 150 years of eutrophication in the northern Adriatic Sea: evidence from a benthic foraminiferal record. *Marine Geology*, 122, 367–384.

Barron, J.A., 1992. Pliocene palaeoclimatic interpretation of DSDP Site 580 (NW Pacific) using diatoms. *Marine Micropaleontology*, 20, 23–44.

Barron, J.A. and Baldauf, J.G., 1989. Tertiary cooling steps and palaeoproductivity as reflected by diatoms and biosiliceous sediments. In W.H. Berger, V.S. Smetecek and G. Wefer (eds), *Productivity of the oceans, past present and future*. Wiley, Chichester, pp. 341–354.

Battarbee, R.W. 1984. Diatom analysis and the acidification of lakes. *Philosophical Transactions of the Royal Society of London*, B305, 451–477.

Battarbee, R.W., 1999. The importance of palaeolimnology to lake restoration. *Hydrobiologia*, 395/396, 149–159.

Battarbee, R.W., Charles, D.F., Sushil, S.D. and Renberg, I., 1999. Diatoms as indicators of lake eutrophication. In F. Stoermer and J.P. Smol (eds), *The Diatoms: Application to the environmental and earth sciences*. Cambridge University Press, Cambridge.

Batten, D.J., 1999. Small palynomorphs. In T.P. Jones and N.P. Rowe (eds), *Fossil Plants and Spores: modern techniques*. Geological Society, London, pp. 15–19.

Båtvik, H., Heimdal, B.R., Fagerbakke, K.M. and Green, J.C., 1997. Effects of unbalanced nutrient regime on coccolith morphology and size in *Emiliania huxleyi* (Prymnesiophyceae). *European Journal of Phycology*, 32, 155–165.

Bé, A.W.H., 1977. An ecological, zoogeographical and taxonomic review of Recent planktonic foraminifera. In A.T.S. Ramsay (ed.), *Oceanic micropalaeontology*, Volume 1, Academic Press, London, pp. 1–100.

Bé, A.W.H. and Tolderlund, D.S., 1971. Distribution and ecology of living planktonic foraminifera in surface waters of the Atlantic and Indian Oceans. In B.M. Funnell and W.R. Riedel (eds), *The micropalaeontology of the oceans*. Cambridge University Press, Cambridge, pp. 105–149.

Bé, A.W., Harrison, S.H. and Lott, L., 1973. *Orbulina universa* d'Orbigny in the Indian Ocean. *Micropaleontology*, 19, 150–192.

Bé, A.W.H., Bishop, J.K.B., Swerdlove, M.S. and Gardner, W.D., 1985. Standing stock, vertical distribution and flux of planktonic foraminifera in the Panama Basin. *Marine Micropaleontology*, 9, 307–333.

Beaufort, L., 1996. Dynamics of the monsoon in the equatorial Indian Ocean over the last 260,000 years. *Quaternary International*, 31, 13–18.

Behre, K.-E. (ed.), 1986. *Anthropogenic indicators in pollen diagrams*. A.A. Balkema, Rotterdam.

Bennett, K.D., 1994a. 'SIMPOLL' Version 2.23: a C program for analysing pollen data and plotting pollen diagrams. *INQUA Working Group on Data-Handling Methods Newsletter*, 11, 4–6.

Bennett, K.E., 1994b. Confidence intervals for age estimates and deposition times in late-Quaternary sediment sequences. *The Holocene*, 4, 337–348.

Bennett, K.R., Whittington, G. and Edwards, K.E., 1994. Recent plant nomenclatural changes and pollen morphology in the British Isles. *Quaternary Newsletter*, 73, 1–6.

Benson, R., 1988. Ostracods and palaeoceanography. In P. De Deckker, J.P. Colin and J.P. Peypouquet (eds), *Ostracoda in the Earth Sciences*. Elsevier, Amsterdam, pp. 1–26.

Benson, R., 1990. Ostracoda and the discovery of global Cainozoic palaeoceanographical events. In R. Whatley and C. Maybury (eds), *Ostracoda and global events*. Chapman and Hall, London, pp. 41–58.

Berger, W.H., 1976. Biogenous deep-sea sediments: production, preservation and interpretation. In J.P. Riley and R. Chester (eds) *Chemical Oceanography*, 5, 265–383.

Berger, W.H. and Diester-Haas, L., 1988. Paleoproductivity: the benthic/planktonic ratio as a productivity index. *Marine Geology*, 81, 15–25.

Berger, W.H. and Wefer, G., 1988. Benthic deep-sea foraminifera: possible consequences of infaunal habitat for paleoceanographic interpretation. *Journal of Foraminiferal Research*, 18, 147–150.

Berger, W.H., Herguera, J.C., Lange, C.B. and Schneider, R., 1994. Paleoproductivity: flux proxies versus nutrient proxies and other problems concerning the Quaternary productivity record. In R. Zahn, T.F. Pedersen, M.A. Kaminski and L. Labeyrie (eds), *Carbon cycling in the glacial ocean: constraints on the ocean's role in global change*. NATO ASI Series, I 17, Springer-Verlag, Berlin, pp. 385–412.

Berglund, B.E. (ed.), 1986. *Handbook of Holocene palaeoecology and palaeohydrology.* Wiley, Chichester.

Berglund, B.E. (ed.), 1991. *The cultural landscape during 6000 years in southern Sweden: the Ystad Project.* Munksgaard, Copenhagen.

Berglund, B.E. and Ralska-Jasiewiczowa, M., 1986. Pollen analysis and pollen diagrams. In B.E. Berglund (ed.), *Handbook of Holocene palaeoecology and palaeohydrology.* Wiley, Chichester, pp. 455–484.

Berglund, B.E., Birks, H.J.B., Ralska-Jasiewiczowa, M. and Wright, H.E. (eds), 1996. *Palaeoecological events during the last 15 000 years.* Wiley, Chichester.

Bernhard, J.M., 1992. Benthic foraminiferal distribution and biomass related to pore-water oxygen content: central California continental slope and rise. *Deep-Sea Research*, 39, 585–605.

Bernhard, J.M., 1993. Experimental and field evidence of Antarctic foraminiferal tolerance to anoxia and hydrogen sulfide. *Marine Micropaleontology*, 20, 203–213.

Bernhard, J.M. and Alve, E., 1996. Survival, ATP pool, and ultrastructural characterization of foraminifera from Drammensfjord (Norway): response to anoxia. *Marine Micropaleontology*, 28, 5–17.

Bernhard, J.M. and Bowser, S.S., 1996. Novel epifluorescence microscopy method to determine life position of foraminifera in sediments. *Journal of Micropalaeontology*, 15, p. 68.

Bernhard, J.M. and Reimers, C.E., 1991. Benthic foraminiferal population fluctuations related to anoxia: Santa Barbara Basin. *Biogeochemistry*, 15, 1127–1149.

Bernhard, J.M. and Sen Gupta, B.K., 1999. Foraminifera of oxygen-depleted environments. In B.K. Sen Gupta (ed.), *Modern foraminifera.* Kluwer Academic Publishers, Dordrecht, The Netherlands, pp. 201–216.

Bernhard, J.M., Sen Gupta, B.K. and Borne, P.F., 1997. Benthic foraminiferal proxy to estimate dysoxic bottom-water oxygen concentrations: Santa Barbara Basin, US Pacific continental margin. *Journal of Foraminiferal Research*, 27, 301–310.

Bickert, T. and Wefer, G., 1996. Late Quaternary deep water circulation in the South Atlantic: reconstruction from carbonate dissolution and benthic stable isotopes. In G. Wefer, W.H. Berger, G. Siedler and D.J. Webb (eds), *The South Atlantic: present and past circulation.* Springer-Verlag, Berlin, pp. 599–620.

Billett, D.S.M., Lampitt, R.S., Rice, A.L. and Mantoura, R.F.G., 1983. Seasonal sedimentation of phytoplankton to the deep-sea benthos. *Nature*, 302, 520–522.

Birks, H.J.B., 1986. Numerical zonation, comparison and correlation of Quaternary pollen-stratigraphical data. In B.E. Berglund (ed.), *Handbook of Holocene Palaeoecology and Palaeohydrology.* Wiley, Chichester, pp. 743–774.

Birks, H.J.B., 1994. The importance of pollen and diatom taxonomic precision in quantitative palaeoecological reconstructions. *Review of Palaeobotany and Palynology*, 83, 107–117.

Birks, H.J.B., 1995. Chapter 6. Quantitative palaeoenvironmental reconstructions. In D. Maddy and J.S. Brew (eds), *Statistical modelling of Quaternary science data.* Technical Guide 5, Cambridge, Quaternary Research Association, pp. 161–254.

Birks, H.J.B. and Birks, H.H., 1980. *Quaternary palaeoecology.* Arnold, London.

Birks, H.J.B. and Gordon, A.D., 1985. *Numerical methods in Quaternary pollen analysis.* Academic Press, London.

Birks, H.J.B. and Line, J.M., 1992. The use of rarefaction analysis for estimating palynological richness from Quaternary pollen-analytical data. *The Holocene*, 2, 1–10.

Birks, H.H., Birks, H.J.B., Kaland, P.E. and Moe, D. (eds), 1988. *The cultural landscape – past, present, future.* Cambridge University Press, Cambridge.

Bjork, S., Persson, T. and Kristersson, I., 1978. Comparison of two concentration methods for pollen in minerogenic sediments. *Geologiska Föreningens i Stockholm Förhandlingar*, 100, 107–111.

Bollmann, J., Baumann, K.H. and Thierstein, H.R., 1998. Global dominance of *Gephyrocapsa* coccoliths in the late Pleistocene: selective dissolution, evolution, or global environmental change? *Paleoceanography*, 13, 517–529.

Boltovskoy, D. and Alder, V.A., 1992. Paleoecological implications of radiolarian distribution and standing stocks versus accumulation rates in the Weddell Sea. *Antarctic Research Series*, 56, pp. 377–384.

Boltovskoy, D. and Jankilevich, S.S., 1985. Radiolarian distribution in east equatorial Pacific plankton. *Oceanologica Acta*, 8, 101–123.

Boltovskoy, D. and Riedel, W.R., 1987. Polycystine radiolaria of the California Current region: seasonal and geographic patterns. *Marine Micropaleontology*, 12, 65–104.

Boltovskoy, D., Alder, V.A. and Abelmann, A., 1993a. Annual flux of radiolaria and other shelled plankters in the eastern equatorial Atlantic at 853m: seasonal variations and polycystine species-specific responses. *Deep-Sea Research I*, 40, 1863–1895.

Boltovskoy, D., Alder, V.A. and Abelmann, A., 1993b. Radiolarian sedimentary imprint in Atlantic equatorial sediments: comparison with the yearly flux at 853m. *Marine Micropaleontology*, 23, 1–12.

Bond, G., Broecker, W., Johnsen, S., *et al.*, 1993. Correlations between climate records from North Atlantic sediments and Greenland ice. *Nature*, 365, 143–147.

Boomer, I. and Eisenhauer, G., 2002. Ostracod faunas as palaeoenvironmental indicators in marginal marine environments. In *The Ostracoda: applications in Quaternary research*, J. Holmes and A. Chivas (eds). American Geophysical Union, Washington, DC.

Boon, J.J., van der Meer, F.W., Schuyl, P.J.W., de Leeuw, J.W., Schenck, P.A. and Burlingame, A.L., 1978. Organic geochemical analyses of core samples from Site 362, Walvis Ridge, DSDP Leg 40. *Initial Reports of the Deep Sea Drilling Project*, 38, 39, 40, 41, 627–637.

Bowen, D.Q., Hughes, S.A., Sykes, G.A. and Miller, G.M., 1989. Land–sea correlations in the Pleistocene based on isoleucine epimerization in non-marine molluscs. *Nature*, 340, 49–51.

Bown, P.R. and Young, J.R., 1998. Techniques. In P. Bown (ed.), *Calcareous nannofossil biostratigraphy*. Chapman and Hall, London, pp. 16–28.

Boyd, P.D.A., 1981. The micropalaeontology and palaeoecology of medieval estuarine sediments from the Fleet and Thames in London. In J.W. Neale and M.D. Brasier (eds), *Microfossils from Recent and fossil shelf seas*. Ellis Horwood, Chichester, pp. 274–292.

Boyle, E.A., 1988. Cadmium: chemical tracer of deepwater paleoceanography. *Paleoceanography*, 3, 471–489.

Boyle, E.A., 1990. Quaternary deepwater paleoceanography. *Science*, 249, 863–870.

Boyle, E.A. and Keigwin, L.D., 1982. Deep circulation of the North Atlantic over the last 200,000 years: geochemical evidence. *Science*, 218, 784–787.

Boyle, E.A. and Keigwin, L.D., 1985/1986. Comparison of Atlantic and Pacific paleochemical records for the last 215,000 years: changes in deep ocean circulation and chemical inventories. *Earth and Planetary Science Letters*, 76, 135–150.

Braarud, T., 1951. Salinity as an ecological factor in marine phytoplankton. *Physiologia Plantarum*, 4, 28–34.

Braarud, T., 1979. The temperature range of the non-motile stage of *Coccolithus pelagicus* in the North Atlantic region. *British Phycological Journal*, 14, 349–352.

Braatz, B.V. and Corliss, B.H., 1987. Calcium carbonate undersaturation of bottom waters in the South Australian Basin during the last 3.2 million years. *Journal of Foraminiferal Research*, 17, 257–271.

Bramlette, M.N., 1958. Significance of coccolithophores in calcium carbonate deposition. *Bulletin of the Geological Society of America*, 69, 121–126.

Bramlette, M.N. and Riedel, W.R., 1954. Stratigraphic value of discoasters and some other microfossils related to recent coccolithophores. *Journal of Paleontology*, 28, 385–403.

Brand, L.E., 1991. Minimum iron requirements of marine phytoplankton and the implications for the biogeochemical control of new production. *Limnology and Oceanography*, 36, 1756–1771.

Brand, L.E., 1994. Physiological ecology of marine coccolithophores. In A. Winter and W.G. Siesser (eds), *Coccolithophores*. Cambridge University Press, Cambridge, pp. 39–49.

Brassell, S.C., Eglinton, G., Marlowe, I.T., Pflaumann, U. and Sarnthein, M., 1986. Molecular stratigraphy: a new tool for climatic assessment. *Nature*, 320, 129–133.

Brand, L.E. and Guillard, R.R.L., 1981. The effects of continuous light and light intensity on the reproduction rates of twenty-two species of marine phytoplankton. *Journal of Experimental Marine Biology and Ecology*, 50, 119–132.

Brasier, M.D. 1980. *Microfossils*. Unwin Hyman, London, 193 pp.

Brassell, S.C., Eglinton, G., Marlowe, I.T., Pflaumann, U. and Sarnthein, M., 1986. Molecular stratigraphy: a new tool for climatic assessment. *Nature*, 320, 129–133.

Bratbak, G., Egge, J.K. and Heldal, M., 1993. Viral mortality of the marine alga *Emiliania huxleyi* (Haptophyceae) and termination of algal blooms. *Marine Ecology Progress Series*, 93, 39–48.

Bratbak, G., Wilson, W. and Heldal, M., 1996. Viral control of *Emiliania huxleyi* blooms? *Journal of Marine Systems*, 9, 75–81.

Broecker, W., Bond, G., Klas, M., Clark, E. and McManus, J., 1992. Origin of the northern Atlantic's Heinrich events. *Climate Dynamics*, 6, 265–273.

Broecker, W.S., Kennett, J.P., Flower, B.P., *et al.*, 1989. Routing of meltwater from the Laurentide Ice Sheet during the Younger Dryas cold episode. *Nature*, 341, 318–321.

Broerse, A.T.C., Brummer, G.-J.A. and Hinte, J.E.V., 2000. Coccolithophore export production in response to monsoonal upwelling of Somalia (northwestern Indian Ocean). *Deep-Sea Research. Part 2, Topical studies in Oceanography*, 47, 2179–2205.

Brookes, D. and Thomas, K.W., 1967. The distribution of pollen grains on microscope slides. I. The non-randomness of the distribution. *Pollen et Spores*, 9, 621–629.

Brouwers, E.M., 1988. Sediment transport detected from the analysis of ostracod population structures: an example from the Alaskan continental shelf, In P.D. Deckker, J.P. Colin and J.P. Peypouquet, *Ostracoda in the earth sciences*. Elsevier, Amsterdam, pp. 231–244.

Bruch, A.A. and Pross, J., 1999. Palynomorph extraction from peat, lignite and coal. In T.P. Jones and N.P. Rowe (eds), *Fossil plants and spores: modern techniques*. Geological Society, London, pp. 26–30.

Bryant, V. and Wrenn, J. (eds), 1998. *New developments in palynomorph sampling, extraction and analysis*. American Association of Stratigraphic Palynologists Contribution Series, vol. 33, 157 pp.

Burnett, J.A., 1998. Upper Cretaceous. In P.R. Bown (ed.), *Calcareous nannofossil biostratigraphy*. Chapman and Hall, London, pp. 132–199.

Butlin, R. and Menozzi, P., 2000. Open questions in evolutionary ecology: do ostracods have the answers? *Hydrobiologia*, 419, 1–14.

Buzas, M.A. and Gibson, T.G., 1969. Species diversity: benthic foraminifera in western North Atlantic. *Science*, 163, 72–75.

Buzas, M.A., Culver, S.J. and Jorissen, F.J., 1993. A statistical evaluation of the microhabitats of living (stained) infaunal benthic foraminifera. *Marine Micropaleontology*, 20, 311–320.

Cameron, N.G., Birks, H.J.B., Jones, V.J., *et al.*, 1999. Surface sediment and epilithic diatom pH calibration sets for remote European Mountain Lakes (AL:PE Project) and their comparison with the surface waters acidification programme (SWAP) calibration set. *Journal of Palaeolimnology*, 22, 291–317.

Caralp, M.-H., 1984. Impact de la matière organique dans des zones de forte productivité sur certains foraminifères benthiques. *Oceanologica Acta*, 7, 509–515.

Caralp, M.-H., 1987. Deep-sea circulation in the northeastern Atlantic over the past 30 000 years: the benthic foraminiferal record. *Oceanologica Acta*, 10, 27–40.

Caralp, M.-H., 1988. Late glacial to Recent deep-sea benthic foraminifera from the northeastern Atlantic (Cadiz Gulf) and western Mediterranean (Alboran Sea): paleoceanographic results. *Marine Micropaleontology*, 13, 265–289.

Caralp, M.-H., 1989. Abundance of *Bulimina exilis* and *Melonis barleeanum*: relationship to the quality and quantity of marine organic matter. *Geo-Marine Letters*, 9, 37–43.

Carbonel, P., 1988. Ostracods and the transition between fresh and saline waters. In P. De Deckker, J.P. Colin and J.P. Peypouquet (eds), *Ostracoda in the earth sciences*. Elsevier, Amsterdam, pp. 157–173.

Carbonel, P. and Hoibian, T., 1988. The impact of organic matter on ostracods from an equatorial deltaic area, the Mahakam Delta – southeastern Kalimantan. In *Evolutionary biology of Ostracoda: its fundamentals and applications*, T. Hanai, N. Ikeya and K. Ishizaki (eds). Kodansha, Tokyo, pp. 353–366.

Casey, R.E., 1971. Radiolarians as indicators of past and present water-masses. In B.M. Funnell and W.R. Riedel (eds), *Micropalaeontology of oceans*. Cambridge University Press, Cambridge, pp. 331–341.

Casey, R.E., 1993. Radiolaria. In J.H. Lipps (ed.), *Fossil prokaryotes and protists*. Blackwell Scientific, Boston, pp. 249–284.

Casey, R.E., Weinheimer, A.L. and Nelson, C.O., 1990. Cenozoic radiolarian evolution and zoogeography of the Pacific. *Bulletin of Marine Science*, 47, 221–232.

Castradori, D., 1993. Calcareous nannofossils and the origin of Eastern Mediterranean sapropels. *Paleoceanography*, 8, 459–471.

Caulet, J.-P., Venec-Peyre, M.-T., Vergnaud-Grazzini, C. and Nigrini, C., 1992. Variation of south Somalian upwelling during the last 160 ka: radiolarian and foraminifera records in core MD 85674. In C.P. Summerhayes, W.L. Prell and K.C. Emeis (eds), *Upwelling systems: evolution since the early Miocene*. Geological Society of London, Special Publication, No. 63, pp. 379–389.

Cayre, O., Beaufort, L. and Vincent, E., 1999. Paleoproductivity in the equatorial Indian Ocean for the last 260,000 yr: a transfer function based on planktonic foraminifera. *Quaternary Science Reviews*, 18, 839–858.

Chambers, F.M., 1995. Climate response, migrational lag, and pollen representation: the problems posed by *Rhododendron* and *Acer*. *Historical Biology*, 9, 243–256.

Chambers, F.M., Mauquoy, D. and Todd, P.A., 1999. Recent rise to dominance of *Molinia caerulea* in Environmentally Sensitive Areas: new perspectives from palaeoecological data. *Journal of Applied Ecology*, 26, 719–733.

Charles, C.D. and Fairbanks, R.G., 1992. Evidence from Southern Ocean sediments for the effect of North Atlantic Deep-Water flux on climate. *Nature*, 335, 416–419.

Charles, C.D., Lynch-Stieglitz, J., Ninnemann, U.S. and Fairbanks, R.G., 1996. Climate connections between the hemispheres revealed by deep sea sediment core/ice core correlation. *Earth and Planetary Science Letters*, 142, 19–27.

Charles, D.F. and Smol, J.P., 1988. New methods for using diatoms and chrysophytes to infer past pH of low alkalinity lakes. *Limnology and Oceanography*, 33, 1451–1462.

Charles, D.F. and Smol, J.P., 1994. Long term chemical changes in lakes: quantitative inferences using biotic remains in the sedimentary record. In L. Baker (ed.), *Environmental chemistry of lakes and reservoirs: advances in chemistry, series 237*. Washington, DC, American Chemical Society, pp. 3–31

Chivas, A.R., De Deckker, P. and Shelley, J.M.G., 1985. Strontium content of ostracods indicates lacustrine palaeosalinity. *Nature*, 316, 251–253.

Chivas, A., De Deckker, P. and Shelley, J., 1986. Magnesium and strontium in non-marine ostracod shells as indicators of palaeosalinity and palaeotemperature. *Hydrobiologia*, 143, 135–142.

Chivas, A.R., De Deckker, P., Cali, J.A., Chapman, A., Kiss, E. and Shelley, J.M.G., 1993. Coupled stable-isotope and trace-element measurements of lacustrine carbonates as paleoclimatic indicators. In P.K Swart, K.C. Lohmann, J. McKenzie and S. Savin, *Climate change in continental isotopic records*. American Geophysical Union, Geophysical Monograph, 78, 113–121.

Clague, J.J., Hutchinson, I., Mathewes, R.W. and Patterson, R.T., 1999. Evidence for late Holocene tsunamis at Catala Lake, British Columbia. *Journal of Coastal Research*, 15, 45–60.

Clark, F.E., Patterson, R.T. and Fishbein, E., 1994. Distribution of Holocene benthic foraminifera from the tropical southwest Pacific Ocean. *Journal of Foraminiferal Research*, 24, 241–267.

Clark, R.B., 1986. *Marine pollution*. Clarendon Press, Oxford, 215 pp.

Clark, R.L., 1982. Point count estimation of charcoal in pollen preparations and thin sections of sediments. *Pollen et Spores*, 24, 523–535.

CLIMAP, 1976. The surface of the ice-age Earth. *Science*, 191, 1131–1137.

CLIMAP, 1981. Seasonal reconstruction of the Earth's surface at the Last Glacial Maximum. *Geological Society of America Map Chart Series*. 36, 1–18.

CLIMAP, 1984. The last interglacial ocean. *Quaternary Research*, 21, 123–224.

Coale, K.H., Johnson, K.S., Fitzwater, S.E. *et al.*, 1996. A massive phytoplankton bloom induced by an ecosystem-scale iron fertilization experiment in the equatorial Pacific Ocean. *Nature*, 383, 495–501.

Coimbra, J., Pinto, I., Würdig, N. and Carmo, D.d., 1999. Zoogeography of Holocene Podocopina (Ostracoda) from the Brazilian Equatorial shelf. *Marine Micropaleontology*, 37, 365–379.

Collins, E.S., Scott, D.B. and Gayes, P.T., 1999. Hurricane records on the South Carolina coast: can they be detected in the sediment record? *Quaternary International*, 56, 15–26.

Conan, S.M. and Brummer, G.J., 2000. Fluxes of planktonic foraminifera in response to monsoonal upwelling on the Somalia Basin margin. *Topical Studies in Oceanography*, 47, 2207–2227.

Conradsen, K. and Heier-Nielsen, S., 1995. Holocene paleoceanography and paleoenvironments of the Skagerrak–Kattegat, Scandinavia. *Paleoceanography*, 10, 801–813.

Conte, M.H., Volkman, J.K. and Eglinton, G., 1994. Lipid biomarkers of the Haptophyta. In J.C. Green and B.S.C. Leadbeater (eds) *The haptophyte algae*. The Systematics Association Special Volume, No. 51. Clarendon Press, Oxford, pp. 351–377.

Corliss, B.H., 1979. Quaternary Antarctic Bottom-Water history: deep-sea benthonic foraminiferal evidence from the southeast Indian Ocean. *Quaternary Research*, 12, 271–289.

Corliss, B.H., 1982. Linkage of North Atlantic and Southern Ocean deep-water circulation during glacial intervals. *Nature*, 298, 458–460.

Corliss, B.H., 1983a. Quaternary circulation of the Antarctic Circumpolar Current. *Deep-Sea Research*, 30, 47–61.

Corliss, B.H., 1983b. Distribution of Holocene deep-sea benthonic foraminifera in the southwest Indian Ocean. *Deep-Sea Research*, 30, 95–117.

Corliss, B.H., 1985. Microhabitats of benthic foraminifera within deep-sea sediments. *Nature*, 314, 435–438.

Corliss, B.H., 1991. Morphology and microhabitat preferences of benthic foraminifera from the northwest Atlantic Ocean. *Marine Micropaleontology*, 17, 195–236.

Corliss, B.H. and Chen, C., 1988. Morphotype patterns of Norwegian Sea deep-sea benthic foraminifera and ecological implications. *Geology*, 16, 716–719.

Corliss, B.H. and Emerson, S., 1990. Distribution of Rose Bengal stained deep-sea benthic foraminifera form the Nova Scotian continental margin and Gulf of Maine. *Deep-Sea Research*, 37, 381–400.

Corliss, B.H. and Fois, E., 1990. Morphotype analysis of deep-sea benthic foraminifera from the Northwest Gulf of Mexico. *Palaios*, 5, 589–605.

Corliss, B.H. and Honjo, S., 1981. Dissolution of deep-sea benthonic foraminifera. *Micropaleontology*, 27, 356–378.

Corliss, B.H., Martinson, D.G. and Keffer, T., 1986. Late Quaternary deep-ocean circulation. *Geological Society of America Bulletin*, 97, 1106–1121.

Corrège, T. and De Deckker, P., 1997. Faunal and geochemical evidence for change in intermediate water temperature and salinity in the western Coral Sea during the late Quaternary. *Palaeogeography, Palaeoclimatology, Palaeoecology*, 131, 185–205.

Coxon, P. and Clayton, G., 1999. Light microscopy of fossil pollen and spores. In T.P. Jones and N.P. Rowe (eds), *Fossil plants and spores: modern techniques*. Geological Society, London, pp. 26–30.

Cronin, T.M. and Raymo, M.E., 1997. Orbital forcing of deep-sea benthic species diversity. *Nature*, 385, 624–627.

Cronin, T., DeMartino, D., Dwyer, G. and Rodriguez-Lazaro, J., 1999. Deep-sea ostracode species diversity: response to late Quaternary climate change. *Marine Micropaleontology*, 37, 231–250.

Cros, L., Kleijne, A., Zeltner, A., Billard, C. and Young, J.R., 2000. New examples of holococcolith–heterococcolith combination coccospheres and their implications for coccolithophorid biology. *Marine Micropaleontology*, 39, 1–34.

Crowley, T.J. and North, G.R. 1991. *Palaeoclimatology*. Oxford University Press, Oxford, 339 pp.

Culver, S.J. and Buzas, M.A., 2000. Global latitudinal species diversity gradient in deep-sea benthic foraminifera. *Deep-Sea Research*, 47, 259–275.

Cunningham, W.L. and Leventer, A. 1998. Distribution of diatom assemblages in surface sediments of the Ross Sea, Antarctica: relationship to modern oceanographic conditions. *Antarctic Science*, 10(2), 134–146.

Cunningham, W.L., Leventer, A., Andrews, J.T., Jennings, A.E. and Licht, K.J., 1999. Late Pleistocene–Holocene marine conditions of the Ross Sea, Antarctica: evidence from the fossil record. *The Holocene*, 9(2), 129–139

Curry, W.B. and Lohmann, G.P., 1982. Carbon isotopic changes in benthic foraminifera from the western South Atlantic: reconstruction of glacial abyssal circulation patterns. *Quaternary Research*, 18, 218–235.

Curry, W.B., Duplessy, J.-C., Labeyrie, L.D., Oppo, D. and Kallel, N., 1988. Quaternary deep-water circulation changes in the distribution of $\delta^{13}C$ of deep water ΣCO_2 between the last glaciation and the Holocene. *Paleoceanography*, 3, 317–342.

Dale, A.L. and Dale, B., 1992. Dinoflagellate contributions to the sediment flux of the Nordic Seas. In S. Honjo (ed.), *Ocean Biocoenosis Series, 5.* Woods Hole Oceanographic Institution Press, Woods Hole, pp. 45–76.

Dale, B., 1983. Dinoflagellate resting cysts: 'benthic plankton'. In G.A. Fryxel (ed.), *Survival strategies of the algae.* Cambridge University Press, Cambridge, pp. 69–136.

Dale, B., 1996. Dinoflagellate cyst ecology: modelling and geological applications. In J. Jansonius and D.G. McGregor (eds), *Palynology: principles and applications.* American Association of Stratigraphic Palynologists Foundation, 3, 1249–1275.

Dale, B., 2001. Marine dinoflagellate cysts as indicators of eutrophication and industrial pollution: a discussion. *The Science of the Total Environment*, 264, 235–240.

Dale B., Dale, A. and Jansen, J.H.F., 2002. Dinoflagellate cysts as environmental indicators in surface sediments from the Congo deep-sea fan and adjacent regions. *Paleogeography, Paleoclimatology, Paleoecology*, in press.

Dale, B., Dale, A.L. and Jansen, J.H.F., in press. Correspondence analysis of dinocyst assemblages to reveal movements in the Angola–Benguela front over the last 200,000 years. *Palaeogeography, Palaeoclimatology, Palaeoecology*.

Dale, B., Thorsen, T.A. and Fjellså, A. 1999. Dinoflagellate cysts as indicators of cultural eutrophication in the Oslofjord, Norway. *Estuarine, Coastal and Shelf Science*, 48, 371–382.

Danielopol, D.L., Carbonel, P. and Colin, J.P., 1990. *Cytherissa* (Ostracoda) – the *Drosophila* of paleolimnology. *Bulletin de l'Institut de Géologie du Bassin d'Aquitaine*, Volume 47/48, 310 pp.

Davies, T.A. and Gorsline, D.S., 1976. Oceanic sediments and sedimentary processes. In J.P. Riley and R. Chester (eds), *Chemical Oceanography* (2nd edn). Academic Press, London, pp. 1–80.

Davis, J.C., 1986. *Statistics and data analysis in geology.* Wiley, New York.

Davis, M.B., 1967. Pollen accumulation rates at Rogers Lake, Connecticut during late- and postglacial time. *Review of Palaeobotany and Palynology*, 2, 219–230.

Davis, M.B., 1976. Pleistocene biogeography of temperate deciduous forests. *Geoscience and Man*, 13, 13–26.

Davis, O.K. (ed.), 1994. *Aspects of archaeological palynology: methodology and applications.* American Association of Stratigraphic Palynologists Contribution Series, vol. 29, 221 pp.

de Beaulieu, J.-L. and Reille, M., 1984. A long Upper Pleistocene record from Les Echets, near Lyon, France. *Boreas*, 13, 111–132.

De Deckker, P., 1977. The distribution of the 'giant' ostracods (Family: Cyprididae Baird, 1845) endemic to Australia. In H. Loffler and D. Danielopol (eds), *Aspects of ecology and zoogeography of Recent and fossil Ostracoda.* Junk, The Hague, pp. 285–294.

De Deckker, P. and Forester, R.M., 1988. The use of ostracods to reconstruct continental records. In P. De Deckker, J.P. Colin and J.P. Peypouquet, *Ostracoda in the earth sciences.* Elsevier, Amsterdam, pp. 175–199.

De Deckker, P., Chivas, A. and Shelley, J., 1999. Uptake of Mg and Sr in the euryhaline ostracod *Cyprideis* determined from in vitro experiments. *Palaeogeography, Palaeoclimatology, Palaeoecology*, 148, 105–116.

De Mendiola, B.R., 1981. Seasonal phytoplankton distribution along the Peruvian Coast. In F.A. Richards (ed.) *Coastal upwelling*, Coastal and Estuarine Sciences 1, American Geophysical Union, Washington, DC, pp. 348–365.

De Menocal, P.B., Oppo, D.W., Fairbanks, R.G. and Prell, W.L., 1992. Pleistocene δ^{13}C variability of North Atlantic Intermediate Water. *Paleoceanography*, 7, 229–250.

De Stigter, H.C., 1996. Recent and fossil benthic foraminifera in the Adriatic Sea: distribution patterns in relation to organic carbon flux and oxygen concentration at the seabed. *Geologica Ultraiectina. Mededelingen van de Faculteit Aardwetenschappen Universiteit Utrecht*, No. 44, 254 pp.

De Vernal, A., Hillaire-Marcel, C. and Bilodeau, G., 1996. Reduced meltwater outflow from the Laurentide ice margin during the Younger Dryas. *Nature*, 381, 774–777.

De Wever, P., Azéma, J. and Fourcade, E. 1994. Radiolaires et radiolarites: production primaire, diagenèse et paléogéographie. *Bullétin des Centres de Recherches Exploration–Production Elf Aquitaine*, 18, 315–379.

Den Dulk, M., Reichardt, G.J., Memon, G.M., Roelofs, E.M.P., Zachariasse, W.J. and van der Zwaan, G.J., 1998. Benthic foraminiferal response to variations in surface water productivity and oxygenation in the northern Arabian Sea. *Marine Micropaleontology*, 35, 43–66.

Denys, L. and Baeteman, C., 1995. Holocene evolution of relative sea-level and local mean high water spring tides in Belgium – a first assessment. *Marine Geology*, 124, 1–19.

Didymus, J.M., Young, J.R. and Mann, S., 1994. Construction and morphogenesis of the chiral ultrastructure of coccoliths from the marine alga *Emiliania huxleyi*. *Proceedings of the Royal Society of London*, B258, 237–245.

Dillion, W.R. and Goldstein, M. 1984. *Multivariate analysis methods and applications*. Wiley, New York.

Dimbleby, G.W., 1985. *The palynology of archaeological sites*. Academic Press, London.

Dingle, R. and Lord, A., 1990. Benthic ostracods and deep water-masses in the Atlantic Ocean. *Palaeogeography, Palaeoclimatology, Palaeoecology*, 80, 213–235.

Dingle, R., Lord, A. and Boomer, I., 1989. Ostracod faunas and water masses across the continental margin off southwestern Africa. *Marine Geology*, 87, 323–328.

do Carmo, D.A., Whatley, R.C. and Timberlake, S., 1999. Variable noding and palaeoecology of a Middle Jurassic limnocytherid ostracod: implications for modern brackish water taxa. *Palaeogeography, Palaeoclimatology, Palaeoecology*, 148, 23–35.

Douglas, R.G. and Woodruff, F., 1981. Deep sea benthic foraminifera. In C. Emiliani (ed.), *The oceanic lithosphere, the sea*, 7. Wiley, New York, pp. 1233–1327.

Dow, R.L., 1978. Radiolarian distribution and the Late Pleistocene history of the southeastern Indian Ocean. *Marine Micropaleontology*, 3, 203–227.

Dransfield, J., Flenley, J.R., King, S.M., Harkness, D.D. and Rapu, S., 1984. A recently extinct palm from Easter Island. *Nature*, 312, 750–752.

Duplessy, J.-C., Shackleton, N.J., Labeyrie, L.D., Oppo, D. and Kallel, N., 1988. Deep water source variations during the last climatic cycle and their impact on the global deep water circulation. *Paleoceanography*, 3, 343–360.

Duplessy, J.-C., Shackleton, N.J. and Matthews, R.K., 1984. ^{13}C record of benthic foraminifera in the last interglacial ocean: implications for the carbon cycle and the global deep water circulation. *Quaternary Research*, 2, 225–243.

Dworetzky, B.A. and Morley, J.J., 1987. Vertical distribution of Radiolaria in the eastern equatorial Atlantic: analysis of a multiple series of closely-spaced plankton tows. *Marine Micropaleontology*, 12, 1–19.

Dwyer, G.S., Cronin, T.M., Baker, P.A., Raymo, M.E., Buzas, J.S. and Correge, T., 1995. North Atlantic deepwater temperature change during late Pliocene and late Quaternary climatic cycles. *Science*, 270, 1347–1351.

Eagar, S., 1999. Distribution of Ostracoda around a coastal sewer outfall: a case study from Wellington, New Zealand. *Journal of the Royal Society of New Zealand*, 29, 257–264.

Eagar, S., 2000. Ostracoda in detection of sewage discharge on a Pacific Atoll. In R. Martin (ed.), *Environmental micropaleontology*. Kluwer, New York, 151–165.

Edwards, L.E., Mudie, P.J. and deVernal, A., 1991. Pliocene paleoclimatic reconstruction using dinoflagellate cysts: comparison of methods. *Quaternary Science Reviews*, 10, 259–274.

Edwards, R.J. and Horton, B.P., 2000. Reconstructing relative sea-level change using UK salt-marsh foraminifera. *Marine Geology*, 169, 41–56.

Egge, J.K. and Heimdal, B.R., 1994. Blooms of phytoplankton including *Emiliania huxleyi* (Haptophyta). Effects of nutrient supply in different N:P ratios. *Sarsia*, 79, 333–348.

Eglinton, G., Bradshaw, S.A., Rosell, A., Sarnthein, M., Pflaumann, U. and Tiedemann, R., 1992. Molecular record of secular sea surface temperature changes on 100-year timescales for glacial terminations I, II and IV. *Nature*, 356, 423–426.

Ellison, R.L., Broome, R. and Ogilvie, R., 1986. Foraminiferal response to trace metal contamination in the Patapsco River and Baltimore Harbor, Maryland. *Marine Pollution Bulletin*, 17, 419–423.

Emiliani, C., 1955. Pleistocene temperatures. *Journal of Geology*, 63, 538–578.

Emiliani, C., 1993. Viral extinctions in deep-sea species. *Nature*, 366, 217–218.

Engstrom, D.R. and Nelson, S.R., 1991. Paleosalinity from trace metals in fossil ostracodes compared with observational records at Devils Lake, North Dakota, USA. *Palaeogeography, Palaeoclimatology, Palaeoecology*, 83, 295–312.

Engstrom, D.R., Fritz, S.C., Almendinger, J.E. and Juggins, S., 2000. Chemical and biological trends during lake evolution in recently deglaciated terrain. *Nature*, 408, 161–166.

Erdtman, G., 1943. *An introduction to pollen analysis*. Chronica Botanica, Waltham, Essex.

Erdtman, G., 1969. *Handbook of palynology*. Munksgaard, Copenhagen.

Erdtman, G., Berglund, B.E. and Praglowski, J., 1961. *An introduction to a Scandinavian pollen flora. I*. Almqvist and Wiksell, Stockholm.

Erdtman, G., Praglowski, J. and Nilsson, S., 1963. *An introduction to a Scandinavian pollen flora. II*. Almqvist and Wiksell, Stockholm.

Estrada, M. and Berdalet, E., 1997. Phytoplankton in a turbulent world. *Scientia Marina*, 61(Suppl.), 125–140.

Faegri, K. and Iversen, J., 1989. *Textbook of pollen analysis, 4th edn*. Wiley, Chichester.

Fariduddin, M. and Loubere, P., 1997. The surface ocean productivity response of deeper water benthic foraminifera in the Atlantic Ocean. *Marine Micropaleontology*, 32, 289–310.

Farrell, J.W., Murray, D.W., McKenna, V.S. and Ravelo, A.C. 1995. Upper ocean temperature and nutrient contrasts inferred from Pleistocene planktonic foraminifera $\delta^{18}O$ and $\delta^{13}C$ in the eastern equatorial Pacific. *Proceedings of the Ocean Drilling Program, Scientific Results*, 138, 289–319.

Farrimond, P., Eglinton, G. and Brassell, S.C., 1986. Alkenones in Cretaceous black shales, Blake–Bahama Basin, western North Atlantic. In J. Leythaeuser and J. Rullkötter (eds), *Advances in organic geochemistry 1985*. Pergamon, Oxford, pp. 897–903.

Fenner, J., Schrader, H.J. and Wiegnik, H., 1976. Diatom phytoplankton studies in the southern Pacific Ocean, composition and correlation to the Antarctic convergence and its palaeoecological significance. *Initial Reports of the Deep Sea Drilling Project*, 35, 757–813.

Fensome, R.A., Taylor, F.J.R., Norris, G., Sarjeant, W.A.S., Wharton, D.I. and Williams, G.L., 1993. A classification of living and fossil dinoflagellates. *Micropaleontology, Special Publication Number 7*, 351 pp.

Fiedler, P.C., Philbrick, V. and Chavez, F.P., 1991. Oceanic upwelling and productivity in the eastern tropical Pacific. *Limnology and Oceanography*, 36, 1834–1850.

Fisher, N.S. and Honjo, S., 1989. Intraspecific differences in temperature and salinity responses in the coccolithophore *Emiliania huxleyi*. *Biological Oceanography*, 6, 355–361.

Fisher, R.A., Corbet, A.S. and Williams, C.B., 1943. The relationship between the number of species and the number of individuals in a random sample of an animal population. *Journal of Animal Ecology*, 12, 42–58.

Flenley, J.R. and King, S.M., 1984. Late Quaternary pollen records from Easter Island. *Nature*, 307, 47–50.

Flenley, J.R., King, S.M., Teller, J.T., Prentice, M.E., Jackson, J. and Chew, C. 1991. The Late Quaternary vegetational and climatic history of Easter Island. *Journal of Quaternary Science*, 6, 85–115.

Flower, B.P., Oppo, D.W., McManus, J.F., Venz, K.A., Hodell, D.A. and Cullen, J.L., 2000. North Atlantic intermediate to deep water circulation and chemical stratification during the past 1 Myr. *Paleoceanography*, 15, 388–403.

Flower, R.J., 1986. The relationship between surface sediment diatom assemblages and pH in 33 Galloway lakes: some regression models for reconstructing pH and their application to sediment cores. *Hydrobiologia*, 143, 93–103.

Foged, N., 1969. Diatoms in a post-glacial core from the bottom of lake Grane Lanso, Denmark. *Bulletin of Geological Society Denmark*, 19, 137–256.

Forester, R., 1983. Relationship of two lacustrine ostracode species to solute composition and salinity: implications for paleohydrochemistry. *Geology*, 11, 435–438.

Forester, R., 1986. Determination of the dissolved anion composition of ancient lakes from fossil ostracodes. *Geology*, 14, 796–798.

Forester, R.M., 1987. Late Quaternary paleoclimate records from lacustrine ostracodes, in North America and adjacent oceans during the last deglaciation. In W.F. Ruddiman and H.E. Wright, *The Geology of North America, v. K-3*. Geological Society of America, Boulder, Colorado, pp. 261–276.

France, I., Duller, A.W.G., Duller, G.A.T. and Lamb, H.F., 2000. A new approach to automated pollen analysis. *Quaternary Science Reviews*, 19, 537–546.

Frodin, D., 2001. *Guide to standard floras of the world, 2nd edn*. Cambridge University Press, Cambridge.

Fryxell, G.A., 1991. Comparison of winter and summer growth stages of the diatom *Eucampia antarctica* from the Kerguelen Convergence Zone. *Proceedings of the Ocean Drilling Program, Scientific Results*, 119, 225–238.

Funnell, B.M., 1995. Global sea-level and the (pen-)insularity of late Cenozoic Britain. In R.C. Preece (ed.), *Island Britain: a Quaternary perspective*. Geological Society of London Special Publication, No. 96, pp. 3–13.

Funnell, B.M., Haslett, S.K., Kennington, K., Swallow, J.E. and Kersley, C.L., 1996. Strangeness of the equatorial ocean during the Olduvai magnetosubchron (1.95 to 1.79 Ma). In A. Moguilevsky and R. Whatley (eds), *Microfossils and oceanic environments*. University of Wales, Aberystwyth Press, pp. 93–109.

Gaby, M.L. and Sen Gupta, B.K., 1985. Late Quaternary benthic foraminifera of the Venezuela Basin. *Marine Geology*, 68, 125–155.

Gage, J.D. and Tyler, P.A., 1991. *Deep-sea biology: a natural history of organisms at the deep-sea floor*. Cambridge University Press, Cambridge, 504 pp.

Ganssen, G.M. and Kroon, D., 2000. The isotopic signature of planktonic foraminifera from NE Atlantic surface sediments: implications for the reconstruction of past oceanic conditions. *Journal of the Geological Society, London*, 157, 693–699.

Garrison, D.L., Buck, K.R. and Fryxell, G.A., 1987. Algal assemblages in Antarctic pack ice and ice-edge plankton. *Journal of Phycology*, 23, 564–572.

Gehrels, W.R., 1994. Determining relative sea-level change from salt-marsh foraminifera and plant zones on the coast of Maine, U.S.A. *Journal of Coastal Research*, 10, 990–1009.

Gehrels, W.R., 1999. Middle and late Holocene sea-level changes in eastern Maine reconstructed from foraminiferal saltmarsh stratigraphy and AMS ^{14}C dates on basal peat. *Quaternary Research*, 52, 350–359.

Gehrels, W.R., 2000. Using foraminiferal transfer functions to produce high-resolution sea-level records from saltmarsh deposits. *The Holocene*, 10, 367–376.

Gehrels, W.R. and Belknap, D.F., 1993. Neotectonic history of eastern Maine evaluated from historic sea-level data and ^{14}C dates on salt-marsh peats. *Geology*, 21, 615–618.

Gehrels, W.R. and Van de Plassche, O., 1999. The use of *Jadammina macrescens* (Brady) and *Balticammina pseudomacrescens* Brönnimann, Lutze and Whittaker (Protozoa: Foraminiferida) as sea-level indicators. *Palaeogeography, Palaeoclimatology, Palaeoecology*, 149, 115–125.

Gehrels, W.R., Belknap, D.F., Pearce, B.R. and Gong, B., 1995. Modeling the contribution of M2 tidal amplification to the Holocene rise of mean high water in the Gulf of Maine and the Bay of Fundy. *Marine Geology*, 124, 71–85.

Gehrels, W.R., Roe, H.M. and Charman, D.J., 2001. Foraminifera, testate amoebae and diatoms as sea-level indicators in UK saltmarshes: a quantitative multiproxy approach. *Journal of Quaternary Science*, 16, 201–220.

Gliozzi, E. and Mazzini, I., 1998. Palaeoenvironmental analysis of Early Pleistocene brackish marshes in the Rieti and Tiberinno intrapenninic basins (Latium and Umbria, Italy) using ostracods (Crustacea). *Palaeogeography, Palaeoclimatology, Palaeoecology*, 140, 325–333.

Godwin, H., 1934a. Pollen analysis: an outline of the problems and potentialities of the method. I. Technique and interpretation. *New Phytologist*, 33, 278–305.

Godwin, H., 1934b. Pollen analysis: an outline of the problems and potentialities of the method. II. General applications of pollen analysis. *New Phytologist*, 33, 325–358.

Godwin, H., 1940. Pollen analysis and forest history of England and Wales. *New Phytologist*, 39, 370–400.

Goldstein, S.T. and Watkins, G.T., 1999. Taphonomy of salt marsh foraminifera: an example from coastal Georgia. *Palaeogeography, Palaeoclimatology, Palaeoecology*, 149, 103–114.

Goll, R.M. and Björklund, K.R., 1971. Radiolaria in surface sediments of the North Atlantic Ocean. *Micropaleontology*, 17, 434–454.

Goll, R.M. and Björklund, K.R., 1974. Radiolaria in surface sediments of the South Atlantic. *Micropaleontology*, 20, 38–75.

Gooday, A.J., 1986. Meiofaunal foraminiferans from the bathyal Porcupine Seabight (northeast Atlantic): size structure, standing stock, taxonomic composition, species diversity and vertical distribution in the sediment. *Deep-Sea Research*, 33, 1345–1373.

Gooday, A.J., 1988. A response by benthic Foraminifera to the deposition of phytodetritus in the deep sea. *Nature*, 332, 70–73.

Gooday, A.J., 1993. Deep-sea benthic foraminiferal species which exploit phytodetritus: characteristic features and controls on distribution. *Marine Micropaleontology*, 22, 187–205.

Gooday, A.J., 1994. The biology of deep-sea foraminifera: a review of some advances and their applications in paleoceanography. *Palaios*, 9, 14–31.

Gooday, A.J., 1996. Epifaunal and shallow infaunal foraminiferal communities at three abyssal NE Atlantic sites subject to differing phytodetritus input regimes. *Deep-Sea Research*, 43, 1395–1421.

Gooday, A.J., 1999. Biodiversity of foraminifera and other protists in the deep sea: scales and patterns. *Belgium Journal of Zoology*, 129, 61–80.

Gooday, A.J. and Alve, E., 2001. Morphological and ecological parallels between sublittoral and abyssal foraminiferal species in the NE Atlantic: a comparison of *Stainforthia fusiformis* and *Stainforthia* sp. *Progress in Oceanography*, 50, 261–283.

Gooday, A.J. and Lambshead, P.J.D., 1989. The influence of seasonally deposited phytodetritus on benthic foraminiferal populations in the bathyal northeast Atlantic. *Marine Ecology Progress Series*, 58, 53–67.

Gooday, A.J. and Rathburn, A. E., 1999. Temporal variability in living deep-sea benthic foraminifera: a review. *Earth-Science Reviews*, 46, 187–212.

Gooday, A.J. and Turley, C.M., 1990. Response by benthic organisms to inputs of organic material to the ocean floor: a review. *Philosophical Transactions of the Royal Society of London*, A331, 119–138.

Gooday, A.J., Bernhard, J.M., Levin, L.A. and Suhr, S.B., 2000. Foraminifera in the Arabian Sea oxygen minimum zone and other oxygen-deficient settings: taxonomic composition, diversity, and relation to metazoan faunas. *Deep-Sea Research*, II, 47, 25–54.

Gooday, A.J., Bett, B.J., Shires, R. and Lambshead, P.J.D., 1998. Deep-sea benthic foraminiferal species diversity in the NE Atlantic and NW Arabian Sea: a synthesis. *Deep-Sea Research*, II, 45, 165–201.

Gooday, A.J., Hughes, J.A. and Levin, L.A., 2001. The foraminiferan macrofauna from three North Carolina (USA) slope sites with contrasting carbon flux: a comparison with the metazoan macrofauna. *Deep-Sea Research I*, 48, 1709–1739.

Gooday, A.J., Levin, L.A., Linke, P. and Heeger, T., 1992. The role of benthic foraminifera in deep-sea food webs and carbon cycling. In G.T. Rowe and V. Pariente (eds), *Deep-sea food chains and the global carbon cycle*. Kluwer Academic Publishers, The Netherlands, pp. 63–91.

Gowing, M.M., 1993. Seasonal radiolarian flux at the VERTEX North Pacific time-series site. *Deep-Sea Research I*, 40, 517–545.

Gowing, M.M. and Coale, S.L., 1989. Fluxes of living radiolarians and their skeletons along a northeast Pacific transect from coastal upwelling to open ocean waters. *Deep-Sea Research I*, 36, 561–576.

Green, D.G. and Dolman, G.S., 1988. Fine resolution pollen analysis. *Journal of Biogeography*, 15, 685–701.

Green, J.C., 1991. Phagotrophy in prymnesiophyte flagellates. In D.J. Patterson and J. Larsen (eds), *The biology of free-living heterotrophic flagellates*. The Systematics Association Special Volume No. 45, Clarendon Press, Oxford, pp. 401–414.

Green, J.C., Heimdal, B.R., Paasche, E. and Moate, R., 1998. Changes in calcification and the dimensions of coccoliths of *Emiliania huxleyi* (Haptophyta) grown at reduced salinities. *Phycologia*, 37, 121–131.

Green, M.A., Aller, R.C. and Aller, J.Y., 1993. Carbonate dissolution and temporal abundances of foraminifera in Long Island Sound sediments. *Limnology and Oceanography*, 38, 331–345.

Grice, K., Klein Breteler, W.C.M., Schouten, S., Grossi, V., de Leeuw, J.W. and Sinninghe Damsté, J.S., 1998. Effects of zooplankton herbivory on biomarker proxy records. *Paleoceanography*, 13, 686–693.

Griffiths, H., 1995. European Quaternary freshwater ostracoda: a biostratigraphic and palaeobiogeographic primer. *Scopolia*, 34, 1–168.

Griffiths, H. and Evans, J., 1995. The late-glacial and early Holocene colonisation of the British Isles by freshwater Ostracoda. In J. Riha (ed.), *Ostracoda and biostratigraphy.* Balkema, Rotterdam, pp. 291–302.

Griffiths, H. and Holmes, J., 2000. Non-marine ostracods and Quaternary palaeoenvironments. *Quaternary Research Association Technical Guide*, 8, 188 pp.

Griffiths, H. and Horne, D., 1998. Fossil distribution of reproductive modes in non-marine ostracods. In K. Martens (ed.), *Sex and parthenogenesis. Evolutionary ecology of reproductive modes in non-marine ostracods.* Backuys Publishers, Leiden, pp. 101–118.

Grigg, U.M. and Siddiqui, Q.A., 1993. Observations on distribution and probable vectors of five cytheracean ostracod species from estuaries and mudflats near Darmouth, Nova Scotia, Canada. In P. Jones and K. McKenzie, *Ostracoda in the earth and life sciences.* Balkema, Rotterdam, pp. 503–514.

Grimm, E.C., 1991. *TILIA and TILIAGRAPH.* Illinois State Museum, Springfield, IL.

Grøssfjeld, K., Larsen, E. and Sejrup, H.P., *et al.*, 1999. Dinoflagellate cysts reflecting surface-water conditions in Voldafjorden, western Norway during the last 11 300 years. *Boreas*, 28, 403–415.

Guilbault, J.-P., Clague, J.J. and Lapointe, M., 1995. Amount of subsidence during a late Holocene earthquake – evidence from fossil tidal marsh foraminifera at Vancouver Island, west coast of Canada. *Palaeogeography, Palaeoclimatology, Palaeoecology*, 118, 49–71.

Guilbault, J.-P., Clague, J.J. and Lapointe, M., 1996. Foraminiferal evidence for the amount of coseismic subsidence during a late Holocene earthquake on Vancouver Island, west coast of Canada. *Quaternary Science Reviews*, 15, 913–937.

Guilderson, T.P., Fairbanks, R.G. and Rubenstone, J.L., 1994. Tropical temperature variations since 20,000 years ago: modulating interhemispheric climate change. *Science*, 263, 663–664.

Guiot, J., 1990. Methods and programs of statistics for paleoclimatology and paleo-ecology. In J. Guiot and L. Labeyrie (eds), *Quantification des changements climatiques: méthode et programmes.* Institut National des Sciences de l'Univers (INSU-France), Monographie No. 1, 253 pp.

Gupta, A.K., 1999. Latest Pliocene through Holocene paleoceanography of the eastern Indian Ocean: benthic foraminiferal evidence. *Marine Geology*, 161, 63–73.

Gupta, A.K. and Srinivasan, M.S., 1990. Response of northern Indian Ocean deep-sea benthic foraminifera to global climates during Pliocene–Pleistocene. *Marine Micropaleontology*, 16, 77–91.

Gupta, S.M. and Fernandes, A.A., 1997. Quaternary radiolarian faunal changes in the tropical Indian Ocean: inferences to paleomonsoonal oscillation of the 10°S hydrographic front. *Current Science*, 72, 965–972.

Hagino, K., Okada, H. and Matsuoka, H., 2000. Spatial dynamics of coccolithophore assemblages in the Equatorial Western–Central Pacific Ocean. *Marine Micropaleontology*, 39, 53–72.

Hallock, P., 2000. Larger foraminifera as indicators of coral-reef vitality. In R. Martin (ed.), *Environmental Micropaleontology.* Kluwer Academic/Plenum Publishers, New York, pp. 121–150.

Hammarlund, D., 1999. Ostracod stable-isotope records from a deglacial isolation sequence in southern Sweden. *Boreas*, 29, 564–574.

Harvey, H.R., Eglinton, G., O'Hara, S.C.M. and Corner, E.D.S., 1987. Biotransformation and assimilation of dietary lipids by *Calanus* feeding on a dinoflagellate. *Geochimica et Cosmochimica Acta*, 51, 3031–3040.

Haslett, S.K., 1992. Early Pleistocene glacial–interglacial radiolarian assemblages from the eastern equatorial Pacific. *Journal of Plankton Research*, 14, 1553–1563.

Haslett, S.K., 1993. Polycystine Radiolaria: geological and environmental applications. *Microscopy and Analysis*, November, 14–16.

Haslett, S.K., 1994. High-resolution radiolarian abundance data through the Late Pliocene Olduvai subchron of ODP Hole 677A (Panama Basin, eastern equatorial Pacific). *Revista Española de Micropaleontologia*, 26, 127–162.

Haslett, S.K., 1995a. Modern and palaeoecological significance of the radiolarian *Spongaster tetras tetras* Ehrenberg in the eastern equatorial Pacific. *P. S. Z. N. I: Marine Ecology*, 16, 273–281.

Haslett, S.K., 1995b. Mapping Holocene upwelling in the eastern equatorial Pacific using Radiolaria. *The Holocene*, 5, 470–478.

Haslett, S.K., 1995c. Pliocene–Pleistocene radiolarian biostratigraphy and palaeoceanography of the North Atlantic. In R.A. Scrutton, M.S. Stoker, G.B. Shimmield and A.W. Tudhope (eds), *The tectonics, sedimentation and palaeoceanography of the North Atlantic region*. Geological Society Special Publication No. 90, pp. 217–225.

Haslett, S.K., 1996. Radiolarian faunal data through the Plio-Pleistocene Olduvai magnetosubchron of ODP Leg 138 sites 847, 850, and 851 (eastern equatorial Pacific). *Revista Española de Micropaleontologia*, 28, 225–256.

Haslett, S.K. and Funnell, B.M., 1996. Sea-surface temperature variation and palaeo-upwelling throughout the Plio-Pleistocene Olduvai subchron of the eastern equatorial Pacific: an analysis of radiolarian data from ODP sites 677, 847, 850 and 851. In A. Moguilevsky and R.C. Whatley (eds), *Microfossils and oceanic environments*. University of Wales, Aberystwyth Press, pp. 155–164.

Haslett, S.K. and Kersley, C.L., 1995. Early Pleistocene planktonic foraminifera from the tropical Indian Ocean. *Microscopy and Analysis*, March, 25–27.

Haslett, S.K. and Kersley, C.L., 1997. Mounting planktonic foraminifera for scanning electron microscopy. *Microscopy and Analysis*, May, 33–34.

Haslett, S.K. and Robinson, P.D., 1991. Detecting Radiolaria in the field. *Journal of Micropalaeontology*, 10, 22.

Haslett, S.K., Davies, P. and Strawbridge, F., 1998. Reconstructing Holocene sea-level change in the Severn Estuary and Somerset Levels: the foraminifera connection. *Archaeology in the Severn Estuary*, 8 (for 1997), 29–40.

Haslett, S.K., Funnell, B.M. and Dunn, C.L. 1994a. Calcite preservation, palaeoproductivity and the radiolarian *Lamprocyrtis neoheteroporos* Kling in Plio-Pleistocene sediments from the eastern equatorial Pacific. *Neues Jahrbuch für Geologie und Paläontologie, Monatschefte*, 1994, 82–94.

Haslett, S.K., Funnell, B.M., Bloxham, K.S. and Dunn, C.L., 1994b. Plio-Pleistocene palaeoceanography of the tropical Indian Ocean (ODP Hole 709C): radiolarian and $CaCO_3$ evidence. *Journal of Quaternary Science*, 9, 199–208.

Haslett, S.K., Strawbridge, F., Martin, N.A. and Davies, C.F.C., 2001. Vertical salt-marsh accretion and its relationship to sea-level in the Severn Estuary, U.K.: an investigation using foraminifera as tidal indicators. *Estuarine, Coastal and Shelf Science*, 52, 143–153.

Hays, J.D. 1965. Radiolaria and late Tertiary and Quaternary history of Antarctic Seas. In G.A. Llano (ed.) *Biology of Antarctic Seas II*. American Geophysical Union (Antarctic Research Series Volume 5), pp. 125–184.

Hays, J.D., Imbrie, J. and Shackleton, N.J. 1976a. Variations in the Earth's orbit: pacemaker of the Ice Ages. *Science*, 194, 1121–1132.

Hays, J.D., Lozano, J.A., Shackleton, N. and Irving, G. 1976b. Reconstruction of the Atlantic and western Indian Ocean sectors of the 18,000 B.P. Antarctic Ocean. *Memoir of the Geological Society of America*, 145, 337–374.

Hayward, B.W., Grenfell, H.R. and Scott, D.B., 1999. Tidal range of foraminifera for determining former sea-level heights in New Zealand. *New Zealand Journal of Geology and Geophysics*, 42, 395–413.

Hecht, A.D., 1976. An ecologic model for test size variation in recent planktonic foraminifera, application to the fossil record. *Journal of Foraminiferal Research*, 6, 295–311.

Hecker, B., 1990. Photographic evidence for the rapid flux of particles to the seafloor and their transport down the continental slope. *Deep-Sea Research*, 37, 1773–1782.

Heinrich, H., 1988. Origin and consequences of cyclic ice rafting in the Northeast Atlantic Ocean during the past 130,000 years. *Quaternary Research*, 29, 142–152.

Hemphill-Haley, E., 1995. Diatom evidence for earthquake induced subsidence and tsunami 300 yr ago in southern coastal Washington. *Geological Society of America Bulletin*, 107(3), 367–378

Henderson, P., 1990. *Freshwater Ostracods. Synopsis of the British Fauna (New Series), No. 4.* Oegstgest, Universal Book Services/Dr W. Backhuys for the Linnean Society of London, 228 pp.

Hendey, N.I., 1937. The planktonic diatoms of the Southern Seas. *Discovery Reports*, 16, 151–364.

Hendey, N.I., 1964. An introductory account of the smaller algae of British coastal waters. Part V. Bacillariophyceae (diatoms). *MAFF, Fisheries Investigation Series iv*, HMSO, London, 317 pp.

Hendy, I.L. and Kennett, J.P., 2000. Dansgaard–Oeschger cycles and the California Current System: planktonic foraminiferal response to rapid climate change in Santa Barbara Basin, Ocean Drilling Program hole 893A. *Paleoceanography*, 15, 30–42.

Henriksson, A.S., 2000. Coccolithophore response to oceanographic changes in the equatorial Atlantic during the last 200,000 years. *Palaeogeography, Palaeoclimatology, Palaeoecology*, 156, 161–173.

Herguera, J.C., 1992. Deep-sea benthic foraminifera and biogenic opal: glacial to postglacial productivity changes in the western equatorial Pacific. *Marine Micropaleontology*, 19, 79–98.

Herguera, J.C., 1994. Nutrient, mixing and export indices: a 250 kyr productivity record from the western equatorial Pacific. In R. Zahn, T.F. Pedersen, M.A. Kaminski and L. Labeyrie (eds), *Carbon cycling in the glacial ocean: constraints on the ocean's role in global change*. NATO ASI Series, I 17, Springer-Verlag, Berlin, 481–520.

Herguera, J.C., 2000. Last glacial paleoproductivity patterns in the eastern equatorial Pacific: benthic foraminifera records. *Marine Micropaleontology*, 40, 259–275.

Herguera, J.C. and Berger, W.H., 1991. Paleoproductivity from benthic foraminifera abundance: glacial to postglacial change in the west-equatorial Pacific. *Geology*, 19, 1173–1176.

Hermelin, J.O.R. and Shimmield, G.B., 1995. Impact of productivity events on the benthic foraminiferal fauna in the Arabian Sea over the last 150,000 years. *Paleoceanography*, 10, 85–116.

Heybroek, H.M., 1963. Diseases and lopping for fodder as possible causes of a prehistoric decline of *Ulmus*. *Acta Botanica Neerlandica*, 12, 1–11.

Hibbert, F.A. and Switsur, V.R., 1976. Radiocarbon dating of Flandrian pollen zones in Wales and Northern England. *New Phytologist*, 77, 793–807.

Hilgen, F.J., Lourens, L.J., Berger, A. and Loutre, M.F., 1993. Evaluation of the astronomically calibrated timescale for the late Pliocene and earliest Pleistocene. *Paleoceanography*, 8, 549–565.

Hill, M.O., 1973. Reciprocal averaging: an eigenvector method of ordination. *J. Ecology*, 61, 237–251.

Hill, M.O. and Gauch, H.G., 1980. Detrended correspondence analysis: an improved ordination technique. *Vegetatio*, 42, 47–58.

Hippensteel, S.P. and Martin, R.E., 1999. Foraminifera as an indicator of overwash deposits, barrier island sediment supply, and barrier island evolution: Folly Island, South Carolina. *Palaeogeography, Palaeoclimatology, Palaeoecology*, 149, 115–125.

Hodell, D.A., 1993. Late Pleistocene paleoceanography of the South Atlantic sector of the Southern Ocean: Ocean Drilling Program Hole 704A. *Paleoceanography*, 8, 47–67.

Hoen, P., 1999. *Glossary of pollen and spore terminology, 2nd revised edn.* Last updated 17/7/99; accessed 17/10/00 at http://www.bio.uu.nl/~palaeo/glossary/glosint.htm

Holcová, K., 1999. Postmortem transport and resedimentation of foraminiferal tests: relations to cyclical changes of foraminiferal assemblages. *Palaeogeography, Palaeoclimatology, Palaeoecology*, 145, 157–182.

Holmes, J.A., 1996. Trace-element and stable-isotope geochemistry of non-marine ostracod shells in Quaternary palaeoenvironmental reconstruction. *Journal of Paleolimnology*, 15, 223–235.

Horne, D.J., 1983. Life-cycles of Podocopid Ostracoda – a review (with particular reference to marine and brackish-water species). In R.F. Maddocks (eds), *Applications of Ostracoda*. University of Houston Geosciences, Houston, pp. 581–590.

Horton, B.P., 1999. The distribution of contemporary intertidal foraminifera at Cowpen Marsh, Tees Estuary, UK: implications for studies of Holocene sea-level changes. *Palaeogeography, Palaeoclimatology, Palaeoecology*, 149, 127–149.

Horton, B.P., Edwards, R.J. and Lloyd, J.M., 1999. UK intertidal foraminiferal distributions: implications for sea-level studies. *Marine Micropaleontology*, 36, 205–223.

Huber, R., Meggers, H., Baumann, K.H., Raymo, M.E. and Henrich, R., 2000. Shell size variation of planktonic foraminifer *Neogloboquadrina pachyderma* sin. in the Norwegian–Greenland Sea during the last 1.3 Myrs: implications for paleoceanographic reconstructions. *Palaeogeography, Palaeoclimatology, Palaeoecology*, 160, 193–212.

Huntley, B., 1992a. Rates of change in the European palynological record of the last 13,000 years and their climatic interpretation. *Climate Dynamics*, 6, 185–191.

Huntley, B., 1992b. Pollen–climate response surfaces and the study of climate change. In J.M. Gray (ed.), *Applications of Quaternary Research*. Quaternary Research Association, Cambridge, pp. 91–99.

Huntley, B. and Birks, H.J.B., 1983. *An atlas of past and present pollen maps for Europe, 0–13,000 years ago*. Cambridge University Press, Cambridge.

Huntley, M., Capon, R., Marin, V., *et al.*, 1983. Acrylic acid in a dinoflagellate suppresses copepod feeding. *EOS*, 64, 1036.

Hurlbert, S.H., 1971. The non-concept of species diversity: a critique and alternative parameters. *Ecology*, 52, 577–586.

Hustedt, F., 1937–1939. Systematische und ökologische Untersuchungen über den Diatomeen-Flora von Java, Bali, Sumatra. *Archiv für Hydrobiologie* (Suppl.) 12 and 16.

Ikehara, M., Kawamura, K., Ohkouchi, N., *et al.*, 1997. Alkenone sea surface temperature in the Southern Ocean for the last two deglaciations. *Geophysical Research Letters*, 24, 679–682.

Imbrie, J. and Kipp, N.G., 1971. A new micropaleontological method for quantitative paleoclimatology: application to a Late Pleistocene Caribbean core. In K.K. Turekian (ed.), *The Late Cenozoic glacial ages*. Yale University Press, New Haven, pp. 71–181.

Inouye, I. and Pienaar, R.N., 1983. Observations on the life cycle and microanatomy of *Thoracosphaera heimii* (Dinophyceae) with special reference to its systematic position. *South African Journal of Botany*, 2, 63–75.

Iversen, J., 1941. Landnam i Danmarks Stenalder. *Danmarks Geologiske Undersøgelse Series II*, 66, 1–25.

Iversen, J., 1944. *Viscum, Hedera* and *Ilex* as climatic indicators. *Geologiska Föreningens i Stockholm Förhandlingar*, 66, 463–483.

Jansen, E., Befring, S., Bugge, T., Eidvin, T., Holtedahl, H., and Sejrup, H.P., 1987. Large submarine slides on the Norwegian continental margins: sediments, transport and timing. *Marine Geology*, 78, 77–107.

Janssen, C.R., 1970. Problems in the recognition of plant communities in pollen diagrams. *Vegetatio*, 20, 187–198.

Jennings, A.E. and Nelson, A.R., 1992. Foraminiferal assemblage zones in Oregon tidal marshes – relation to marsh floral zones and sea level. *Journal of Foraminiferal Research*, 22, 13–29.

Jennings, A.E., Nelson, A.R., Scott, D.B. and Aravena, J.C., 1995. Marsh foraminiferal assemblages in the Valdivia Estuary, south-central Chile, relative to vascular plants and sea level. *Journal of Coastal Research*, 11, 107–123.

Jessen, K., 1949. Studies in Late Quaternary deposits and flora – history of Ireland. *Proceedings of the Royal Irish Academy*, 52 B6, 85–290 and Pl. III–XVI.

Jian, Z. and Wang, L., 1997. Late Quaternary benthic foraminifera and deep-water paleoceanography in the South China Sea. *Marine Micropaleontology*, 32, 127–154.

Jian, Z., Li, B., Huang, B. and Wang, J., 2000. *Globorotalia truncatulinoides* as indicator of upper-ocean thermal structure during the Quaternary: evidence from the South China Sea and Okinawa Trough. *Palaeogeography, Palaeoclimatology, Palaeoecology*, 162, 287–298.

Jian, Z., Wang, L., Kienast, M., *et al.*, 1999. Benthic foraminiferal paleoceanography of the South China Sea over the last 40,000 years. *Marine Geology*, 156, 159–186.

Johnson, D.A. and Knoll, A.H., 1974. Radiolaria as paleoclimatic indicators: Pleistocene climatic fluctuations in the equatorial Pacific Ocean. *Quaternary Research*, 4, 206–216.

Johnson, D.A. and Nigrini, C., 1980. Radiolarian biogeography in surface sediments of the western Indian Ocean. *Marine Micropaleontology*, 5, 111–152.

Johnson, D.A. and Nigrini, C. 1982. Radiolarian biogeography in surface sediments of the eastern Indian Ocean. *Marine Micropaleontology*, 7, 237–281.

Jones, H.L.J., Leadbeater, B.S.C. and Green, J.C., 1994. Mixotrophy in haptophytes. In J.C. Green and B.S.C. Leadbeater (eds) *The haptophyte algae*. The Systematics Association Special Volume No. 51, Clarendon Press, Oxford, pp. 247–263.

Jones, R.A., 1998. Focused microwave digestion and the oxidation of palynological samples. *Review of Palaeobotany and Palynology*, 103, 17–22.

Jones, R.W. and Charnock, M.A., 1985. 'Morphogroups' of agglutinating foraminifera, their life positions and feeding habits and potential applicability in (paleo)ecological studies. *Revue de Paléobiologie*, 4, 311–320.

Jones, R.L. and Keen, D.H., 1993. *Pleistocene environments in the British Isles.* Chapman and Hall, London.

Jones, R.L., Whatley, R.C., Cronin, T.M. and Dowsett, H.J., 1999. Reconstructing late Quaternary deep-water masses in the eastern Arctic Ocean using benthonic Ostracoda. *Marine Micropaleontology*, 37, 251–272.

Jones, T.P. and Row, N.P. (eds), 1999. *Fossil plants and spores: modern techniques.* Geological Society, London.

Jongman, R.H.G., Ter Braak, C.J.F., van Tongeren, O.F.R. (eds), 1995. *Data Analysis in Community and Landscape Ecology* (reprinted edition). Cambridge University Press, Cambridge.

Jordan, R.W. and Chamberlain, A.H.L., 1997. Biodiversity among haptophyte algae. *Biodiversity and Conservation*, 6, 131–152.

Jordan, R.W. and Green, J.C., 1994. A check-list of the extant Haptophyta of the world. *Journal of the Marine Biological Association of the United Kingdom*, 74, 149–174.

Jordan, R.W. and Winter, A., 2000. Assemblages of coccolithophorids and other living microplankton off the coast of Puerto Rico during January–May 1995. *Marine Micropaleontology*, 39, 113–130.

Jordan, R.W., Kleijne, A., Heimdal, B.R. and Green, J.C., 1995. A glossary of the extant Haptophyta of the world. *Journal of the Marine Biological Association of the United Kingdom*, 75, 769–814.

Jordan, R.W., Zhao, M., Eglinton, G. and Weaver, P.P.E., 1996. Coccolith and alkenone stratigraphy and palaeoceanography at an upwelling site off NW Africa (ODP 658C) during the last 130,000 years. In A. Moguilevsky and R.C. Whatley (eds) *Microfossils and Oceanic Environments*. Aberystwyth Press, University of Wales, UK, pp. 111–130.

Jorissen, F.J., 1999a. Benthic foraminiferal microhabitats below the sediment–water interface. In B.K. Sen Gupta (ed.), *Modern Foraminifera*. Kluwer Academic Publishers, Dordrecht, The Netherlands, pp. 161–179.

Jorissen, F.J., 1999b. Benthic foraminiferal successions across Late Quaternary Mediterranean sapropels. *Marine Geology*, 153, 91–101.

Jorissen, F.J., De Stigter, H.C. and Widmark, J.G.V., 1995. A conceptual model explaining benthic foraminiferal microhabitats. *Marine Micropaleontology*, 26, 3–15.

Jorissen, F.J., Wittling, I., Peypouquet, J.P., Rabouille, C. and Relexans, J.C., 1998. Live benthic foraminiferal faunas off Cape Blanc, NW Africa: community structure and microhabitats. *Deep-Sea Research*, I, 45, 2157–2188.

Juggins, S. and Ter Braak, C.J.F., 1998. *CALIBRATE Version 0.82. A Computer Program for the Graphical Display and Analysis of Species/Environment Relationships by Weighted Averaging, [Weighted Averaging] Partial Least Squares and Principal Components Analysis.* Department of Geography, University of Newcastle, Newcastle-upon-Tyne.

Kaczmarska, I., Barbrick, N.E., Ehrman, J.M. and Cant, G.P., 1993. *Eucampia* Index as an indicator of the late Pleistocene oscillations of the winter sea-ice extent at the ODP Leg 119 Site 745B at the Kerguelen Plateau. *Hydrobiologia*, 269/270, 103–112.

Kaiho, K., 1991. Global changes of Paleogene aerobic/anaerobic benthic foraminifera and deep-sea circulation. *Palaeogeography, Palaeoclimatology, Palaeoecology*, 83, 65–85.

Kaiho, K., 1994. Benthic foraminiferal dissolved oxygen index and dissolved oxygen levels in the modern ocean. *Geology*, 22, 719–722.

Kaiho, K., 1999. Effect of organic carbon flux and dissolved oxygen on the benthic foraminiferal oxygen index (BFOI). *Marine Micropaleontology*, 37, 67–76.

Kanaya, T. and Koizumi, I., 1966. Interpretation of diatom thanatocoenoses from the north Pacific applied to a study of core V20–130. (Studies of a deep sea core V20–130 Part iv). *Scientific Report of Tohoku University Sendai, Second Series (Geology)*, 37, 89–130.

Kang, S.H. and Fryxell, G.A., 1992. *Fragilariopsis cylindrus* (Grunow) Krieger: the most abundant diatom in water column assemblages of Antarctic marginal ice edge zones. *Polar Biology*, 12, 609–627.

Karwath, B., Janofske, D., Tietjen, F. and Willems, H., 2000. Temperature effects on growth and cell size in the marine calcareous dinoflagellate *Thoracosphaera heimii*. *Marine Micropaleontology*, 39, 43–51.

Kawachi, M. and Inouye, I., 1995. Functional roles of the haptonema and the spine scales in the feeding process of *Chrysochromulina spinifera* (Fournier) Pienaar et Norris (Haptophyta = Prymnesiophyta). *Phycologia*, 34, 193–200.

Kawachi, M., Inouye, I., Maeda, O. and Chihara, M., 1991. The haptonema as a food-capturing device: observations on *Chrysochromulina hirta* (Prymnesiophyceae). *Phycologia*, 30, 563–573.

Kellam, S.J. and Walker, J.M., 1989. Antibacterial activity from marine microalgae in laboratory culture. *British Phycological Journal*, 24, 191–194.

Keller, M.D., 1988/1989. Dimethyl sulfide production and marine phytoplankton: the importance of species composition and cell size. *Biological Oceanography*, 6, 375–382.

Kemp, A.E.S and Baldauf, J.G., 1993. Vast Neogene laminated diatom mat deposits from the eastern equatorial Pacific. *Nature*, 362, 141–143.

Kennington, K., Haslett, S.K. and Funnell, B.M., 1999. Offshore transport of neritic diatoms as indicators of surface current and trade wind strength in the Plio-Pleistocene eastern equatorial Pacific. *Palaeogeography, Palaeoclimatology, Palaeoecology*, 149, 151–171.

Kilenyi, T.I., 1972. Transient and balanced genetic polymorphism as an explanation of variable noding in the ostracode genus *Cyprideis torosa*. *Micropaleontology*, 18, 47–63.

Kincaid, E., Thunnell, R.C., Le, J., Lange, C.B., Weinheimer, A.L. and Reid, F.M., 2000. Planktonic foraminiferal fluxes in the Santa Barbara Basin: response to seasonal and interannual hydrographic changes. *Topical Studies in Oceanography*, 47, 1157–1176.

Kinkel, H., Baumann, K.-H. and Cepek, M., 2000. Coccolithophores in the equatorial Atlantic Ocean: response to seasonal and Late Quaternary surface water variability. *Marine Micropaleontology*, 39, 87–112.

Kitazato, H. and Ohga, T., 1995. Seasonal changes in deep-sea benthic foraminiferal populations: results of long-term observations at Sagami Bay, Japan. In H. Sakai and Y. Nozaki (eds), *Biogeochemical processes and ocean flux in the western Pacific*. Terra Scientific Publishing Company, Tokyo, pp. 331–342.

Kitazato, H., Shirayama, Y., Nakatsuka, T., *et al.*, 2000. Seasonal phytodetritus deposition and responses of bathyal benthic foraminiferal populations in Sagami Bay, Japan: preliminary results from 'Project Sagami 1996–1999'. *Marine Micropaleontology*, 40, 135–149.

Kleijne, A., 1990. Distribution and malformation of extant calcareous nannoplankton in the Indonesian Sea. *Marine Micropaleontology*, 16, 293–316.

Kleijne, A., 1991. Holococcolithophorids from the Indian Ocean, Red Sea, Mediterranean Sea and North Atlantic Ocean. *Marine Micropaleontology*, 17, 1–76.

Kling, S.A., 1978. Radiolaria. In B.U. Haq and A. Boersma (eds), *Introduction to marine micropaleontology*. Elsevier, New York, pp. 203–244.

Kling, S.A. 1979. Vertical distribution of polycystine radiolarians in the central North Pacific. *Marine Micropaleontology*, 4, 295–318.

Kling, S.A. and Boltovskoy, D., 1995. Radiolarian vertical-distribution patterns across the southern California Current. *Deep-Sea Research*, 42, 191–231.

Klovan, J.E. and Imbrie, J., 1971. An algorithm and Fortran IV program for large-scale Q-mode factor analysis and calculation of factor scores. *Mathematical Geology*, 3, 61–77.

Koc Karpuz, N. and Schrader, H.J., 1990. Surface sediment diatom distribution and Holocene paleotemperature variations in the Greenland, Iceland and Norwegian Sea. *Palaeoceanography*, 5(4), 557–580.

Koizumi, I., 1989. Holocene pulses of diatom growth in the warm Tsushima Current in the Japan Sea. *Diatom Research*, 4, 55–68.

Koutsoukos, E.A.M. and Hart, M.B., 1990. Cretaceous foraminiferal morphogroup distribution patterns, palaeocommunities and trophic structures: a case study from the Sergipe Basin, Brazil. *Transactions of the Royal Society of Edinburgh: Earth Sciences*, 81, 221–246.

Kovach, W.L., 1995. *MVSP – A MultiVariate Statistical Package for IBM PC's, ver. 2.2.* Kovach Computing Services, Pentraeth, Wales, UK, 71 pp.

Kovach, W.L., 1998. *MVSP – A MultiVariate Statistical Package, for Windows, ver. 3.0.* Kovach Computing Services, Pentraeth, Wales, UK, 127 pp.

Kristensen, P., Heier-Nielsen, S. and Hylleberg, J., 1995. Late-Holocene salinity fluctuations in Bjørnsholm Bay, Limfjorden, Denmark, as deduced from micro- and macrofossil analysis. *The Holocene*, 5, 313–322.

Kristensen, P., Knudsen, K.L. and Sejrup, H.P., 1998. A Middle Pleistocene glacial–interglacial succession in the Inner Silver Pit, southern North Sea: foraminiferal stratigraphy and amino acid geochronology. *Quaternary Science Reviews*, 17, 901–911.

Kroon, D. and Ganssen, E., 1988. Northern Indian Ocean upwelling cells and the stable isotope composition of living planktonic foraminifers. In G.J.A. Brummer and D. Kroon (eds), *Planktonic foraminifers as tracers of ocean climate history*. Free University Press, Amsterdam, pp. 299–319.

Ku, T.-L. and Luo, S., 1992. Carbon isotopic variations on glacial-to-interglacial time scales in the ocean: modeling and implications. *Paleoceanography*, 7, 543–562.

Kuhnt, W., Hess, S. and Jian, Z., 1999. Quantitative composition of benthic foraminiferal assemblages as a proxy indicator for organic carbon flux rates in the South China Sea. *Marine Geology*, 156, 123–157.

Kutzbach, J.E. and Wright, H.E., 1985. Simulation of the climate of 18,000 years BP. Results for the North American/North Atlantic/European sector and comparison with the geologic record of North America. *Quaternary Science Reviews*, 4, 147–187.

Lambeck, K., Smither, C. and Johnston, P., 1998. Sea level change, glacial rebound and mantle viscosity of northern Europe. *Geophysical Journal International*, 134, 102–144.

Lea, D.W., 1993. Constraints on the alkalinity and circulation of glacial Circumpolar Deep Water from benthic foraminiferal barium. *Global Biogeochemical Cycles*, 7, 695–710.

Lea, D.W., 1999. Trace elements in foraminiferal calcite. In B.K. Sen Gupta (ed.), *Modern Foraminifera*. Kluwer Academic Publishers, Dordrecht, The Netherlands, pp. 259–277.

Lea, D.W. and Boyle, E.A., 1989. Barium content of benthic foraminifera controlled by bottom water composition. *Nature*, 338, 751–753.

Lea, D.W. and Boyle, E.A., 1990a. A 210,000-year record of barium variability in the deep northwest Atlantic Ocean. *Nature*, 347, 269–272.

Lea, D.W. and Boyle, E.A., 1990b. Foraminiferal reconstruction of barium distributions in water masses of the glacial oceans. *Paleoceanography*, 5, 719–742.

Lea, D.W., Mashiotta, T.A. and Spero, H.J., 1999. Controls on magnesium and strontium uptake in planktonic foraminifera determined by live culturing. *Geochimica et Cosmochimica Acta*, 63, 2369–2379.

Lee, J.J. and Anderson, O.R., 1991. *Biology of foraminifera*. Academic Press, London, 368 pp.

Lee, J.J. and Hallock, P. (eds), 2000. Advances in the biology of foraminifera. *Micropaleontology*, Suppl. 46, 1–198.

Lentfer, C.J. and Boyd, W.E., 2000. Simultaneous extraction of phytoliths, pollen and spores from sediments. *Journal of Archaeological Science*, 27, 363–372.

Levasseur, M., Michaud, S., Egge, J., *et al.*, 1996. Production of DMSP and DMS during a mesocosm study of an *Emiliania huxleyi* bloom: influence of bacteria and *Calanus finmarchicus* grazing. *Marine Biology*, 126, 609–618.

Leventer, A., Dunbar, R.B. and DeMaster, D.J., 1993. Diatom evidence for late Holocene climatic events in Granite Harbor, Antarctica. *Paleoceanography*, 8(3), 373–386.

Levin, L.A. and Gage, J.D., 1998. Relationships between oxygen, organic matter and the diversity of bathyal macrofauna. *Deep-Sea Research*, II, 45, 129–163.

Libes, S.M., 1992. *An introduction to marine biogeochemistry*. Wiley, New York, 734 pp.

Linke, P. and Lutze, G.F., 1993. Microhabitat preferences of benthic foraminifera – a static concept or a dynamic adaptation to optimize food acquisition? *Marine Micropaleontology*, 20, 215–234.

Lisitzin, A.P., 1971. Distribution of siliceous microfossils in suspension and in bottom sediments. In B.M. Funnell and W.R. Riedel (eds), *The Micropalaeontology of the Oceans*. Cambridge University Press, Cambridge, pp. 173–195.

Lister, G., 1988. Stable-isotopes from lacustrine Ostracoda as tracers for continental palaeoenvironments. In P. De Deckker, J.P. Colin and J.P. Peypouquet, *Ostracoda in the earth sciences*. Elsevier, Amsterdam, pp. 201–218.

Lloyd, J., 2000. Combined foraminiferal and thecamoebian environmental reconstruction from an isolation basin in NW Scotland: implications for sea-level studies. *Journal of Foraminiferal Research*, 30, 294–305.

Locker, S., 1996. Quantitative radiolarian slides prepared from soft marine sediments. *Micropaleontology*, 42, 407–411.

Loeblich, A.R. and Tappan, H., 1987. *Foraminiferal genera and their classification*. Van Nostrand Reinhold Company, New York, 2 volumes.

Lohmann, G.P., 1978. Abyssal benthonic foraminifera as hydrographic indicators in the western South Atlantic. *Journal of Foraminiferal Research*, 8, 6–34.

Lombari, G. and Boden, G., 1985. *Modern radiolarian global distributions*. Cushman Foundation for Foraminiferal Research, Special Publication No. 16A, 125 pp.

Lotter, A.F., 1988. Paläoökologische und Paläolimnologische Studie des Rotsees bei Luzern. Pollen-, grossrest-, diatomeen- und sedimentanalytische Untersuchungen. *Dissertationes Botanicae*, 124, 187 pp.

Loubere, P., 1991. Deep-sea benthic foraminiferal assemblage response to a surface ocean productivity gradient: a test. *Paleoceanography*, 6, 193–204.

Loubere, P., 1994. Quantitative estimation of surface ocean productivity and bottom water oxygen concentration using benthic foraminifera. *Paleoceanography*, 9, 723–737.

Loubere, P., 1996. The surface ocean productivity and bottom water oxygen signals in deep water benthic foraminiferal assemblages. *Marine Micropaleontology*, 28, 247–261.

Loubere, P., 1998. The impact of seasonality on the benthos as reflected in the assemblages of deep-sea foraminifera. *Deep-Sea Research, I*, 45, 409–432.

Loubere, P., 1999. A multiproxy reconstruction of biological productivity and oceanography in the eastern equatorial Pacific for the past 30,000 years. *Marine Micropaleontology*, 37, 173–198.

Loubere, P. and Fariduddin, M., 1999a. Quantitative estimation of global patterns of surface ocean biological productivity and its seasonal variation on timescales from centuries to millenia. *Global Biogeochemical Cycles*, 13, 115–133.

Loubere, P. and Fariduddin, M., 1999b. Benthic foraminifera and the flux of organic carbon to the seabed. In B.K. Sen Gupta (ed.), *Modern Foraminifera*. Kluwer Academic Publishers, Dordrecht, The Netherlands, pp. 181–199.

Loubere, P. and Qian, H., 1997. Reconstructing palaeoecology and paleoenvironmental variables using factor analysis and regression: some limitations. *Marine Micropaleontology*, 31, 205–217.

Lowe, J.J. and Walker, M.J.C., 1997. *Reconstructing Quaternary environments, 2nd edn.* Longman, London.

Lozano, J.A. and Hays, J.D., 1976. Relationship of radiolarian assemblages to sediment types and physical oceanography on the Atlantic and western Indian Ocean sectors of the Antarctic Ocean. *Memoir of the Geological Society of America*, 145, 303–336.

Lund, D.C. and Mix, A.C., 1998. Millennial-scale deep water oscillations: reflections of the North Atlantic in the deep Pacific from 10 to 60 ka. *Paleoceanography*, 13, 10–19.

Lutze, G.F. and Coulbourn, W.T., 1984. Recent benthic from the continental margin of northwest Africa: community structure and distribution. *Marine Micropaleontology*, 8, 361–401.

Lutze, G.F. and Thiel, H., 1989. Epibenthic foraminifera from elevated microhabitats: *Cibicidoides wuellerstorfi* and *Planulina ariminensis*. *Journal of Foraminiferal Research*, 19, 153–158.

Lutze, G.F., Pflaumann, U. and Weinholz, P., 1986. Jungquatäre Fluktuationen der benthischen Foraminiferenfaunen in Tiefsee-Sedimenten vor NW-Afrika: Eine Reaktion auf Productivitätsänderungen im Oberflächenwasser. *'Meteor' Forschungs-Ergebnisse, Reihe C*, 40, 163–180.

Lyle, M.W., Prahl, F.G. and Sparrow, M.A., 1992. Upwelling and productivity changes inferred from a temperature record in the central equatorial Pacific. *Nature*, 355, 812–815.

MacDonald, G.M., 1990. Palynology. In B.H. Warner (ed.) *Methods in Quaternary Ecology*. Geoscience Canada Reprint Series 5, pp. 37–52.

Mackensen, A., 1992. Neogene benthic foraminifers from the southern Indian Ocean (Kerguelen Plateau): biostratigraphy and paleoecology. In S.J. Wise, R. Schlich, *et al.* (eds), *Proceedings of the Ocean Drilling Program, Scientific Results*, 120. College Station, TX (Ocean Drilling Program), pp. 649–673.

Mackensen, A., Hubberten, H.-W., Bickert, T., Fischer, G. and Fütterer, D.K., 1993a. The $\delta^{13}C$ in benthic foraminiferal tests of *Fontbotia wuellerstorfi* (Schwager) relative to the $\delta^{13}C$ of dissolved inorganic carbon in Southern Ocean deep water: implications for glacial ocean circulation models. *Paleoceanography*, 8, 587–610.

Mackensen, A., Fütterer, D.K., Grobe, H. and Schmiedl, G., 1993b. Benthic foraminiferal assemblages from the eastern South Atlantic Polar Front region between 35° and 57°S: distribution, ecology and fossilization potential. *Marine Micropaleontology*, 22, 33–69.

Mackensen, A., Grobe, H., Hubberten, H.-W. and Kuhn, G., 1994. Benthic foraminiferal assemblages and the $\delta^{13}C$-signal in the Atlantic sector of the Southern Ocean: glacial-to-interglacial contrasts. In R. Zahn, T.F. Pedersen, M.A. Kaminski and L. Labeyrie, (eds), *Carbon cycling in the glacial ocean: constraints on the ocean's role in global change*. NATO ASI Series, I 17, Springer-Verlag, Berlin, pp. 105–144.

Mackensen, A., Grobe, H., Kuhn, G. and Fütterer, D.K., 1990. Benthic foraminiferal assemblages from the eastern Weddell Sea between 68 and 73°S: distribution, ecology and fossilization potential. *Marine Micropaleontology*, 16, 241–283.

Mackensen, A., Schmiedl, G., Harloff, J. and Giese, M., 1995. Deep-sea foraminifera in the South Atlantic Ocean: ecology and assemblage generation. *Micropaleontology*, 41, 342–358.

Mackensen, A., Schumacher, S., Radke, J. and Schmidt, D.N., 2000. Microhabitat preferences and stable carbon isotopes of endobenthic foraminifera: clue to quantitative reconstruction of oceanic new production? *Marine Micropaleontology*, 40, 233–258.

Maddocks, R.F., 1992. Ostracoda. In A.G. Humes (ed.), *Microscopic anatomy of invertebrates*. Wiley-Liss, pp. 415–441.

Maddy, D. and Brew, J. (eds), 1995. *Statistical modelling of Quaternary science data*. Quaternary Research Association, Cambridge.

Maestrini, S.Y. and Granéli, E., 1991. Environmental conditions and ecophysiological mechanisms which led to the 1988 *Chrysochromulina polylepis* bloom: an hypothesis. *Oceanologica Acta*, 14, 397–413.

Majoran, S. and Nordberg, K., 1997. Late Weichselian ostracod assemblages from the southern Kattegat, Scandinavia: a palaeoenvironmental study. *Boreas*, 26, 181–200.

Malin, G., Turner, S., Liss, P., Holligan, P. and Harbour, D., 1993. Dimethylsulphide and dimethylsulphoniopropionate in the Northeast Atlantic during the summer coccolithophore bloom. *Deep-Sea Research*, 40, 1487–1508.

Malmgren, B.A. and Haq, B.U., 1982. Assessment of quantitative techniques in palaeobiogeography. *Marine Micropaleontology*, 7, 213–236.

Malmgren, B.A. and Kennett, J.P., 1978. Late Quaternary paleoclimatic applications of mean size variations in *Globigerina bulloides* d'Orbigny in the southern Indian Ocean. *Journal of Paleontology*, 52, 1195–1207.

Marchant, M., Hebbeln, D. and Wefer, G., 1999. High resolution planktic foraminiferal record of the last 13,300 years from the upwelling area off Chile. *Marine Geology*, 161, 115–128.

Marchitto, T.M., Jr, Curry, W.B. and Oppo, D.W., 2000. Zinc concentrations in benthic foraminifera reflect seawater chemistry. *Paleoceanography*, 15, 299–306.

Margalef, R., 1978. Life-forms of phytoplankton as survival alternatives in an unstable environment. *Oceanologica Acta*, 1, 493–509.

Marlowe, I.T., Green, J.C., Neal, A.C., Brassell, S.C., Eglinton, G. and Course, P.A., 1984. Long-chain (n-C_{37}–C_{39}) alkenones in the Prymnesiophyceae. Distribution of alkenones and other lipids and their taxonomic significance. *British Phycological Journal*, 19, 203–216.

Marlowe, I.T., Brassell, S.C., Eglinton, G. and Green, J.C., 1990. Long-chain alkenones and alkyl alkenoates and the fossil coccolith record of marine sediments. *Chemical Geology*, 88, 349–375.

Martens, K., 1998. *Sex and parthenogenesis. Evolutionary ecology of reproductive modes in non-marine ostracods*. Leiden, Backhuys Publishers, 336 pp.

Martin, R. (ed.), 2000. *Environmental micropaleontology*. Kluwer Academic/Plenum Publishers, New York, 481 pp.

Martinez, J.I., De Decker, P. and Barrows, T.T., 1999. Palaeoceanography of the last glacial maximum in the eastern Indian Ocean: planktonic foraminiferal evidence. *Palaeogeography, Palaeoclimatology, Palaeoecology*, 147, 73–99.

Mashiotta, T.A., Lea, D.W. and Spero, H.J., 1997. Experimental determination of cadmium uptake in shells of the planktonic foraminifera *Orbulina universa* and

Globigerina bulloides: implications for surface water paleoreconstructions. *Geochimica et Cosmochimica Acta*, 61, 4053–4065.

Matsumoto, K. and Lynch-Stieglitz, J., 1999. Similar glacial and Holocene deep water circulation inferred from southeast Pacific benthic foraminiferal carbon isotope composition. *Paleoceanography*, 14, 149–163.

Matsuoka, A. and Anderson, O.R., 1992. Experimental and observational studies of radiolarian physiological ecology: 5. Temperature and salinity tolerance of *Dictyocoryne truncatum*. *Marine Micropaleontology*, 19, 299–313.

Matsuoka, K., 1999. Eutrophication process recorded in dinoflagellate cyst assemblages – a case of Yokohama Port, Tokyo Bay, Japan. *The Science of the Total Environment*, 231, 17–35.

Mazzini, I., Aanadon, P. and Barbieri, M., *et al.*, 1999. Late Quaternary sea-level changes along the Tyrrhenian coast near Orbetello (Tuscany, central Italy): palaeoenvironmental reconstruction using ostracods. *Marine Micropaleontology*, 37, 289–312.

McCorkle, D.C., Keigwin, L.D., Corliss, B.H. and Emerson, S.R., 1990. The influence of microhabitats on the carbon isotopic composition of deep-sea benthic foraminifera. *Paleoceanography*, 5, 161–185.

McCorkle, D.C., Corliss, B.H. and Farnham, C.A., 1997. Vertical distributions and stable isotopic compositions of live (stained) benthic foraminifera from the North Carolina and California continental margin. *Deep-Sea Research*, 44, 983–1024.

McCorkle, D.C., Heggie, D.T. and Veeh, H.H., 1998. Glacial and Holocene stable isotope distributions in the southeastern Indian Ocean. *Paleoceanography*, 13, 20–34.

McCoy, W., 1988. Amino acid racemization in fossil non-marine ostracod shells: a potential tool for the study of Quaternary stratigraphy, chronology, and paleotemperature. In P. De Deckker, J.P. Colin and J.P. Peypouquet, *Ostracoda in the earth sciences*. Elsevier, Amsterdam, pp. 219–229.

McIntyre, A. and Bé, A.W.H., 1967. Modern Coccolithophoridae of the Atlantic Ocean – I. Placoliths and cyrtoliths. *Deep-Sea Research*, 14, 561–597.

McIntyre, A. and Molfino, B., 1996. Forcing of Atlantic equatorial and subpolar millennial cycles by precession. *Science*, 274, 1867–1870.

McIntyre, A., Kipp, N.G. and Bé, A.W.H., *et al.*, 1976. Glacial North Atlantic 18,000 years ago: a CLIMAP reconstruction. *Memoir, Geological Society of America*, 145, 43–76.

Medlin, L.K. and Priddle, J., 1990. *Polar marine diatoms*. British Antarctic Survey, Cambridge, 214 pp.

Meisch, C., 2000. Freshwater Ostracoda of Western and Central Europe. In J. Schwoerbel and P. Zwick, *Suesswasserfauna von Mitteleuropa 8/3*. Spektrum Akademischer Verlag, Heidelberg, p. 522.

Meriläinen, J., 1967. The diatom flora and the hydrogen ion concentration of the water. *Annales Botanici Fennici*, 4, 51–58.

Merrett, M.J., Dong, L.F. and Nimer, N.A., 1993. Nitrate availability and calcite production in *Emiliania huxleyi* Lohmann. *European Journal of Phycology*, 28, 243–246.

Mezquita, F., Hernández, R. and Rueda, J., 1999. Ecology and distribution of ostracods in a polluted Mediterranean river. *Palaeogeography, Palaeoclimatology, Palaeoecology*, 148, 87–103.

Michaels, A.F., Caron, D.A., Swanberg, N.R., Howse, F.A. and Michaels, C.M., 1995. Planktonic sarcodines (Acantharia, Radiolaria, Foraminifera) in surface waters near Bermuda – abundance, biomass and vertical flux. *Journal of Plankton Research*, 17, 131–163.

Milankovitch, M., 1941. Kankron der Erdbestrahlung und Seine andwendung auf das eiszeeiten problem. *Royal Serbian Academy, Special Edition*, 132 (translated by the Israel program for scientific translations, Jerusalem).

Miller, K.G. and Lohmann, G.P., 1982. Environmental distribution of Recent benthic foraminifera on the northeast United States continental slope. *Geological Society of America Bulletin*, 93, 200–206.

Mitchell, G.F., 1951. Studies in Irish Quaternary deposits: no. 7. *Proceedings of the Royal Irish Academy*, 53B, 113–206 and pl. IV–VIII.

Mjaaland, G., 1956. Some laboratory experiments on the coccolithophorid *Coccolithus huxleyi. Oikos*, 7, 251–255.

Moestrup, Ø., 1994. Economic aspects: 'blooms', nuisance species, and toxins. In J.C. Green and B.S.C. Leadbeater (eds), *The haptophyte algae*. The Systematics Association Special Volume No. 51, Clarendon Press, Oxford, pp. 265–285.

Molfino, B. and McIntyre, A., 1990a. Precessional forcing of nutricline dynamics in the equatorial Atlantic. *Science*, 249, 766–769.

Molfino, B. and McIntyre, A., 1990b. Nutricline variation in the equatorial Atlantic coincident with the Younger Dryas. *Paleoceanography*, 5, 997–1008.

Molina-Cruz, A., 1977a. Radiolarian assemblages and their relationship to the oceanography of the subtropical southeastern Pacific. *Marine Micropaleontology*, 2, 315–352.

Molina-Cruz, A., 1977b. The relation of the southern trade winds to upwelling processes during the last 75,000 years. *Quaternary Research*, 8, 324–338.

Molina-Cruz, A., 1984. Radiolaria as indicators of upwelling processes: the Peruvian connection. *Marine Micropaleontology*, 9, 53–75.

Molina-Cruz, A., 1988. Late Quaternary oceanography of the mouth of the Gulf of California: the Polycystine connection. *Paleoceanography*, 3, 447–459.

Molina-Cruz, A. and Martinez-López, M., 1994. Oceanography of the Gulf of Tehuantepec, Mexico, indicated by Radiolaria remains. *Palaeogeography, Palaeoclimatology, Palaeoecology*, 110, 179–195.

Moodley, L., Schaub, B.E.M., Van der Zwaan, G.J. and Herman, P.M.J., 1998. Tolerance of benthic foraminifera (Protista: Sarcodina) to hydrogen sulphide. *Marine Ecology Progress Series*, 169, 77–86.

Moodley, L., Van der Zwaan, G.J., Herman, P.M.J., Kempers, A.J. and Van Breugel, P., 1997. Differential response of benthic meiofauna to anoxia with special reference to foraminifera (Protista: Sarcodina). *Marine Ecology Progress Series*, 158, 151–163.

Moore, P.D., Webb, J.A. and Collinson, M.E., 1991. *Pollen Analysis*. Blackwell, Oxford.

Moore, R.C. (ed.), 1961. *Treatise on Invertebrate Paleontology. Part Q: Arthropoda 3; Crustacea, Ostracoda*. Geological Society of America and University of Kansas Press, 442 pp.

Moore, T.C., Jr, 1973. Late Pleistocene–Holocene oceanographic changes in the northeastern Pacific. *Quaternary Research*, 3, 99–109.

Moore, T.C., Jr, 1978. The distribution of radiolarian assemblages in the modern and ice-age Pacific. *Marine Micropaleontology*, 3, 229–266.

Moore, T.C., Jr., Burckle, L.H., Geitzenauer, K., *et al.* 1980. The reconstruction of sea surface temperatures in the Pacific Ocean of 18 000 B.P. *Marine Micropaleontology*, 5, 215–247.

Mudie, P.J. and Harland, R., 1996. Aquatic Quaternary, Chapter 21. In J. Jansonius and D.C. McGregor (eds), *Palynology: principles and applications*. American Association of Stratigraphic Palynologists Foundation, 2, pp. 843–877.

Mullineaux, L.S. and Lohmann, G.P., 1981. Late Quaternary stagnations and recirculation of the eastern Mediterranean: changes in the deep water recorded by fossil benthic foraminifera. *Journal of Foraminiferal Research*, 11, 20–39.

Murray, J.W., 1973. *Distribution and ecology of living benthic foraminiferids*. Heinemann, London, 274 pp.

Murray, J.W., 1986. Benthic foraminifers and Neogene bottom-water masses at Deep Sea Drilling Project Leg 94 North Atlantic sites. In W.F. Ruddiman, R.B. Kidd, E. Thomas, *et al*. (eds), *Initial Reports of the Deep Sea Drilling Project*, 94. US Government Printing Office, Washington, DC, pp. 965–979.

Murray, J.W., 1988. Neogene bottom-water masses and benthic foraminifera in the NE Atlantic Ocean. *Journal of the Geological Society, London*, 145, 125–132.

Murray, J.W., 1991. *Ecology and palaeoecology of benthic foraminifera*. Longman, Harlow, Essex, 397 pp.

Murray, J.W., 1995. Microfossil indicators of ocean water masses, circulation and climate. In D.W.J. Bosence and P.A. Allison (eds), *Marine palaeoenvironmental analysis from fossils*. Geological Society of London, Special Publication, 83, pp. 245–264.

Murray, J.W., 2000. Revised taxonomy, *An Atlas of British Recent Foraminiferids*. *Journal of Micropalaeontology*, 19, 44.

Murray, J.W., 2001. The niche of benthic foraminifera, critical thresholds and proxies. *Marine Micropaleontology*, 41, 1–7.

Murray, J.W. and Alve, E., 1999. Natural dissolution of modern shallow water benthic foraminifera: taphonomic effects on the palaeoecological record. *Palaeogeography, Palaeoclimatology, Palaeoecology*, 146, 195–209.

Murray, J.W. and Hawkins, A.B., 1976. Sediment transport in the Severn Estuary during the past 8000–9000 years. *Journal of the Geological Society*, 132, 385–398.

Murray, J.W., Weston, J.F., Haddon, C.A. and Powell, A.D.J., 1986. Miocene to Recent bottom water masses of the north-east Atlantic: an analysis of benthic foraminifera. In C.P. Summerhayes and N.J. Shackleton (eds), *North Atlantic palaeoceanography*. Geological Society of London, Special Publication, 21, pp. 219–230.

Naidu, P.D. and Malmgren, B.A., 1995a. Do benthic foraminifer records represent a productivity index in oxygen mimimum zone areas? An evaluation from the Oman Margin, Arabian Sea. *Marine Micropaleontology*, 26, 49–55.

Naidu, P.D. and Malmgren, B.A., 1995b. Monsoon upwelling effects on test size of some planktonic foraminiferal species from the Oman Margin, Arabian Sea. *Paleoceanography*, 10, 117–122.

Naish, T. and Kamp, P.J.J., 1997. Foraminiferal depth palaeoecology of Late Pliocene shelf sequences and systems tracts, Wanganui Basin, New Zealand. *Sedimentary Geology*, 110, 237–255.

Nakagawa, T., Brugiapaglia, E., Digerfeldt, G., Reille, M., De Beaulieu, J.L. and Yasuda, Y., 1998. Dense-media separation as a more efficient pollen extraction method for use with organic sediment/deposit samples: comparison with the conventional method. *Boreas*, 27, 15–24.

Nanninga, H.J. and Tyrrell, T., 1996. Importance of light for the formation of algal blooms by *Emiliania huxleyi*. *Marine Ecology Progress Series*, 136, 195–203.

Neale, J.W., 1988. Ostracods and palaeosalinity reconstruction. In P. De Deckker, J.P. Colin and J.P. Peypouquet (eds), *Ostracoda in the earth sciences*. Elsevier, Amsterdam, pp. 125–155.

Needham, H.D., Habib, D. and Heezen, B.C., 1969. Upper Carboniferous palynomorphs as a tracer of red sediment dispersal patterns in the northwest Atlantic. *Journal of Geology*, 77, 113–120.

Nees, S., 1997. Late Quaternary palaeoceanography of the Tasman Sea: the benthic foraminiferal view. *Palaeogeography, Palaeoclimatology, Palaeoecology*, 131, 365–389.

Nees, S., Altenbach, A.V., Kassens, H. and Thiede, J., 1997. High-resolution record of foraminiferal response to late Quaternary sea-ice retreat in the Norwegian–Greenland Sea. *Geology*, 25, 659–662.

Nees, S., Armand, L., De Deckker, P., Labracherie, M. and Passlow, V., 1999. A diatom and benthic foraminiferal record from the South Tasman Rise (southeastern Indian Ocean): implications for palaeoceanographic changes for the last 200,000 years. *Marine Micropaleontology*, 38, 69–89.

Nigrini, C., 1967. Radiolaria in pelagic sediments from the Indian and Atlantic Oceans. *Bulletin of the Scripps Institution of Oceanography*, 11, 1–125.

Nigrini, C., 1968. Radiolaria from eastern tropical Pacific sediments. *Micropaleontology*, 14, 51–63.

Nigrini, C., 1970. Radiolarian assemblages in the North Pacific and their application to a study of Quaternary sediments in core V20–130. *Memoir of the Geological Society of America*, 126, 139–183.

Nigrini, C. and Caulet, J.-P., 1992. Late Neogene radiolarian assemblages characteristic of Indo-Pacific areas of upwelling. *Micropaleontology*, 38, 139–164.

Nigrini, C. and Moore, T.C., Jr, 1979. *A guide to modern radiolaria*. Cushman Foundation for Foraminiferal Research, Special Publication, No. 16, 260 pp.

Nimer, N.A. and Merrett, M.J., 1993. Calcification rate in *Emiliania huxleyi* Lohmann in response to light, nitrate and availability of inorganic carbon. *New Phytologist*, 123, 673–677.

Nolet, G.J. and Corliss, B.H., 1990. Benthic foraminiferal evidence for reduced deep-water circulation during sapropel deposition. *Marine Geology*, 94, 109–130.

Nordberg, K., Gustafsson, M. and Krantz, A.L., 2000. Decreasing oxygen concentrations in the Gullmar Fjord, Sweden, as confirmed by benthic foraminifera, and the possible association with NAO. *Journal of Marine Systems*, 23, 303–316.

Nurnberg, D., Muller, A. and Schneider, R.R., 2000. Paleo-sea surface temperature calculations in the equatorial east Atlantic from Mg/Ca ratios in planktic foraminifera: a comparison to sea surface estimates from U^K_{37}, oxygen isotopes, and foraminiferal transfer function. *Paleoceanography*, 15, 124–134.

Nygaard, G., 1956. Ancient and recent flora of diatoms and chysophyceae in Lake Gribsö. Studies on the humic acid Lake Gribsö. *Folia Limnologica Scandinavica*, 8, 32–94.

Nygaard, K. and Tobiesen, A., 1993. Bacterivory in algae: a survival strategy during nutrient limitation. *Limnology and Oceanography*, 38, 273–279.

Oggioni, E. and Zandini, L., 1987. Response of benthic foraminifera to stagnant episodes – a quantitative study of Core Ban 81–23, eastern Mediterranean. *Marine Geology*, 75, 241–261.

Ohga, T. and Kitazato, H., 1997. Seasonal changes in bathyal foraminiferal populations in response to the flux of organic matter (Sagami Bay, Japan). *Terra Nova*, 9, 33–37.

Ohkouchi, N., Kawamura, K., Nakamura, T. and Taira, A., 1994. Small changes in the sea surface temperature during the last 20,000 years: molecular evidence from the western tropical Pacific. *Geophysical Research Letters*, 21, 2207–2210.

Ohkushi, K., Thomas, E. and Kawahata, H., 2000. Abyssal benthic foraminifera from the northwestern Pacific (Shatsky Rise) during the last 298 kyr. *Marine Micropaleontology*, 38, 119–147.

Okada, H., 1983. Modern nannofossil assemblages in sediments of coastal and marginal seas along the western Pacific Ocean. In J.E. Meulenkamp (ed.), *Reconstruction of marine paleoenvironments*. Utrecht Micropaleontological Bulletin, 30, pp. 171–187.

Okada, H. and Honjo, S., 1973. The distribution of oceanic coccolithophorids in the Pacific. *Deep-Sea Research*, 20, 355–374.

Okada, H. and Honjo, S., 1975. Distribution of coccolithophores in marginal seas along the western Pacific Ocean and in the Red Sea. *Marine Biology*, 31, 271–285.

Okada, H. and Matsuoka, M., 1996. Lower-photic nannoflora as an indicator of the late Quaternary monsoonal palaeo-record in the tropical Indian Ocean. In A. Moguilevsky and R.C. Whatley (eds), *Microfossils and oceanic environments*. Aberystwyth Press, University of Wales, UK, pp. 231–245.

Okada, H. and McIntyre, A., 1979. Seasonal distribution of modern coccolithophores in the Western North Atlantic Ocean. *Marine Biology*, 54, 319–328.

Okada, H. and Wells, P., 1997. Late Quaternary nannofossil indicators of climate change in two deep-sea cores associated with the Leeuwin current off Western Australia. *Palaeogeography, Palaeoclimatology, Palaeoecology*, 131, 413–432.

Oppo, D.W. and Fairbanks, R.G., 1987. Variability in the deep and intermediate water circulation of the Atlantic Ocean during the past 25,000 years: Northern Hemisphere modulation of the Southern Ocean. *Earth and Planetary Science Letters*, 86, 1–15.

Oppo, D.W. and Rosenthal, Y., 1994. Cd/Ca changes in a deep Cape Basin core over the past 730,000 years: response of circumpolar deepwater variability to northern hemisphere ice sheet melting? *Paleoceanography*, 9, 661–675.

Oppo, D.W., Fairbanks, R.G., Gordon, A.L. and Shackleton, N.J., 1990. Late Pleistocene Southern Ocean δ^{13}C variability. *Paleoceanography*, 5, 43–54.

Paasche, E. and Brubak, S., 1994. Enhanced calcification in the coccolithophorid *Emiliania huxleyi* (Haptophyceae) under phosphorus limitation. *Phycologia*, 33, 324–330.

Paasche, E., Brubak, S., Skattebøl, S., Young, J.R. and Green, J.C., 1996. Growth and calcification in the coccolithophorid *Emiliania huxleyi* (Haptophyceae) at low salinities. *Phycologia*, 35, 394–403.

Paasche, E., in press. A review of the coccolithophorid *Emiliania huxleyi* (Prymnesiophyceae), with particular reference to growth, coccolith formation, and calcification–photosynthesis interactions. *Phycologia*.

Palmer, A.J.M. and Abbot, W.H., 1986. Diatoms as indicators of sea level change. In O. van de Plassche (ed.), *Sea-level research, a manual for the collection and evaluation of data*. Geo-Books, Norwich, UK, pp. 457–489.

Parke, M. and Adams, I., 1960. The motile (*Crystallolithus hyalinus* Gaarder & Markali) and non-motile phases in the life history of *Coccolithus pelagicus* (Wallich) Schiller. *Journal of the Marine Biological Association of the United Kingdom*, 39, 263–274.

Parker, W.C. and Arnold, A.J., 1999. Quantitative methods of data analysis in foraminiferal ecology. In B.K. Sen Gupta (ed.), *Modern Foraminifera*. Kluwer Academic Publishers, Dordrecht, The Netherlands, pp. 71–89.

Patrick, R. and Reimer, C.W., 1966–1975. The diatoms of the United States. 2 vols. *Monograph of the Academy of Natural Sciences of Philadelphia*, 13 pp.

Patterson, R.T. and Fishbein, E., 1989. Re-examination of the statistical methods used to determine the number of point counts needed for micropalaeontological quantitative research. *Journal of Paleontology*, 63, 245–248.

Patterson, R.T., Guilbault, J.-P. and Clague, J.J., 1999. Taphonomy of tidal marsh foraminifera: implications of surface sample thickness for high-resolution sea-level studies. *Palaeogeography, Palaeoclimatology, Palaeoecology*, 149, 199–211.

Paul, M.A. and Barras, B.F., 1998. A geotechnical correction for post-depositional sediment compression: examples from the Forth Valley, Scotland. *Journal of Quaternary Science*, 13, 171–176.

Pawlowski, J., 1991. Distribution and taxonomy of some benthic tiny foraminifers from the Bermuda Rise. *Micropaleontology*, 37, 163–172.

Pedersen, T.F., Pickering, M., Vogel, J.S., Southon, J.N. and Nelson, D.E., 1988. The response of benthic foraminifera to productivity cycles in the eastern equatorial Pacific: faunal and geochemical constraints on glacial bottom water oxygen levels. *Paleoceanography*, 3, 157–168.

Peglar, S.M. and Birks, H.J.B., 1993. The mid-Holocene *Ulmus* decline at Diss Mere, south-east England – disease and human impact? *Vegetation History and Archaeobotany*, 2, 61–68.

Pennington, W., 1984. Long term natural acidification of upland sites in Cumbria: evidence from post glacial lake sediments. *Freshwater Biological Association Annual Report*, 52, 28–46.

Perch-Nielsen, K., 1985a. Silicoflagellates. In H.M. Bolli, J.B. Saunders and K. Perch-Nielsen (eds), *Plankton stratigraphy*. Cambridge University Press, pp. 811–846.

Perch-Nielsen, K., 1985b. Mesozoic calcareous nannofossils. In H.M. Bolli, J.B. Saunders and K. Perch-Nielsen (eds), *Plankton Stratigraphy*, Cambridge University Press, Cambridge, pp. 329–426.

Peypouquet, J.-P., 1975. Les variation des caractères morphologiques internes chez les Ostracodes des genres *Krithe* et *Parakrithe*: relation possible avec la teneur en dissous dans l'eau. *Bullétin de l'Institut Géologique du Bassin d'Aquitaine*, 17, 81–88.

Peypouquet, J.-P., Ducasse, O., Gayet, J. and Pratviel, L., 1980. Agradation et dégradation des tests d'ostracodes. Intérêt pour la connaissance de l'évolution paléohydrologique des domaines margino-littoraux. In *Cristallisation, déformation, dissolution des carbonates*, Réunion spéciale, Bordeaux III, 357–369.

Peypouquet, J.-P., Carbonel, P., Ducasse, O., Tolderer-Farmer, M. and Lété, C., 1988. Environmentally cued polymorphism of ostracods. In T. Hanai, N. Ikeya and K. Ishizaki (eds), *Evolutionary biology of Ostracoda: its fundamentals and applications*. Kodansha, Tokyo, pp. 1003–1018.

Pflaumann, U. and Jian, Z., 1999. Modern distribution patterns of planktonic foraminifera in the South China Sea and western Pacific: a new transfer technique to estimate regional sea-surface temperatures. *Marine Geology*, 156, 41–83.

Phleger, F.B., 1960. *Ecology and distribution of Recent foraminifera*. Johns Hopkins Press, Baltimore, 297 pp.

Phleger, F.B. and Soutar, A., 1973. Production of benthic foraminifera in three east Pacific oxygen minima. *Micropaleontology*, 19, 110–115.

Phleger, F.B. and Walton, W.R., 1950. Ecology of marsh and bay foraminifera, Barnstable, Mass. *American Journal of Science*, 148, 174–194.

Pichon, J.J., Labracherie, M., Labeyrie, L.D. and Duprat, J., 1987. Transfer functions between diatom assemblages and surface hydrology of the Southern Ocean. *Palaeogeography, Palaeoclimatology, Palaeoecology*, 61, 79–95.

Pike, J. and Kemp, A.E., 1999. Diatom mats in Gulf of California sediments: implications for the palaeoenvironmental interpretation of laminated sediments and silica burial. *Geology*, 27, 311–314.

Pilcher, J.R., 1993. Radiocarbon dating and the palynologist: a realistic approach to precision and accuracy. In F.M. Chambers (ed.), *Climate change and human impact on the landscape*. Chapman and Hall, London, pp. 23–32.

Pisias, N.G., 1978. Paleoceanography of the Santa Barbara Basin during the last 8000 years. *Quaternary Research*, 10, 366–384.

Pisias, N.G., 1979. Model for paleoceanographic reconstructions of the California Current during the last 8000 years. *Quaternary Research*, 11, 373–386.

Pisias, N.G., 1986. Vertical water mass circulation and the distribution of Radiolaria in surface sediments of the Gulf of California. *Marine Micropaleontology*, 10, 189–205.

Pisias, N.G., Roelofs, A. and Weber, M., 1997. Radiolarian-based transfer functions for estimating mean surface ocean temperatures and seasonal range. *Paleoceanography*, 12, 365–379.

Pizzuto, J.E. and Schwendt, A.E., 1997. Mathematical modeling of autocompaction of a Holocene transgressive valley-fill deposit, Wolfe Glade, Delaware. *Geology*, 25, 57–60.

Pokras, E.M., 1987. Diatom record of late Quaternary climatic change in the eastern Equatorial Atlantic and tropical Africa. *Palaeoceanography*, 2(3), 273–286.

Prahl, F.G. and Wakeham, S.G., 1987. Calibration of unsaturation patterns in long-chain ketone compositions for palaeotemperature assessment. *Nature*, 330, 367–369.

Prahl, F.G., Muehlhausen, L.A. and Zahnle, D.L., 1988. Further evaluation of long-chain alkenones as indicators of paleoceanographic conditions. *Geochimica et Cosmochimica Acta*, 52, 2303–2310.

Prahl, F.G., Collier, R.B., Dymond, J., Lyle, M. and Sparrow, M.A., 1993. A biomarker perspective on prymnesiophyte productivity in the northeast Pacific Ocean. *Deep-Sea Research I*, 40, 2061–2076.

Preece, R.C. and Robinson, J.E., 1984. Late Devensian and Flandrian environmental history of the Ancholme Valley, Lincolnshire: molluscan and ostracod evidence. *Journal of Biogeography*, 11, 319–352.

Preece, R.C., Bennett, K.D. and Robinson, J.E., 1984. The biostratigraphy of an early Flandrian tufa at Inchrory, Glen Avon, Banffshire. *Scottish Journal of Geology*, 20, 143–159.

Preece, R.C., Coxon, P. and Robinson, J.E., 1986. New biostratigraphic evidence of the post-glacial colonization of Ireland and for Mesolithic forest disturbance. *Journal of Biogeography*, 13, 487–509.

Prentice, C., 1986a. Multivariate methods for data analysis. In B.E. Berglund (ed.), *Handbook of Holocene palaeoecology and palaeohydrology*. Wiley, Chichester, pp. 775–798.

Prentice, C., 1986b. Forest-composition calibration of pollen data. In B.E. Berglund (ed.) *Handbook of Holocene palaeoecology and palaeohydrology*. John Wiley, Chichester, pp. 799–816.

Prezelin, B.B., Samuelsson, G. and Matlick, H.A., 1986. Photosystem II photoinhibition and altered kinetics of photosynthesis during nutrient-dependent high-light photo-adaption in *Gonyaulax polyhedra*. *Marine Biology*, 93, 1–12.

Proctor, L.M. and Fuhrman, J.A., 1990. Viral mortality of marine bacteria and cyanobacteria. *Nature*, 343, 60–62.

Punt, W. and Clarke, G.C.S., 1976–. *The Northwest European Pollen Flora I–*. Elsevier, Amsterdam.

Rathburn, A.E. and Corliss, B.H., 1994. The ecology of living (stained) deep-sea benthic foraminifera from the Sulu Sea. *Paleoceanography*, 9, 87–150.

Raymo, M.E., Ruddiman, W.F. and Froelich, P.N., 1988. Influence of late Cenozoic mountain building on ocean geochemical cycles. *Geology*, 16, 649–653.

Raymo, M.E., Ruddiman, W.F., Shackleton, N.J. and Oppo, D.W., 1990. Evolution of Atlantic–Pacific $\delta^{13}C$ gradients over the last 2.5 m.y. *Earth and Planetary Science Letters*, 5, 353–368.

Raymo, M.E., Oppo, D.W. and Curry, W.B., 1997. The mid-Pleistocene climate transition: a deep sea carbon isotopic perspective. *Paleoceanography*, 12, 546–559.

Reinhardt, E.G., Patterson, R.T. and Schröder-Adams, C.J., 1994. Geoarchaeology of the ancient harbor site of Caesarea Maritima, Israel: evidence from sedimentology and paleoecology of benthic foraminifera. *Journal of Foraminiferal Research*, 24, 37–48.

Renberg, I. and Hellberg, T., 1982. The pH history of lakes in southwestern Sweden, as calculated from the subfossil diatom flora of the sediments. *Ambio*, 11, 30–33.

Renz, G.W., 1976. The distribution and ecology of Radiolaria in the central Pacific: plankton and surface sediments. *Bulletin of the Scripps Institution of Oceanography*, 22, 1–267.

Reille, M., 1992. *Pollen et spores d'Europe et d'Afrique du Nord*. Laboratoire de Botanique Historique et Palynologie, Marseilles.

Rhiel, E., Krupinskak, K. and Wehrmeyer, W., 1986. Effect of nitrogen starvation on the function and organisation of the photosynthetic membranes in *Crytomonas maculata* (Cryptophyceae). *Planta*, 169, 361–369.

Rice, A.L., Billett, D.S.M., Fry, J., John, A.W.G., Lampitt, R.S., Mantoura, R.F.G. and Morris, R.J., 1986. Seasonal deposition of phytodetritus to the deep-sea floor. *Proceedings of the Royal Society of Edinburgh B*, 88, 265–279.

Rice, A.L., Thurston, M.H. and Bett, B.J., 1994. The IOSDL DEEPSEAS programme: introduction and photographic evidence for the presence and absence of a seasonal input of phytodetritus at contrasting abyssal sites in the northeastern Atlantic. *Deep-Sea Research*, 41, 1305–1320.

Rickaby, R.E.M. and Elderfield, H., 1999. Planktonic foraminiferal Cd/Ca: paleo-nutrients or paleotemperature? *Paleoceanography*, 14, 293–303.

Riedel, W.R. 1958. Radiolaria in Antarctic sediments. *B.A.N.Z. Antarctic Research Expedition Reports, series B*, 6, 217–255.

Rind, D. and Peteet, D., 1985 Terrestrial conditions at the last glacial maximum and CLIMAP sea surface temperatures: are they consistent? *Quaternary Research*, 24, 1–22.

Robinson, P.D. and Haslett, S.K., 1995. A radiolarian dated sponge microsclere assemblage from the Miocene Dos Bocas Formation of Ecuador. *Journal of South American Earth Sciences*, 8, 195–200.

Roelofs, A.K. and Pisias, N.G., 1986. Revised technique for preparing quantitative radiolarian slides from deep-sea sediments. *Micropaleontology*, 32, 182–185.

Rohling, E.J., 1994. Review and new aspects concerning the formation of eastern Mediterranean sapropels. *Marine Geology*, 122, 1–28.

Rohling, E.J. and Cooke, S., 1999. Stable oxygen and carbon isotopes in foraminiferal carbonate shells. In B.K. Sen Gupta (ed.) *Modern Foraminifera*. Kluwer Academic Publishers, Dordrecht, The Netherlands, pp. 239–258.

Romine, K. and Moore, T.C. Jr, 1981. Radiolarian assemblage distributions and paleoceanography of the eastern equatorial Pacific Ocean during the last 127 000 years. *Palaeogeography, Palaeoclimatology, Palaeoecology*, 35, 281–314.

Rosenfeld, A. and Vesper, B., 1977. The variability of the sieve-pores in recent and fossil species of *Cyprideis torosa* (Jones, 1850) as an indicator for salinity and palaeosalinity. In H. Loffler and D. Danielopol (eds), *Aspects of Ecology and Zoogeography of Recent and Fossil Ostracoda*. Junk, The Hague, pp. 55–67.

Rosenfeld, A., Ortal, R. and Honigstein, A., 2000. Ostracods as indicators of river pollution in Northern Israel. In R. Martin (ed.), *Environmental Micropaleontology*. Kluwer, New York, pp. 167–180.

Rosenthal, Y., Boyle, E.A. and Labeyrie, L., 1997a. Last glacial maximum paleo-chemistry and deepwater circulation in the Southern Ocean: evidence from foraminiferal cadmium. *Paleoceanography*, 12, 787–796.

Rosenthal, Y., Boyle, E.A. and Slowey, N., 1997b. Temperature control on the incorporation of magnesium, strontium, fluorine, and cadmium into benthic foraminiferal shells from Little Bahama Bank: prospects for thermocline paleoceanography. *Geochimica et Cosmochimica Acta*, 61, 3633–3643.

Rosoff, D.B. and Corliss, B.H., 1992. An analysis of Recent deep-sea benthic foraminiferal morphotypes from the Norwegian and Greenland seas. *Palaeogeography, Palaeoclimatology, Palaeoecology*, 91, 13–20.

Ross, C.R. and Kennett, J.P., 1983. Late Quaternary paleoceanography as recorded by benthonic foraminifera in Strait of Sicily sediment sequences. *Marine Micropaleontology*, 8, 315–336.

Ross, R., Cox, E.J., Kareyava, N.I., Simonsen, R. and Sims, P.A., 1979. An amended terminology for the siliceous components of the diatom cell. *Nova Hedwigia Beiheft*, 64, 513–533.

Round, F.E., 1990. The effect of liming on the benthic diatom populations in three upland Welsh lakes. *Diatom Research*, 5, 129–140.

Round, F.E., Crawford, R.M. and Mann, D.G., 1990. *The diatoms: biology and morphology of the genera*. Cambridge University Press, Cambridge, UK.

Rowe, G.T., 1983. Biomass and production of the deep-sea macrobenthos. In G.T. Rowe (ed.), *The Sea, 8*. Wiley, New York, pp. 97–121.

Rowe, P.J., Atkinson, T.C. and Turner, C., 1999. U-series dating of Hoxnian interglacial deposits at Marks Tey, Essex, England. *Journal of Quaternary Science*, 14, 693–702.

Sachs, H.M., 1973a. North Pacific radiolarian assemblages and their relationship to oceanographic parameters. *Quaternary Research*, 3, 73–88.

Sachs, H.M., 1973b. Late Pleistocene history of the North Pacific: evidence from quantitative study of Radiolaria in core V21–173. *Quaternary Research*, 3, 89–98.

Sætre, M.L.L., Dale, B., Abdullah, M.I. and Sætre, G.-P., 1997. Dinoflagellate cysts as possible indicators of industrial pollution in a Norwegian fjord. *Marine Environmental Research*, 44(20), 167–189.

Sakshaug, E., Andresen, K., Myklestad, S. and Olsen, Y., 1982. Nutrient status of phytoplankton communities in Norwegian waters (marine, brackish, and fresh) are revealed by their chemical composition. *Journal of Plankton Research*, 5, 175–196.

Sancetta, C.A., 1995. Diatoms in the Gulf of California: seasonal flux patterns and the sediment record for the last 15,000 years. *Palaeoceanography*, 10, 67–84.

Sanders, H.L., 1968. Marine benthic diversity: a comparative study. *American Naturalist*, 102, 243–282.

Sanfilippo, A., 1987. Stratigraphy and evolution of tropical Cenozoic Radiolaria. *Meddelanden från Stockholms Universitets Geologiska Institution*, 270, 44 pp.

Sanfilippo, A., Westberg-Smith, M.J. and Riedel, W.R., 1985. Cenozoic Radiolaria. In H.M. Bolli, J.B. Saunders and K. Perch-Nielsen (eds), *Plankton stratigraphy*. Cambridge University Press, Cambridge, pp. 631–712.

Sanyal, A., Hemming, N.G., Hanson, G.N. and Broecker, W.S., 1995. Evidence for a higher pH in the glacial ocean from boron isotopes in foraminifera. *Nature*, 373, 234–236.

Sarnthein, M., Winn, K., Jung, S.J.A., Duplessy, J.-C., Labeyrie, L., Erlenkeuser, H. and Ganssen, G., 1994. Changes in east Atlantic deepwater circulation over the last 30,000 years: eight time slice reconstructions. *Palaeoceanography*, 9, 209–267.

Sautter, L.R. and Thunnell, R.C., 1991a. Planktonic foraminiferal response to upwelling and seasonal hydrographic conditions: sediment trap results from San Pedro Basin, southern California Bight. *Journal of Foraminiferal Research*, 21, 347–363.

Sautter, L.R. and Thunnell, R.C., 1991b. Seasonal variability in the $\delta^{18}O$ and $\delta^{13}C$ of planktonic foraminifera from an upwelling environment: sediment trap results from the San Pedro Basin, southern California Bight. *Paleoceanography*, 6, 307–334.

Sawada, K., Handa, N. and Nakatsuka, T., 1998. Production and transport of long-chain alkenones and alkyl alkenoates in a sea water column in the northwestern Pacific off central Japan. *Marine Chemistry*, 59, 219–234.

Schafer, C.T., Collins, E.S. and Smith, J.N., 1991. Relationship of Foraminifera and thecamoebian distributions to sediments contaminated by pulp mill effluent: Saguenay Fiord, Quebec, Canada. *Marine Micropaleontology*, 17, 255–283.

Schafer, C.T., Winters, G.V., Scott, D.B., Pocklington, P., Cole, F.E. and Honig, C., 1995. Survey of living foraminifera and polychaete populations at some Canadian aquaculture sites: potential for impact mapping and monitoring. *Journal of Foraminiferal Research*, 25, 236–259.

Schmiedl, G., 1995. Rekonstruktion der spätquartären Tiefenwasserzirkulation und Produktivität im östlichen Südatlantik anhand von benthischen Foraminiferen vergesellschaftungen. *Berichte zur Polarforschung*, 160.

Schmiedl, G. and Mackensen, A., 1997. Late Quaternary paleoproductivity and deep water circulation in the eastern South Atlantic Ocean: evidence from benthic foraminifera. *Palaeogeography, Palaeoclimatology, Palaeoecology*, 130, 43–80.

Schmiedl, G., Mackensen, A. and Müller, P.J., 1997. Recent benthic foraminifera from the eastern South Atlantic Ocean: dependence on food supply and water masses. *Marine Micropaleontology*, 32, 249–287.

Schmiedl, G., Hemleben, C., Keller, J. and Segl, M., 1998. Impact of climatic changes on the benthic foraminiferal fauna in the Ionian Sea during the last 330,000 years. *Paleoceanography*, 13, 447–458.

Schmuker, B., 2000. The influence of shelf vicinity on the distribution of planktic foraminifera south of Puerto Rico. *Marine Geology*, 166, 125–143.

Schnitker, D., 1974. West Atlantic abyssal circulation during the past 120,000 years. *Nature*, 248, 385–387.

Schnitker, D., 1979. The deep waters of the western North Atlantic during the past 24,000 years, and the re-initiation of the Western Boundary Undercurrent. *Marine Micropalaeontology*, 4, 265–280.

Schnitker, D., 1980. Quaternary deep-sea benthic foraminifers and bottom water masses. *Annual Review of Earth and Planetary Science*, 8, 343–370.

Schnitker, D., 1982. Climatic variability and deep ocean circulation: evidence from the North Atlantic. *Palaeogeography, Palaeoclimatology, Palaeoecology*, 40, 213–234.

Schnitker, D., 1994. Deep-sea benthic foraminifers: food and bottom water masses. In R. Zahn, T.F. Pedersen, M.A. Kaminski and L. Labeyrie (eds), *Carbon cycling in the glacial ocean: constraints on the ocean's role in global change*. NATO ASI Series, I 17, Springer-Verlag, Berlin, pp. 539–554.

Schoning, K. and Wastegård, S., 1999. Ostracod assemblages in late Quaternary varved glaciomarine clay of the Baltic Sea Yoldia stage in eastern middle Sweden. *Palaeogeography, Palaeoclimatology, Palaeoecology*, 148, 313–325.

Schrader, H.J., 1992a. Peruvian coastal primary palaeo-productivity during the last 200,000 years. In C.P. Summerhays, W.L. Prell and C. Emeis (eds), *Upwelling systems: evolution since the early Miocene*. Geological Society, Special Publication 64, pp. 391–409.

Schrader, H.J., 1992b. Coastal upwelling and atmospheric CO_2 changes over the last 400,000 years: Peru. *Marine Geology*, 107, 239–248.

Schrader, H.J., 1992c. Comparison of Quaternary coastal upwelling proxies off central Peru. *Marine Micropalaeontology*, 19, 29–47.

Schrader, H.J. and Koc Karpuz, N., 1990. Norwegian–Iceland seas: transfer functions between marine planktic diatoms and surface water temperature. In U. Bleil and J. Thiede (eds), *Geological history of the polar oceans: Arctic versus Antarctic.* Kluwer, Amsterdam, pp. 337–361.

Schrader, H.J. and Sorkness, R. 1990. Spatial and temporal variation of Peruvian coastal upwelling during the latest Quaternary. In E. Suess *et al., Proceedings of the ODP, Scientific Results,* 112, pp. 1179–1232.

Schrader, H.J. and Sorkness, R., 1991. Peruvian coastal upwelling: Late Quaternary productivity changes revealed by diatoms. *Marine Geology,* 97, 233–249.

Schrader, H.J., Swanberg, N., Lycke, A.K., Paetzel, M. and Schrader, T., 1993. Diatom inferred productivity changes in the eastern equatorial Pacific: the Quaternary record of ODP Leg 111, Site 677. *Hydrobiologia,* 269, 137–151.

Schramm, C.T., 1985. Implications of radiolarian assemblages for the Late Quaternary palaeoceanography of the eastern equatorial Pacific. *Quaternary Research,* 24, 204–218.

Schröder, C.J., Scott, D.B. and Medioli, F.S., 1987. Can smaller benthic foraminifera be ignored in paleoenvironmental analyses? *Journal of Foraminiferal Research,* 17, 101–105.

Schwalb, A., Burns, S. and Kelts, K., 1999. Holocene environments from stable-isotope stratigraphy of ostracods and authigenic carbonate in Chilean Altiplano lakes. *Palaeogeography, Palaeoclimatology, Palaeoecology,* 148, 133–152.

Scott, D.B., 1976. Brackish-water foraminifera from southern California and description of *Polysaccammina ipohalina* n. gen., n. sp. *Journal of Foraminiferal Research,* 6, 312–321.

Scott, D.B. and Medioli, F.S., 1978. Vertical zonation of marsh foraminifera as accurate indicators of former sea-levels. *Nature,* 272, 538–541.

Scott, D.B. and Medioli, F.S., 1980. Quantitative studies of marsh foraminiferal distributions in Nova Scotia: implications for sea-level studies. *Cushman Foundation for Foraminiferal Research Special Publication,* 17, 1–57.

Scott, D.B., Schnack, E.J., Ferrero, L., Espinosa, M. and Barbosa, C.F., 1990. Recent marsh foraminifera from the east coast of South America: comparison to the northern hemisphere. In C. Hemleben, C. Kaminski, W. Kuhnt and D.B. Scott (eds), *Paleoecology, biostratigraphy and taxonomy of agglutinated foraminifera.* NATO ASI Series C-327, Kluwer, Dordrecht, pp. 717–737.

Scott, D.B., Collins, E.S., Duggan, J., Asioli, A., Saito, T. and Hasegawa, S., 1996. Pacific Rim marsh foraminiferal distributions: implications for sea-level studies. *Journal of Coastal Research,* 12, 850–861.

Scott, D.B., Medioli, F.S. and Schafer, C.T., 2001. *Monitoring in Coastal Environments using Foraminifera and Thecamoebian Indicators.* Cambridge University Press, Cambridge, 177 pp.

Scott, D.K. and Leckie, R.M., 1990. Foraminiferal zonation of Great Sippewissett salt marsh (Falmouth, Massachusetts). *Journal of Foraminiferal Research,* 20, 248–266.

Sen Gupta, B.K. (ed.), 1999. *Modern foraminifera.* Kluwer Academic Publishers, Dordrecht, 371 pp.

Sen Gupta, B.K. and Machain-Castillo, M.L., 1993. Benthic foraminifera in oxygen-poor habitats. *Marine Micropaleontology,* 20, 183–201.

Sen Gupta, B.K., Lee, R.F. and May, M.S. 1981. Upwelling and an unusual assemblage of benthic foraminifera on the northern Florida continental slope. *Journal of Paleontology,* 55, 853–857.

Sen Gupta, B.K., Temples, T.J. and Dallmeyer, M.D.G., 1982. Late Quaternary benthic foraminifera of the Grenada Basin: stratigraphy and paleoceanography. *Marine Micropaleontology,* 7, 297–309.

Sen Gupta, B.K., Shin, I.C. and Wendler, S.T., 1987. Relevance of specimen size in distribution studies of deep-sea benthic foraminifera. *Palaios,* 2, 332–338.

Shackleton, N.J. and Hall, M.A., 1989. Stable isotope history of the Pleistocene at ODP Site 677. *Proceedings of the Ocean Drilling Program, Scientific Results*, 111, 295–316.

Shackleton, N.J. and Opdyke, N.D., 1973. Oxygen isotope and palaeomagnetic stratigraphy of equatorial Pacific Core V28–238: oxygen isotope temperatures and ice volumes on a 105 year and 106 year scale. *Quaternary Research*, 3, 39–55.

Shackleton, N.J. and Opdyke, N.D., 1976. Oxygen isotope and palaeomagnetic stratigraphy of equatorial Pacific core V28–239. Late Pliocene to early Pleistocene. In R.M. Cline and J.D. Hays (eds), *Investigations of Late Quaternary palaeoceanography and palaeoclimatology*. Geological Society of America Memoirs, 145, pp. 449–464.

Shackleton, N.J. and Opdyke, N.D., 1977. Oxygen isotope and palaeomagnetic evidence for early northern hemisphere glaciation. *Nature*, 270, 216–219.

Shackleton, N.J., Berger, A. and Peltier, W.R., 1991. An alternative astronomical calibration of the lower Pleistocene timescale based on ODP Site 677. *Transactions of the Royal Society of Edinburgh, Earth Sciences*, 81, 251–261.

Shackleton, N.J., Hall, M.A. and Pate, D., 1995a. Pliocene stable isotope stratigraphy of Site 846. *Proceedings of the Ocean Drilling Program, Scientific Results*, 138, 337–355.

Shackleton, N.J., Crowhurst, S., Hagelberg, T., Pisias, N.G. and Schneider, D.A., 1995b. A new Late Neogene time scale: application to ODP Leg 138 sites. *Proceedings of the Ocean Drilling Program, Scientific Results*, 138, 73–101.

Shannon, C.E. and Weaver, W., 1963. *The mathematical theory of communication*. University of Illinois Press, Urbana.

Sharifi, A.R., Croudace, I.W. and Austin, R.L., 1991. Benthic foraminiferids as pollution indicators in Southampton Water, southern England, U.K. *Journal of Micropalaeontology*, 10, 109–113.

Shennan, I., Innes, J.B., Long, A.J. and Zong, Y., 1993. Late Devensian and Holocene sea-level changes at Rumach, near Arisaig, northwest Scotland. *Norsk Geologisk Tidsskrift*, 73, 161–174.

Shennan, I., Innes, J.B., Long, A.J. and Zong, Y., 1994. Late Devensian and Holocene sea-level changes at Loch nan Eala, near Arisaig, northwest Scotland. *Journal of Quaternary Science*, 9, 261–283.

Shennan, I., Innes, J.B., Long, A.J. and Zong, Y. 1995a. Holocene relative sea-level changes and coastal vegetation history at Kentra Moss, Argyll, northwest Scotland. *Marine Geology*, 124, 43–59.

Shennan, I., Innes, J.B., Long, A.J. and Zong, Y., 1995b. Late Devensian and Holocene relative sea level changes in northwestern Scotland: new data to test existing models. *Quaternary International*, 26, 97–123.

Shennan, I., Tooley, M. and Green, F., 1999. Sea level, climate change and coastal evolution in Morar, northwest Scotland. *Geologie en Mijnbouw*, 77, 247–262.

Shennan, I., Lambeck, K., Horton, B. M. *et al.*, 2000. Late Devensian and Holocene records of relative sea-level changes in northwest Scotland and their implications for glacial-hydro-isostatic modelling. *Quaternary Science Reviews*, 19, 1103–1135.

Shi, G.R., 1993. Multivariate data analysis in palaeoecology and palaeobiogeography. *Palaeogeography, Palaeoclimatology, Palaeoecology*, 105, 199–234.

Shirayama, Y., 1984. Vertical distribution of meiobenthos in the sediment profile in bathyal, abyssal and hadal deep sea systems of the western Pacific. *Oceanologica Acta*, 7, 123–129.

Sibuet, M., Lambert, C.E., Chesselet, R. and Laubier, L., 1989. Density of the major size groups of benthic fauna and trophic input in deep basins of the Atlantic Ocean. *Journal of Marine Research*, 47, 851–867.

Sicre, M.-A., Ternois, Y., Miquel, J.-C. and Marty, J.-C., 1999. Alkenones in the Northwestern Mediterranean sea: interannual variability and vertical transfer. *Geophysical Research Letters*, 26, 1735–1738.

Sieburth, J.M., 1960. Acrylic acid, an 'antibiotic' principle in *Phaeocystis* blooms in Antarctic waters. *Science*, 132, 676–677.

Sikes, E.L., Farrington, J.W. and Keigwin, L.D., 1991. Use of the alkenone unsaturation ratio U^K_{37} to determine past sea surface temperatures: core-top SST calibrations and methodology considerations. *Earth and Planetary Science Letters*, 104, 36–47.

Simmons, I.G., 1993. Vegetation change during the Mesolithic in the British Isles: some amplifications. In F.M. Chambers (ed.), *Climate Change and human impact on the landscape*. Chapman and Hall, London, pp. 109–118.

Sirocko, F., 1996. Classics in physical geography revisited – Emiliani, C. (1955) Pleistocene temperatures. *Journal of Geology*, 63, 538–578. *Progress in Physical Geography*, 20, 447–452.

Slack, J.M., Kaesler, R.L. and Kontrovitz, M., 2000. Trend, signal and noise in the ecology of Ostracoda: information from rare species in low-diversity assemblages. *Hydrobiologia*, 419, 181–189.

Sloan, D., 1995. Use of foraminiferal biostratigraphy in mitigating pollution and seismic problems, San Francisco, California. *Journal of Foraminiferal Research*, 25, 260–266.

Smart, C.W. and Gooday, A.J., 1997. Recent benthic foraminifera in the abyssal northeast Atlantic Ocean: relation to phytodetrital inputs. *Journal of Foraminiferal Research*, 27, 85–92.

Smart, C.W. and Murray, J.W., 1994. An early Miocene Atlantic-wide foraminiferal/palaeoceanographic event. *Palaeogeography, Palaeoclimatology, Palaeoecology*, 108, 139–148.

Smart, C.W., King, S.C., Gooday, A.J., Murray, J.W. and Thomas, E., 1994. A benthic foraminiferal proxy of pulsed organic matter paleofluxes. *Marine Micropaleontology*, 23, 89–99.

Smetacek, V.S., 1985. Role of sinking in diatom life-history cycles: ecological, evolutionary and geological significance. *Marine Biology*, 84, 239–251.

Smith, A.G., 1981. The Neolithic. In I.G. Simmons and M.J. Tooley (eds), *The environment in British prehistory*. Duckworth, London.

Smith, A.G. and Pilcher, J.R., 1973. Radiocarbon dates and vegetational history of the British Isles. *New Phytologist*, 72, 903–914.

Smith, A.J. and Forester, R.M., 1994. Estimating past precipitation and temperature from fossil ostracodes. *ASCE and ANS, 5th International Conference, Las Vegas, Nevada, International High Level Radioactive Waste Management Proceedings*, pp. 2545–2552.

Smith, A.J., Donovan, J.J., Ito, E. and Engstrom, D.R., 1997. Ground-water processes controlling a prairie lake's response to middle Holocene drought. *Geology*, 25, 391–394.

Smith, C.R., 1994. Tempo and mode in deep-sea benthic ecology: punctuated equilibrium revisited. *Palaios*, 9, 3–13.

Smith, C.R., Hoover, D.J., Doan, S.E., *et al.*, 1996. Phytodetritus at the abyssal seafloor across 10° of latitude in the central equatorial Pacific. *Deep-Sea Research*, 43, 1309–1338.

Smith, K.L., Kaufmann, R.S. and Baldwin, R.J., 1994. Coupling of near-bottom pelagic processes at abyssal depths in the eastern North Pacific Ocean. *Limnology and Oceanography*, 39, 1101–1118.

Smith, R. and Martens, K., 2000. The ontogeny of the cyprid ostracod *Eucypris virens* (Jurine, 1820) (Crustacea, Ostracoda). *Hydrobiologia*, 419, 31–63.

Snoeijs, P., 1993. *Intercalibration and distribution of diatom species in the Baltic Sea*. Vol. 1. Opulus Press, Uppsala.

Snoeijs, P. and Balashova, N., 1998. *Intercalibration and distribution of diatom species in the Baltic Sea.* Vol. 5. Opulus Press, Uppsala.

Snoeijs, P. and Kasperoviciene, J., 1996. *Intercalibration and distribution of diatom species in the Baltic Sea.* Vol. 4. Opulus Press, Uppsala.

Snoeijs, P. and Potapova, M., 1995. *Intercalibration and distribution of diatom species in the Baltic Sea.* Vol. 3. Opulus Press, Uppsala..

Snoeijs, P. and Vilbaste, S., 1994. *Intercalibration and distribution of diatom species in the Baltic Sea*, Vol. 2. Opulus Press, Uppsala.

Sournia, A., 1982. Is there a shade flora? *Journal of Plankton Research*, 4, 391–399.

Stevenson, A.C., Birks, H.J.B., Anderson, D.S., *et al.*, 1991. *The surface waters acidification programme: modern diatom/lake water chemistry data-set.* Ensis, London.

Stockmarr, J., 1973. Tablets with spores used in absolute pollen analysis. *Pollen et Spores*, 13, 615–621.

Stockner, J.G., 1971. Preliminary characterisation of lakes in the experimental lakes area, north-western Ontario, using diatom occurrence in lake sediments. *Journal of the Fisheries Board of Canada*, 28, 265–275.

Stockner, J.B. and Benson, W.W., 1967. The succession of diatom assemblages in the recent sediments of Lake Washington. *Limnology and Oceanography*, 12, 513–532.

Stoermer, E.F., Emmert, G., Julius, M.L. and Schelske, C.L., 1996. Paleolimnologic evidence of rapid change in Lake Erie's trophic status. *Canadian Journal of Fisheries and Aquatic Sciences*, 53, 1451–1458.

Stott, L.D., Hayden, T.P. and Griffith, J., 1996. Benthic foraminifera at the Los Angeles County Whites Point outfall revisited. *Journal of Foraminiferal Research*, 26, 357–368.

Stouff, V., Debenay, J.-P. and Lesourd, M., 1999. Origin of double and multiple tests in benthic foraminifera: observations in laboratory cultures. *Marine Micropaleontology*, 36, 189–204.

Stover, L.E., Brinkhuis, H. and Damassa, S.P., 1996. Mesozoic–Tertiary dinoflagellates, acritarchs and prasinophytes. In J. Jansonius and D.G. McGregor (eds), *Palynology: principles and applications.* American Association of Stratigraphic Palynologists Foundation, 2, pp. 641–750.

Streeter, S.S., 1973. Bottom water and benthonic foraminifera in the North Atlantic – glacial–interglacial contrasts. *Quaternary Research*, 3, 131–141.

Streeter, S.S. and Shackleton, N.J., 1979. Paleocirculation of the deep North Atlantic: 150,000-year record of benthic foraminifera and oxygen-18. *Science*, 203, 168–171.

Struck, U., 1995. Stepwise postglacial migration of benthic foraminifera into the abyssal northeastern Norwegian Sea. *Marine Micropaleontology*, 26, 207–213.

Stute, M., Forster, M. and Frischkorn, H., 1995. Cooling of tropical Brazil (5°C) during the last glacial maximum. *Science*, 269, 379–383.

Sugita, S., Gaillard, M.-J. and Broström, A., 1999. Landscape openness and pollen records: a simulation approach. *The Holocene*, 9, 409–421.

Sugiyama, K. and Anderson, O.R., 1997. Experimental and observational studies of radiolarian physiological ecology. 6. Effects of silicate-supplemented seawater on the longevity and weight gain of spongiose radiolarians *Spongaster tetras* and *Dictyocoryne truncatum. Marine Micropaleontology*, 29, 159–172.

Suttle, C.A. and Chan, A.M., 1995. Virus infecting the marine prymnesiophyte *Chrysochromulina* spp.: isolation, preliminary characterization and natural abundance. *Marine Ecology Progress Series*, 118, 275–282.

Suttle, C.A., Chan, A.M. and Cottrell, M.T., 1990. Infection of phytoplankton by viruses and reduction of primary productivity. *Nature*, 347, 467–469.

Sywula, T., Glazewska, I., Whatley, R.C. and Moguilevsky, A., 1995. Genetic differentiation in the brackish-water ostracod *Cyprideis torosa*. *Marine Biology*, 121, 647–653.

Takahashi, K., 1991. *Radiolaria: flux, ecology, and taxonomy in the Pacific and Atlantic*. Woods Hole Oceanographic Institution, Ocean Biocoenosis Series No. 3, 303 pp.

Takahashi, K. and Okada, H., 2000. The paleoceanography for the last 30,000 years in the southeastern Indian Ocean by means of calcareous nannofossils. *Marine Micropaleontology*, 40, 83–103.

Tangen, K., Brand, L.E., Blackwelder, P.L. and Guillard, R.R., 1982. *Thoracosphaera heimii* (Lohmann) Kamptner is a dinophyte: observations on its morphology and life cycle. *Marine Micropaleontology*, 7, 193–212.

Ter Braak, C.J.F., 1990. *CANOCO – A FORTRAN program for CANOnical Community Ordination by [partial] [canonical] correspondence analysis, principal components analysis and redundancy analysis (version 3.10)*. Microcomputer Power, Ithaca, NY.

Ter Braak, C.J.F. and Smilauer, P., 1998. *CANOCO Reference Manual and User's Guide to CANOCO for Windows: Software for Canonical Community Ordination (version 4)*. Microcomputer Power, Ithaca, NY, USA, 352 pp.

Thiel, H., Pfannkuche, O. and Schriever, G., 1990. Phytodetritus on the deep-sea floor in a central oceanic region of the northeast Atlantic. *Biological Oceanography*, 6, 203–239.

Thomas, E., 1986. Late Oligocene to Recent benthic foraminifers from Deep Sea Drilling Project Sites 608 and 610, northeastern North Atlantic. In W.F. Ruddiman, R.B. Kidd and E. Thomas *et al.* (eds), *Initial Reports of the Deep Sea Drilling Project*, 94. US Government Printing Office, Washington, DC, pp. 997–1031.

Thomas, E., 1990. Late Cretaceous through Neogene deep-sea benthic foraminifers (Maud Rise, Weddell Sea, Antarctica). In P.F. Barker and J.P. Kennett *et al.* (eds), *Proceedings of the Ocean Drilling Program, Scientific Results*, 113. College Station, TX (Ocean Drilling Program), pp. 571–594.

Thomas, E., 1992. Cenozoic deep-sea circulation: evidence from deep-sea benthic foraminifera. *Antarctic Research Series*, 56, 141–165.

Thomas, E. and Gooday, A.J., 1996. Cenozoic deep-sea benthic foraminifers: tracers for changes in oceanic productivity? *Geology*, 24, 355–358.

Thomas, E., Booth, L., Maslin, M. and Shackleton, N.J., 1995. Northeastern Atlantic benthic foraminifera during the last 45,000 years: changes in productivity seen from the bottom up. *Paleoceanography*, 10, 545–562.

Thomsen, H.A., Ostergaard, J.B. and Hansen, L.E., 1991. Heteromorphic life histories in Arctic coccolithophorids (Prymnesiophyceae). *Journal of Phycology*, 27, 634–642.

Thorsen, T.A. and Dale, B., 1997. Dinoflagellate cysts as indicators of pollution and past climate in a Norwegian fjord. *The Holocene*, 7(4), 433–446.

Thunell, R.C., 1976. Optimum indices of calcium carbonate dissolution in deep-sea sediments. *Geology*, 4, 525–528.

Thunell, R.C., Anderson, D., Gellar, D. and Qingmin, M., 1994. Sea-surface temperature estimates for the tropical western Pacific during the last glaciation and their implications for the Pacific Warm Pool. *Quaternary Research*, 41, 255–264.

Tooley, M.J., 1982. Sea level changes in northern England. *Proceedings of the Geologists' Association*, 93, 43–51.

Tortell, P.D., Maldonado, M.T. and Price, N.M., 1996. The role of heterotrophic bacteria in iron-limited ocean ecosystems. *Nature*, 383, 330–332.

Turner, C., 1970. The Middle Pleistocene Deposits at Marks Tey, Essex. *Philosophical Transactions of the Royal Society of London*, B257, 373–440.

Turner, C. and West, R.G., 1968. The subdivision and zonation of interglacial periods. *Eiszeitalter und Gegenwart*, 19, 93–101.

Tyrrell, T., 1999. The relative influences of nitrogen and phosphorus on oceanic primary production. *Nature*, 400, 525–531.

Tyrrell, T., Holligan, P.M. and Mobley, C.D., 1999. Optical impacts of oceanic coccolithophore blooms. *Journal of Geophysical Research*, 104(C2), 3223–3241.

Ufkes, E., Jansen, J.H.F. and Schneider, R.R., 2000. Anomalous occurrences of *Neogloboquadrina pachyderma* (left) in a 420-ky upwelling record from Walvis Ridge (SE Atlantic). *Marine Micropaleontology*, 40, 23–42.

Ujiié, H., Tanaka, Y. and Ono, T., 1991. Late Quaternary paleoceanographic record from the Middle Ryukyu trench slope, Northwest Pacific. *Marine Micropaleontology*, 18, 115–128.

Underdal, B., Skulberg, O.M., Dahl, E. and Aune, T., 1989. Disastrous bloom of *Chrysochromulina polylepis* (Prymnesiophyceae) in Norwegian coastal waters 1988 – mortality in marine biota. *Ambio*, 18, 265–270.

van Bleijswijk, J.D.L., Kempers, R.S., van der Wal, P., Westbroek, P., Egge, J.K. and Lukk, T., 1994. Standing stocks of PIC, POC, PON and *Emiliania huxleyi* coccospheres and liths in sea water enclosures with different phosphate loadings. *Sarsia*, 79, 307–317.

Vance, D. and Burton, K., 1999. Neodymium isotopes in planktonic foraminifera: a record of the response of continental weathering and ocean circulation rates to climate change. *Earth and Planetary Science Letters*, 173, 365–379.

Van der Zwaan, G.J., 1982. Paleoecology of Late Miocene Mediterranean foraminifera. *Utrecht Micropaleontological Bulletin*, 25, 5–193.

Van der Zwaan, G.J., Jorissen, F. and De Stigter, H., 1990. The depth dependency of planktonic/benthic foraminiferal ratios: constraints and applications. *Marine Geology*, 95, 1–16.

Van der Zwaan, G.J., Duijnstee, I.A.P., Den Dulk, M., *et al.*, 1999. Benthic foraminifers: proxies or problems? A review of paleoecological concepts. *Earth-Science Reviews*, 46, 213–236.

van Geel, B., 1986. Application of fungal and algal remains and other microfossils in palaynological analyses. In B.E. Berglund (ed.), *Handbook of Holocene palaeoecology and palaeohydrology*. Wiley, Chichester, pp. 497–506.

van Harten, D., 1996. *Cyprideis torosa* revisited. Of salinity, nodes and shell size. In M.C. Keen (ed.), *Proceedings of the 2nd European Ostracodologists Meeting, London*, pp. 226–230.

van Harten, D., 2000. Variable noding in *Cyprideis torosa* (Ostracoda, Crustacea): an overview, experimental results and a model from Catastrophe Theory. *Hydrobiologia*, 419, 131–139.

Vénec-Peyré, M.-T., Caulet, J.P. and Grazzini, C.V., 1995. Paleohydrographic changes in the Somali Basin (5°N upwelling and equatorial areas) during the last 160 kyr, based on correspondence analysis of foraminiferal and radiolarian assemblages. *Paleoceanography*, 10, 473–491.

Vergnaud-Grazzini, C., Venec-Petre, M.T., Caulet, J.P. and Lerasle, N., 1995. Fertility tracers and monsoon forcing at an equatorial site of the Somali Basin (northwest Indian Ocean). *Marine Micropaleontology*, 26, 137–152.

Vesper, B., 1972. Zum problem der Buckelbildung bei *Cyprideis torosa* (Jones, 1850) (Crustacea, Ostracoda, Cytheridae). *Mitteilungen aus dem Hamburgischen Zoologisches Museum und Institut*, 68, 79–94.

Vesper, B., 1975. To the problem of noding on *Cyprideis torosa* (Jones, 1850). *Bulletin of American Paleontology*, 65, 205–216.

Vidal, L., Labeyrie, L. and van Weering, T.C.E., 1998. Benthic $\delta^{18}O$ records in the North Atlantic over the last glacial period (60–10 kyr): evidence for brine formation. *Paleoceanography*, 13, 245–251.

Villanueva, J., Grimalt, J.O., Cortijo, E., Vidal, L. and Labeyrie, L., 1997. A biomarker approach to the organic matter deposited in the North Atlantic during the last climatic cycle. *Geochimica et Cosmochimica Acta*, 61, 4633–4646.

Villanueva, J., Grimalt, J.O., Cortijo, E., Vidal, L. and Labeyrie, L., 1998. Assessment of sea surface temperature variations in the central North Atlantic using the alkenone unsaturation index ($U^{K'}_{37}$). *Geochimica et Cosmochimica Acta*, 62, 2421–2427.

Villereal, T.A., 1988. Positive buoyancy in the oceanic diatom *Rhizosolenia debyana* H. Peragallo. *Deep Sea Research*, 35, 1037–1045.

Volkman, J.K., Eglinton, G., Corner, E.D.S. and Forsberg, T.E.V., 1980a. Long-chain alkenes and alkenones in the marine coccolithophorid *Emiliania huxleyi*. *Phytochemistry*, 19, 2619–2622.

Volkman, J.K., Eglinton, G., Corner, E.D.S. and Sargent, J.R., 1980b. Novel unsaturated straight-chain C_{37}–C_{39} methyl and ethyl ketones in marine sediments and a coccolithophore *Emiliania huxleyi*. In A.G. Douglas and J.R. Maxwell (eds), *Advances in Organic Geochemistry, 1979. Physics and Chemistry of the Earth, 12*, Pergamon Press, Oxford, pp. 219–227.

Volkman, J.K., Barrett, S.M., Blackman, S.I. and Sikes, E.L., 1995. Alkenones in *Gephyrocapsa oceanica*: implications for studies of paleoclimate. *Geochimica et Cosmochimica Acta*, 59, 513–520.

von Grafenstein, U., Erlenkeuser, H., Brauer, A., Jouzel, J. and Johnsen, S.J., 1999a. A mid-European decadal isotope-climate record from 15,500 to 5000 years B.P. *Science*, 284, 1654–1657.

von Grafenstein, U., Erlernkeuser, E. and Trimborn, P., 1999b. Oxygen and carbon isotopes in modern fresh-water ostracod valves: assessing vital offsets and autecological effects. *Palaeogeography, Palaeoclimatology, Palaeoecology*, 148, 117–132.

Von Post, L., 1916. Om Skogsträdspollen i sydsvenska torfmosselagerföljder (föredragsreferat). *Geologiska Föreningens i Stockholm Förhandlingar*, 38, 384–394.

Von Post, L., 1946. The prospect for pollen analysis in the study of the Earth's climatic history. *New Phytologist*, 45, 193–217.

Vos, P. and de Wolf, H., 1988. Methodological aspects of paleo-ecological diatom research in coastal areas of the Netherlands. *Geologie en Mijnbouw*, 67, 31–40.

Walker, S.E. and Goldstein, S.T., 1999. Taphonomic tiering: experimental field taphonomy of molluscs and foraminifera above and below the sediment–water interface. *Palaeogeography, Palaeoclimatology, Palaeoecology*, 149, 227–244.

Wall, D. and Dale, B., 1968. Modern dinoflagellate cysts and evolution of the Peridiniales. *Micropaleontology*, 14, 265–304.

Wang, P. and Murray, J.W., 1983. The use of foraminifera as indicators of tidal effects in estuarine deposits. *Marine Geology*, 51, 239–250.

Wansard, G., 1996. Quantification of paleotemperature changes during isotopic stage 2 in the La Draga continental sequence (NE Spain) based on the Mg/Ca ratio of freshwater ostracods. *Quaternary Science Reviews*, 15, 237–245.

WDC, 2000. *NOAA Paleoclimatology Program – Free Software*. Page dated 2 March 2000, accessed 29/9/2000 at http://www.ngdc.noaa.gov/paleo/softlib.html

Wei, W. and Peleo-Alampay, A., 1993. Updated Cenozoic nannofossil magnetobiochronology. *International Nannoplankton Association Newsletter*, 15, 15–17.

Weinheimer, A.L. and Cayan, D.R., 1997. Radiolarian assemblages from Santa Barbara Basin sediments: recent interdecadal variability. *Paleoceanography*, 12, 658–670.

Welling, L.A. and Pisias, N.G., 1993. Seasonal trends and preservation biases of Polycystine Radiolaria in the Northern California Current system. *Paleoceanography*, 8, 351–372.

Welling, L.A. and Pisias, N.G., 1995. A new settling method for preparing quantitative radiolarian slides from plankton, sediment trap and deep-sea sediment samples. *Micropaleontology*, 41, 375–380.

Welling, L.A. and Pisias, N.G., 1998a. Radiolarian fluxes, stocks, and population residence times in surface waters of the central equatorial Pacific. *Deep-Sea Research*, 45, 639–671.

Welling, L.A. and Pisias, N.G., 1998b. How do radiolarian sediment assemblages represent surface ocean ecology in the central equatorial Pacific? *Paleoceanography*, 13, 131–149.

Welling, L.A., Pisias, N.G. and Roelofs, A.K., 1992. Radiolarian microfauna in the northern California Current System: indicators of multiple processes controlling productivity. In C.P. Summerhayes, W.L. Prell and K.C. Emeis (eds), *Upwelling systems: evolution since the Early Miocene*. Geological Society of London, Special Publication, No. 63, pp. 177–195.

Wells, P., Wells, G., Cali, J. and Chivas, A., 1994. Response of deep-sea benthic foraminifera to Late Quaternary climate changes, southeast Indian Ocean, offshore Western Australia. *Marine Micropaleontology*, 23, 185–229.

West, R.G., 1956. The Quaternary deposits at Hoxne, Suffolk. *Philosophical Transactions of the Royal Society of London*, B239, 265–356.

West, R.G., 1980. Pleistocene forest history in East Anglia. *New Phytologist*, 85, 571–622.

West, R.G., 2000. *Plant life of Quaternary cold stages: evidence from the British Isles*. Cambridge University Press, Cambridge.

Weston, J.F. and Murray, J.W., 1984. Benthic foraminifera as deep-sea water-mass indicators. In H.J. Oertli (ed.) *Benthos '83: Second International Symposium on Benthic Foraminifera (Pau, 1983)*. Elf-Aquitaine, Esso REP and Total CFP, Pau, France, pp. 605–610.

Whatley, R.C., 1988. Population structure of ostracods: some general principles for the recognition of palaeoenvironments. In *Ostracoda in the earth sciences*, P. De Deckker, J.P. Colin and J.P. Peypouquet (eds). Elsevier, Amsterdam, pp. 245–256.

Whatley, R.C., 1991. The platycopid signal: a means of detecting kenoxic events using Ostracoda. *Journal of Micropalaeontology*, 10, 181–184.

Whatley, R.C., 1993. Ostracoda as biostratigraphical indices in Cenozoic deep-sea sequences. In E.A. Hailwood and R.B. Kidd (eds), *High resolution stratigraphy*. Geological Society of London Special Publication, pp. 155–167.

Whatley, R.C., 1996. The bonds unloosed: the contribution of Ostracoda to our understanding of deep-sea events and processes. In A. Moguilevsky and R.C. Whatley, *Microfossils and oceanic environments*. University of Wales, Aberystwyth Press, Aberystwyth, pp. 3–25.

Whatley, R.C. and Ayress, M., 1988. Pandemic and endemic distribution patterns in Quaternary deep-sea ostracods. In T. Hanai, N. Ikeya and K. Ishizaki (eds), *Evolutionary biology of Ostracoda: its fundamentals and applications*. Kodansha, Tokyo, pp. 739–755.

Whatley, R.C. and Coles, G., 1987. The late Miocene to Quaternary Ostracoda of Leg 94, Deep Sea Drilling Project. *Revista Española de Micropaleontologia*, 19, 33–97.

Whatley, R.C. and Jones, R., 1999. The marine podocopid Ostracoda of Easter Island: a paradox in zoogeography and evolution. *Marine Micropaleontology*, 37, 327–344.

Whatley, R.C. and Quanhong, Z., 1993. The *Krithe* problem: a case history of the distribution of *Krithe* and *Parakrithe* (Crustacea, Ostracoda) in the South China Sea. *Palaeogeography, Palaeoclimatology, Palaeoecology*, 103, 281–297.

Whatley, R.C. and Roberts, R., 1995. Marine Ostracoda from Pitcairn, Oeno and Henderson Islands. *Biological Journal of the Linnean Society*, 56, 359–364.

Wilbur, K.M. and Watabe, N., 1963. Experimental studies on calcification in molluscs and the alga *Coccolithus huxleyi*. *Annals of the New York Academy of Sciences*, 109, 82–112.

Wilbur, K.M. and Watabe, N., 1967. Mechanisms of calcium carbonate deposition in coccolithophorids and molluscs. *Studies in Tropical Oceanography*, 5, 133–154.

Williams, H.F.L., 1994. Intertidal benthic foraminiferal biofacies on the central Gulf Coast of Texas: modern distribution and application to sea level reconstruction. *Micropaleontology*, 40, 169–183.

Williams, M., Dunkerley, D., De Deckker, P. and Chappell, J., 1998. *Quaternary Environments* (2nd edn). Arnold, London, 329 pp.

Winter, A., Jordan, R.W. and Roth, P.H., 1994. Biogeography of living coccolithophores in ocean waters. In A. Winter and W.G. Siesser (eds), *Coccolithophores*. Cambridge University Press, Cambridge, pp. 161–177.

Woillard, G.M., 1978. Grande Pile peat bog: a continuous pollen record for the last 140,000 years. *Quaternary Research*, 9, 1–21.

Wolfe, G.V., Steinke, M. and Kirst, G.O., 1997. Grazing-activated chemical defence in a unicellular marine alga. *Nature*, 387, 894–897.

Wollenburg, J.E. and Kuhnt, W., 2000. The response of benthic foraminifers to carbon flux and primary production in the Arctic Ocean. *Marine Micropaleontology*, 40, 189–231.

Wood, G.D., Gabriel, A.M. and Lawson, J.C., 1996. Palynological techniques – processing and microscopy. In J. Jansonius and D.G. McGregor (eds.), *Palynology: principles and applications*. American Association of Stratigraphic Palynologists Foundation, 1, pp. 29–50.

Wood, R.A., Keen, A.B., Mitchell, J.F.B. and Gregory, J.M., 1999. Changing spatial structure of the thermohaline circulation in response to atmospheric CO_2. *Nature*, 399, 572–575.

Yoder, J.A., Ackleson, S., Barber, R. and Flament, P., 1994. A line in the sea. *Nature*, 371, 689–692.

Young, J.R., 1992(1993). The description and analysis of coccolith structure. *Knihovnicka zemniho plynu a nafty*, 14a, 35–71.

Young, J.R., 1994. Functions of coccoliths. In A. Winter and W.G. Siesser (eds), *Coccolithophores*. Cambridge University Press, Cambridge, pp. 63–82.

Young, J.R., 1998. Neogene. In P.R. Bown (ed.), *Calcareous nannofossil biostratigraphy*. British Micropalaeontological Society Publications Series, Chapman and Hall, London, pp. 225–265.

Young, J.R. and Bown, P.R., 1997. Cenozoic calcareous nannoplankton classification. *Journal of Nannoplankton Research*, 19, 36–47.

Young, J.R. and Ziveri, P., 2000. Calculation of coccolith volume and its use in calibration of carbonate flux estimates. *Deep-Sea Research, Part 2, Topical Studies in Oceanography*, 47, 1679–1700.

Young, J.R., Didymus, J.M., Bown, P.R., Prins, B. and Mann, S., 1992. Crystal assembly and phylogenetic evolution in heterococcoliths. *Nature*, 356, 516–518.

Young, J.R., Bergen, J.A. and Bown, P.R. (eds), 1997. Guidelines for coccolith and calcareous nannofossil terminology. *Palaeontology*, 40, 875–912.

Young, J.R., Davis, S.A., Bown, P.R. and Mann, S., 1999. Coccolith ultrastructure and biomineralisation. *Journal of Structural Biology*, 126, 195–215.

Zagwijn, W.H., 1960. Aspects of the Pliocene and Early Pleistocene vegetation in the Netherlands. *Mededelingen van de Geologische Stichtung*, CIII(5), 1–78.

Zahn, R. and Mix, A.C., 1991. Benthic foraminiferal $\delta^{18}O$ in the ocean's temperature–salinity–density field: constraints on ice age thermohaline circulation. *Paleoceanography*, 6, 1–20.

Zahn, R., Sarnthein, M. and Erlenkeuser, H., 1987. Benthic isotope evidence for changes of the Mediterranean outflow during the late Quaternary. *Paleoceanography*, 2, 543–559.

Zhao, M., Rosell, A. and Eglinton, G., 1993. Comparison of two U^{K}_{37}-sea surface temperature records for the last climatic cycle at ODP Site 658 from the sub-tropical Northeast Atlantic. *Palaeogeography, Palaeoclimatology, Palaeoecology*, 103, 57–65.

Zhao, M., Eglinton, G., Haslett, S., Jordan, R.W., Sarnthein, M. and Zhang, Z., 2000. Marine and terrestrial biomarker records for the last 35,000 years at ODP site 658C off NW Africa. *Organic Geochemistry*, 31, 919–930.

Zielinski, U. and Gersonde, R., 1997. Diatom distributions in Southern Ocean surface sediments (Atlantic sector): implications for palaeoenvironmental reconstructions. *Palaeogeography, Palaeoclimatology, Palaeoecology*, 129, 213–250.

Zielinski, U., Gersonde, R., Sieger, R. and Fütterer, D., 1998. Quaternary surface water temperature estimations: calibration of a diatom transfer function for the Southern Ocean. *Palaeoceanography*, 13(4), 365–383.

INDEX

Note: Page numbers in *italics* refer to tables, page numbers in **bold** refer to figures.